BIOMAT 2006

International Symposium on
Mathematical and Computational Biology

BIOMAT 2006

International Symposium on
Mathematical and Computational Biology

Manaus, Brazil 27 – 30 November 2006

edited by

Rubem P Mondaini

Universidade Federal do Rio de Janeiro, Brazil

Rui Dilão

Instituto Superior Técnico, Portugal

World Scientific

NEW JERSEY • LONDON • SINGAPORE • BEIJING • SHANGHAI • HONG KONG • TAIPEI • CHENNAI

Published by

World Scientific Publishing Co. Pte. Ltd.
5 Toh Tuck Link, Singapore 596224
USA office: 27 Warren Street, Suite 401-402, Hackensack, NJ 07601
UK office: 57 Shelton Street, Covent Garden, London WC2H 9HE

British Library Cataloguing-in-Publication Data
A catalogue record for this book is available from the British Library.

BIOMAT 2006
International Symposium on Mathematical and Computational Biology

ISBN-13 978-981-270-768-0
ISBN-10 981-270-768-9

Printed in Singapore by World Scientific Printers (S) Pte Ltd

Preface

This book contains the selected works of the BIOMAT 2006 International Symposium on Mathematical and Computational Biology. This series of symposia started in 2001, in Rio de Janeiro, Brazil, being the oldest interdisciplinary series of conferences in Latin America in the area of biomathematics. A successful realization every year is due to the expertise of the members of the BIOMAT Consortium as well as to the members of the BIOMAT Editorial Board, its Referees, and Scientific Program Committees.

The BIOMAT 2006 Symposium was held in the city of Manaus in the Brazilian Equatorial Rain forest, from November the 25th to December the 1st. We had fifteen Keynote Speakers from Europe and Americas and an impressive number of contributed works presented by scientists and research students from Brazil and abroad. The BIOMAT tutorials, which are already traditional in the BIOMAT symposia, and are lectured on the first two days of these conferences, are a source of motivation for future researchers in these interdisciplinary topics.

The topics of the BIOMAT 2006 Symposium were a combination of state of the art research and review approaches. They range from cell dynamics and surface reaction models of protocells, to the study of collective steady states of cells, to the modelling of infectious diseases like HIV epidemiology, molecular genetic mechanisms of hepatitis B virus, and the dynamics of tuberculosis. Models of physiological disorders like tumor growth and 3D reconstruction of objects were also analyzed. Topics on the modelling of DNA and proteins by using de novo structure prediction, substitution matrices and Steiner trees were discussed. Other subjects covered in the BIOMAT 2006 Symposium were studies in population dynamics like insect sociality, multistability on predator-prey models, and techniques of impulsive differential equations in bio-economics.

We are indebted to the Board of Trustees of two Brazilian Federal sponsoring agencies: Coordination for the Improvement of Higher Education Personnel — CAPES, and National Council for Scientific and Technological Development — CNPq. We thank also PETROBRAS, the Brazilian Oil company and the world leader of oil research on deep sea waters, and the PETROBRAS-CENPES Research Centre. We have received financial support from SUFRAMA — Superintendency of the Manaus Free Trade zone, and from UNINORTE University Centre. The National Institute for Research of the Amazon (INPA) has provided an excursion to its "Science

Park". We thank specially the directors and representatives of these institutions: Prof. José Fernandes de Lima from CAPES, Dr. Mª. de Lourdes Queirós from CNPq, Dr. Gina Vasquez from PETROBRAS-CENPES, Dr. Auxiliadora Tupinambá from SUFRAMA, and Dr. Wanderli Tadei from INPA.

Our warmest thanks are due to the representatives of two host institutions of the BIOMAT 2006 Symposium at Manaus: Prof. Mª. Hercília Tribuzzy, Dr. Isa Leal and Dr. Tania Castelo Branco, from UNINORTE University Centre. Dr. Andrea Waichman, Dr. Marta Gusmão and Dr. José Pedro Cordeiro from UFAM — Federal University of Amazonas.

On behalf of the Editorial Board of the BIOMAT Consortium, we thank all the authors, participants and sponsors of the BIOMAT 2006 Symposium for their continuous support to the scientific activities and administrative tasks of this successful conference.

Rubem P. Mondaini and Rui Dilão

Manaus, December 2006

Contents

DNA and proteins

Population dynamics

SYSTEMS *STEM CELL* BIOLOGY*

INGO ROEDER
University of Leipzig
Institute for Medical Informatics, Statistics and Epidemiology
Haertelstr 16-18, D-04107 Leipzig, Germany
E-mail: ingo.roeder@imise.uni-leipzig.de

Within the last decade, our modeling attempts in stem cell biology have considerably evolved. Starting from the cellular level, our models now comprise a broad spectrum of phenomena on different scales, ranging from the molecular to the tissue level. Such a scale-bridging description of biological processes does exactly match the intentions of the newly emerging field of systems biology with its central objective to understand biological complexity from molecular scales to ecosystems by a joint application of experimental and theoretical techniques. This work is an attempt to illustrate our systems biological perspective on tissue stem cell organization. Herein, I will describe the general principles of a new concept that understands stem cell organization as a dynamic, self-organizing process rather than as a pre-defined sequence of discrete developmental steps, as classically proposed.

The suitability of these principles to explain a broad variety of experimental results is illustrated for hematopoietic stem cells (HSC). For this system, the general stem cell concept has been translated into a stochastic model that comprises the processes of stem cell self-renewal and differentiation as well as lineage specification on the cellular level. Starting from this model, I will describe one possible way to extend the description towards the intra-cellular level. To do so, we considered a simple transcription factor network as the underlying mechanism controlling lineage specification decisions of HSC and analyzed its dynamical properties applying a system of ordinary differential equations. Finally, a clinical application of the proposed single-cell based model of HSC will shortly be outlined. This application again extends the description level of the model, now also incorporating systemic effects of therapeutic interventions. Based on the assumption that chronic myeloid leukemia can be modeled by a clonal competition process of normal and malignant cells, we analyzed the potential dynamic treatment effects of the tyrosine kinase inhibitor *imatinib*. Our results suggest a selective activity of *imatinib* on proliferating cells which implies the hypothesis that the therapeutic efficiency might benefit from a combination of *imatinib* with drugs promoting the cell cycle activation of primitive stem cells.

---

*Work partially supported by grants LO942/9-1,2 and RO3500/1-1 of the German Research Foundation DFG.

1. Introduction

Systems biology is an emerging new field of science, covering the complexity of biology from molecular scales to ecosystems. Particularly, it is intended to contribute to a *systemic* understanding of biological processes and regulatory principles applying experimental *and* theoretical (i.e. modeling) techniques. Also with respect to stem cell biology, the enormous amount of emerging data (e.g. by the application of high-throughput techniques in the field of genomics, proteomics or transcriptomics) as well as the recognizable complexity of these systems, apparently call for the application of mathematical and computational methods. This clearly shows that a systems biological approach can and even has to be applied to the field of stem cell biology to achieve a comprehensive understanding, which itself is a prerequisite for the successful application of stem cells in therapeutic setting.

Using the example of hematopoietic stem cells (HSC), this work is illustrating how the application of a simple theoretical concept (approximating biological mechanisms) and its mathematical implementation (providing quantitative results) can be used to understand a broad variety of experimental phenomena. Starting from a description of stem cell self-renewal and differentiation as well as lineage specification on the cellular level, it will be described how the proposed generic regulatory principles might be explained by interactions of molecular regulators. Additionally, the possibility to apply quantitative stem cell modeling to the description of hematologic disorders, including the possibility to predict effects of different treatment options is illustrated.

1.1. *Defining tissue stem cells*

At the begin I would like to point to the fact that there are different types of stem cells. The most prominent classification is into embryonic stem cells (ESC) and adult tissue stem cells (TSC). Whereas ESC are derived by particular culturing methods from the inner cell mass of an blastocyst (i.e. an very early state of an embryo), TSC can be found in different tissues throughout the life of an organism[1]. Although both types are denoted as *stem cells*, ESC and TSC do considerably differ with respect to their potential and their functionality. In the following I will exclusively consider tissue stem cells and whenever the term *stem cell* is used, it refers to this cell type.

To analyze stem cells experimentally it is necessary to characterize and select these cells for further investigation. Such a prospective selection process imposes the question: *Is this particular cell a stem cell?* This question implies the idea that one can indeed decide about the nature of a selected cell without relating it to other cells and without testing its capabilities. This, however, is a rather unrealistic point of view. To explain this, let us start by looking at the definition of tissue stem cells. Although there are a number of definitions around (see e.g.[1,2,3]), there is a consensus that its *functional* attributes, and not an explicitly observable characteristic, has to be considered as the gold standard for a stem cell characterization.

The following definition, which explicitly refers to the functional capabilities, has been formulated by Loeffler and Roeder[4] on the basis of a definition previously given by Potten and Loeffler[2]: *Stem cells of a particular tissue are a (potentially heterogeneous) population of functionally undifferentiated cells, capable (i) of homing to an appropriate growth environment, (ii) of proliferation, (iii) of production of a large number of differentiated progeny, (iv) of self-renewing their population, (v) of regenerating the functional tissue after injury, and (vi) with a flexibility and reversibility in the use of these options.* This choice of a functional definition is inherently consistent with the biological role of a stem cell, particularly linked to the functional tissue regeneration feature. However, it imposes difficulties since in order to identify whether or not a cell is a stem cell, its function has to be tested. This requires an experimental manipulation, subjecting the cell to a functional bioassay, which inevitably alters its properties. In analogy to the Heisenberg's uncertainty principle in quantum physics[5] this phenomenon is sometimes called *uncertainty principle of stem cell biology*[2]. Simply speaking, this principle states that the very act of measuring the functional system properties always changes the characteristics of the system, therefore, giving rise to a certain degree of uncertainty in the characterization of the system. If this analogy holds true for tissue stem cells - and there is strong evidence for that - all prospective statements that can be made about stem cells will be necessarily probabilistic statements about the future behavior under particular conditions.

1.2. *Conceptual challenges in tissue stem cell biology*

An essential issue of the above given definition of TSC is the flexibility criterion. There is increasing evidence for flexibility and reversibility of stem cells which will be highlighted by a few examples, preferably related to the hematopoietic system. Tissue stem cells are heterogeneous with respect to functional properties such as cycling activity, engraftment potential or differentiation status, and to the expression of specific markers such as adhesion molecules or cell surface antigens. However, experimental evidence is accumulating that these properties are able to reversibly change[6,7,8,9]. Many authors have described the variability in the proliferative status of hematopoietic stem cells. One important finding in this respect is the fact that primitive cells may leave the cell cycle for many days and even months, but that almost all re-enter cycling activity from time to time[10,11]. Experimental evidence is also provided for reversible changes of the stem cell phenotypes involving differentiation profiles, adhesion protein expression and engraftment/homing behavior associated with the cell cycle status or the point in the circadian rhythm[8]. Also the expression of cell surface markers (e.g. CD34) on hematopoietic stem cells is not constant but may fluctuate. The property can be gained and lost without affecting the stem cell quality[7,12]. Furthermore, there is a lot of indirect evidence for fluctuations in the stem cell population based on the clonal composition of functional cells. Chimerism induced by bone marrow transplantations in animal models has been shown to fluctuate with time[13,14,15] indicating variations in the composition of active and inactive tissue stem cells. Similar observations were made following retroviral marking of individual stem cell clones which highlight the relative differences of inheritable cellular properties between stem cell clones and their impact on the competitive potential[16,17,18]. Another level of flexibility was found for lineage specification within the hematopoietic tissue. It is possible to bias the degree of erythroid, granuloid, or lymphoid lineage commitment by several maneuvers altering the growth-conditions in different culture systems[6,?]. The present concept to explain the fluctuations observed in lineage specification is based on a dynamic network of interacting transcription factors[20,21]. Some authors put forward the concept of fluctuating levels of transcription factors with threshold dependent commitment[22]. Moreover, tissue stem cells specified for one type of tissue (e.g. hematopoiesis) can be manipulated in such a way that they can act as tissue stem cells of another tissue[23,24,25,26]. As suggested by experimental observations on these tissue plasticity phenomena, microenvironmental effects seem to play an essential role in directing cellular development. Clearly, tissue plasticity represents a particular degree of flexibility consistent with the above definition.

Because classical stem cell concepts are not able to explain all these experimental findings consistently, new conceptual approaches are required. To be validated, such concepts need a rigorous examination by quantitative and predictive modeling. In the following, some general ideas on how to proceed with such theories are presented and illustrated by a worked model of hematopoietic stem cells.

1.3. *Predictive theories and quantitative models*

Within the natural sciences a model is understood as a simplifying abstraction of a more complex construct or process. Theoretical models in biology include qualitative concepts (i.e. descriptive representations) and quantitative models (i.e. mathematical representations) of a biological process. In contrast to qualitative concepts, quantitative models allow for an analytical, numerical, or simulation analysis. The more we realize that we cannot prospectively determine stem cells directly, the more we need theoretical approaches to cope with the complexity. There is a tremendous need for general and specific theoretical concepts of tissue stem cell organization, as well as for related quantitative models to validate the concept by comparison of model predictions and experimental results.

Such a theoretical framework of tissue stem cell functioning will have several advantages: Model predictions can assist biologists to select and design experimental strategies and they help to anticipate the impact of manipulations to a system and its response. Modeling is able to discriminate similar and to link different phenomena. Specifically, models originating from the same principles adapted to different systems (i.e. tissues or cell types) may help to understand generic construction and regulation principles. Furthermore, they contribute to the understanding of latent mechanisms or crucial parameters of biological processes and may predict new phenomena.

The following list represents a summary of general requirements which quantitative models should fulfill in order to be suitable to serve as the basis for a theoretical framework of tissue stem cell organization: The model cells must fulfill the criteria listed in the definition of tissue stem cells consistently. This has the following implications: (i) The models must be based on populations of individual cells to follow clonal development, to conform with the uncertainty principle, and to enable considerations of population fluctuations. (ii) They must consider growth environments and the interactions between the cells. (iii) The system has to be dynamic in time and possibly space. (iv) The system requires assumptions on mechanism to regulate proliferation, cellular differentiation, and cell-cell / cell-growth environment interactions. (v) The model concept must be comprehensive in the sense of being applicable to the normal unperturbed *in vivo* homeostasis as well as to any *in vivo* or *in vitro* assay/intervention procedure. This criterion requests that system-measurement interactions must be consistently considered.

2. A new perspective on stem cell systems

As discussed above, the basic concept of a functional definition of TSC has widely been accepted. Such a functional definition implies that one does not require *stemness* as an explicit attribute of cells, but rather considers it as a functional endpoint. Therefore, any concept on TSC has to specify assumptions about the mechanisms that potentially control the regenerative and proliferative potential of these cells, such as proliferation, differentiation, maturation, lineage specification, and homing. Hence, the task is to design a dynamic process that drives and controls the cellular attributes. Central points herein are aspects of capabilities (i.e. actual and potential expression of cellular properties), of flexibility, and of reversibility. Apparently, all these aspects are determined by the genetic and epigenetic status of the cells and by the activity of the signal transduction pathways including the transcription factor networks. Presently, it is impossible to describe these processes in any reasonable detail. However, it is possible to propose a simplified basic scheme of the generic principles underlying the cellular dynamics.

2.1. *Phenotypic reversibility as a generic principle of stem cell systems*

Classical concepts of tissue stem cell organization are almost exclusively based on the assumption of a strict unidirectional developmental hierarchy. However, as mentioned above, many current experimental results are challenging these concepts and show the necessity of new conceptual approaches to understand TSC organization. Flexibility and reversibility of tissue/lineage specification and of cellular properties within a tissue (summarized under the term *phenotypic reversibility*) have major implications with regard to concepts of stem cell function. Therefore, our group proposed to give up the view of tissue stem cells as being entities with a pre-programmed development and to replace it by a concept that makes the capabilities for flexible and regulated tissue self-organization the new paradigm[4]. Such a concept permits to incorporate context-dependent phenotypic reversibility and generation of stem cell heterogeneity as the result of a dynamically regulated process and it strictly avoids assumptions that end up with direct or indirect labeling of particular cells as stem cells *a priori*. All model cells are characterized only by functional properties (e.g. proliferating or not, having an affinity for homing to a particular growth environment, sensitivity to particular growth factors etc.) and request that the system behavior changes these properties such that the population fulfill the functional criteria of the stem cell definition.

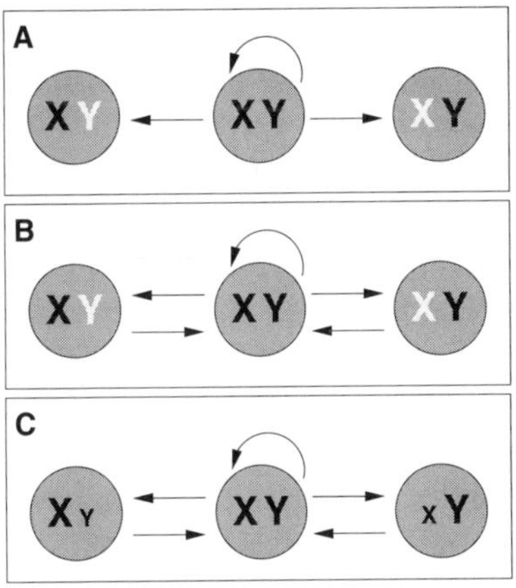

Figure 1. Examples of state transition graphs according to level 1 and 2 dynamics. X and Y illustrate certain genes or functionally related gene clusters. Whereas the color is coding for the level 1 dynamics status (black: sensitive, white: insensitive), the font size illustrates the quantitative expression level according to level 2 dynamics. (A) Shows irreversible loss of cellular properties due to permanent level 1 inactivation. Only self-maintenance of XY state possible. (B) Due to reversible changes (plasticity) with respect to level 1 dynamics (sensitive, insensitive) true self-renewal of XY state possible. (C) Reversibility (plasticity) of XY state due to changes with respect to quantitative level 2 dynamics.

To explain this concept, let us consider the activity of genes[a] determining the behavior of TSC. Because there might be circumstances when particular gene sets are insensitive to activation by regulatory molecules, e.g. if epigenetic constellations prevent accessibility or if key regulator molecules are lacking[23], two levels of gene activity control are conceptually distinguished: Level 1 is qualitative and decides whether a gene is accessible for activation or not (sensitive or insensitive). Level 2 is quantitative

[a]here, genes are used as examples for regulatory units, neglecting any post-transcriptional regulation and identifying genes with their products, e.g. regulator proteins.

and describes the degree of gene expression in a sensitive gene. Within this concept, a gene may not be expressed for two very different reasons. It may either not be sensitive (level 1 dynamics) or it may be sensitive, but it is not activated due to lack of challenge (level 2 dynamics). State-transition graphs can be used to characterize this two level dynamics. If they contain only self-maintaining and irreversible acyclic transitions between states, a population can be self-maintaining but not self-renewing (Figure 1A). In contrast, figures 1B and C illustrate state transition graphs which are characterized by reversible transitions. This would imply the property of true self-renewal, in the sense that cellular properties can be reestablished even if they had been lost/locked (level 1 dynamics) or down regulated (level 2 dynamics) before. We, furthermore, assume that the preferred direction of cellular development is dependent on growth environment specific signals. Therefore, alternating homing to various growth environments would yield a rather fluctuating development. In such a setting not only the influence of the environments, but particularly the frequency of transitions between them would be important. Figure 2 illustrates how signals from different growth environments can influence the cellular fate. Although only explained for level 2 dynamics, growth environmental signals could also affect transient or permanent inactivation of genes, i.e. the level 1 dynamics.

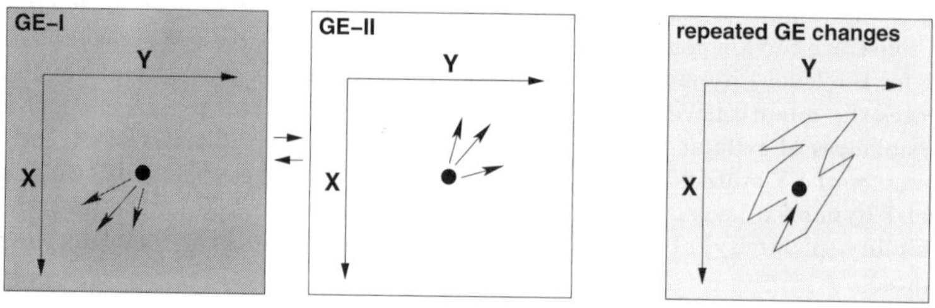

Figure 2. Dependency of cellular development on growth environment. This figure illustrates the actual position of a cell (black dot) and the preferred developmental directions (arrows) with respect to level 2 dynamics of cellular properties X and Y (e.g. gene expression) depending on the actual growth environment (GE). Alternation between different growth environments can induce fluctuating expression of cellular properties (quantitative plasticity), as illustrated in the rightmost panel by one possible example trajectory.

Taken together, such a general concept of growth environment dependent dynamics of reversibly changing cellular properties is a possibility to explain processes of self-renewal and differentiation in tissue stem cell systems. This general framework has been translated into a stochastic, single-cell based model for HSC[28] which is summarized in the next section.

3. Modeling hematopoietic stem cell organization on the cellular level

3.1. *Modeling self-renewal and differentiation*

To set up a mathematical description describing stem cell self-renewal and differentiation in the hematopoietic system, the following assumptions are made: Cellular properties of HSC can reversibly change within a range of potential options. The direction of cellular development and the decision whether a certain property is actually expressed, depends on the internal state of the cell and on signals from its growth environment. Individual cells are considered to reside in one of two possible growth-environments (denoted as GE-A and GE-Ω). It should be noted that these two growth-environments represent different signaling contexts of the cells and that they are in general not spatially defined. However, regarding the system of HSC, the two growth environments could be interpreted as signaling contexts in or outside a specific stem cell niche environment, as suggested biologically[29,30].

On the basis of these general assumptions, the state of each cell in the model is characterized by a property a which describes its affinity to reside in GE-A, by its position in the cell cycle c, and by its actual growth environment membership m. Therefore, the state z of an individual cell at time t can be written as the vector: $z(t) = \Big(a(t), c(t), m(t)\Big)^T$. with the following components:

(1) the affinity $a \in [a_{min}, a_{max}]$,
(2) the position in the cell cycle c with $0 \leq c < \tau_c$, with τ_c describing the cell cycle time,
(3) a growth environment membership indicator $m \in \{A, \Omega\}$.

At discrete time steps of length Δt ($\Delta t \ll \tau_c$) the cells are forced to decide whether they will stay in the actual growth-environment or whether they will change to the other. The transition of a cell from GE-Ω to GE-A, which is assumed to be possible during G_1-phase of the cell cycle[b] only,

[b]cell cycle sequence: G_1 (gap 1), S (DNA-synthesis), G_2 (gap 2), M (mitosis)

occurs with probability α, i.e.

$$P\Big(m(t+\Delta t) = A \,|\, m(t) = \Omega\Big) = \begin{cases} \alpha & \text{for } \; c(t) \leq \tau_{g_1} \\ 0 & \text{for } \; c(t) > \tau_{g_1} \end{cases},$$

$$\text{with} \quad \alpha = \frac{a(t)}{a_{max}} \cdot f_\alpha(N_A). \tag{1}$$

N_A denotes the total number of cells in GE-A and τ_{g_1} the length of the G_1-phase. The function f_α models the capacity of GE-A to assimilate cells. This capacity decreases monotonically with rising numbers of cells in GE-A representing a limited resource of binding sites.

The transition from GE-A to GE-Ω happens with probability ω, i.e.

$$P\Big(m(t+\Delta t) = \Omega \,|\, m(t) = A\Big) = \omega,$$

$$\text{with} \quad \omega = \frac{a_{min}}{a(t)} \cdot f_\omega(N_\Omega). \tag{2}$$

N_Ω is denoting the total number of cells in GE-Ω. This transition (reactivation into cell cycle) is assumed to occur always at the check point between G_1 and S-phase. f_ω represents the cell production demand with low numbers of proliferating cells inducing an activation of dormant cells into cell cycle.

Because no detailed knowledge is available about the underlying biological and chemical mechanisms of the growth environment transition processes and to allow for flexibility in the analysis of the system properties, the GE-transition characteristics f_α, f_ω are modeled by a general class of sigmoid functions of the form

$$f(N) = \frac{1}{\nu_1 + \nu_2 \cdot exp\left(\nu_3 \cdot \dfrac{N}{\tilde{N}}\right)} + \nu_4.$$

The parameters ν_1, ν_2, ν_3, and ν_4 determine the shape of f, and $\tilde{N}$ is a scaling factor for N. It is possible to uniquely determine ν_1, ν_2, ν_3, and ν_4 by fixing the more intuitive values $f(0)$, $f(\frac{\tilde{N}}{2})$, $f(\tilde{N})$, and $f(\infty) := lim_{N\to\infty} f(N)$:

$$\begin{aligned}
\nu_1 &= (h_1 h_3 - h_2^2)/(h_1 + h_3 - 2h_2) \\
\nu_2 &= h_1 - \nu_1 \\
\nu_3 &= \ln((h_3 - \nu_1)/\nu_2) \\
\nu_4 &= f(\infty)
\end{aligned}$$

with the dummy variables

$$
\begin{aligned}
h_1 &= 1 \,/\, [f(0) - f(\infty)] \\
h_2 &= 1 \,/\, \left[f\left(\tilde{N}/2\right) - f(\infty) \right] \\
h_3 &= 1 \,/\, [f(\tilde{N}) - f(\infty)].
\end{aligned}
$$

To obtain the probabilities α and ω, the transition characteristics f_α and f_ω are modulated by the individual affinity a in the sense that cells with high a are more likely to change to GE-A whereas cells with low a tend to reside in GE-Ω (see 1 and 2).

If cells do not undertake a transition from one growth-environment to the other at the time points $t_0 + k \cdot \Delta t$, with $k \in N$ (probabilities $1 - \alpha$ and $1 - \omega$, respectively) they develop inside the actual growth-environment according to the following deterministic rules:

$$
\begin{aligned}
& m(t) = \Omega : && (3) \\
& \quad a(t + \Delta t) := \begin{cases} a(t)/d & \text{if} \quad a(t)/d \geq a_{min}\,, \quad \text{with} \quad d > 1 \\ 0 & \text{else} \end{cases} \\
& \quad c(t + \Delta t) := \begin{cases} c(t) + \Delta t & \text{if} \quad c(t) + \Delta t < \tau_c \\ 0 & \text{else} \end{cases} \\
& m(t) = A : && (4) \\
& \quad a(t + \Delta t) := \begin{cases} a(t) \cdot r & \text{if} \quad a(t) \cdot r < a_{max}\,, \quad \text{with } r > 1 \\ a_{max} & \text{else} \end{cases} \\
& \quad c(t + \Delta t) \equiv \tau_{g_1}\,, \text{ with } \tau_{g_1} \text{ describing the length of cell cycle phase } G_1
\end{aligned}
$$

Cell division (in GE-Ω) at $c(t) \geq \tau_c$ is realized by the replacement of the mother cell by two daughter cells, inheriting the actual state vector $z(t)$. If the affinity a of a cells has fallen below the critical value a_{min}, it is set to 0 and the cell is assumed to be the founder of a maturing clone producing differentiated cells. All these differentiated clones are assumed to have a fixed life time λ_c. Figure 3 provides a schematic representation of the model as well as examples of the transition characteristics f_α and f_ω for three different values of a, respectively.

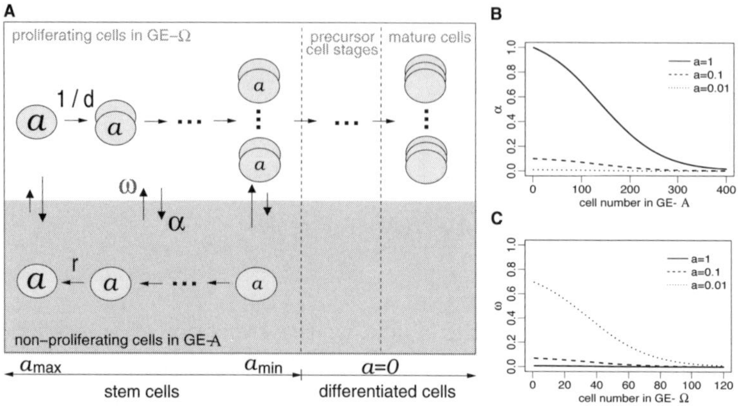

Figure 3. Schematic representation of the model. (A) The lower part represents growth environment GE-A and the upper part GE-Ω. Cell amplification due to proliferation in GE-Ω is illustrated by growing cell numbers (cell groups separated by vertical dots represent large cell numbers). Whereas the growth environment affinity a of the cells decreases by factor $1/d$ per time step in GE-Ω, it increases by factor r per time step in GE-A. The actual quantity of the affinity a is sketched by different font sizes. If a fell below a critical threshold a_{min}, the cell lost its potential to switch to GE-A and a is set to 0 (represented by empty cells). Transition between GE-A and Ω occurs with intensities $\alpha = (a/a_{max})f_\alpha$ and $\omega = (a_{min}/a)f_\omega$, which depend on the value of a (represented by the differently scaled vertical arrows) and on the cell numbers in the target GE. Typical profiles of the cell number dependent transition characteristics f_α and f_ω for different values of affinity a shown in panels (B) and (C).

We showed that the proposed concept is fully compatible with the functional definition of tissue stem cells and consistently explains a broad variety of macroscopic phenomena including e.g. heterogeneity of repopulation or colony-forming potential, *in vitro* colony-growth kinetics, *in vivo* repopulation after different system disturbances or fluctuating clonal contributions[28]. As one example let us consider the coexistence of cells from two different mouse strain backgrounds (DBA/2 and C57BL/6) in one common host. As shown in figure 4, the chimerism in lethally irradiated host mice transplanted with mixtures of DBA/2 (D2) and C57BL/6 (B6) bone marrow cells is not constant. It shows a typical peak of D2 contribution shortly after transplantation, followed by a slow decline in the contribution of this cell type. This dynamic behavior can be reproduced after a secondary (and even tertiary - data not shown) transplantation. It could be demonstrated by the model that such a behavior can be consistently explained by small quantitative differences in the cell-micro environment interaction (modeled by the transition characteristics f_α, f_ω) between the two strains. These parameter changes are also consistent with the experimental observation that both mouse strains equally reconstitute ablated hosts in a non-competitive situation. For a detailed description of experimental data and simulation results the reader is referred to Roeder et al. 2005[31].

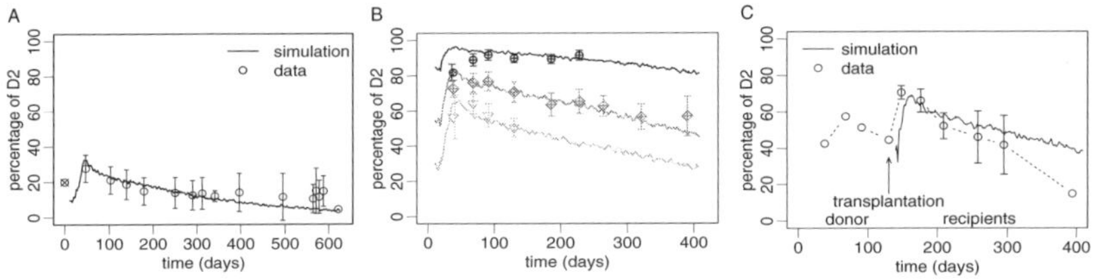

Figure 4. Simulation results on chimerism development. (A) Data points (open circles) represent the observed chimerism levels (mean ± 1 SD) in primary radiation chimeras. The solid line shows the simulated chimerism of mature model leukocytes (average of 100 simulation runs). (B) Effect of the initial D2:B6 ratio: Data points represent the results (mean ± 1 SD) from three independent experiments using different D2 proportions of the transplant. Solid lines represent corresponding average simulation results. These are based on identical parameter sets, but different initial D2 proportions: 85% - black, 50% - dark gray, 30% - light gray. (C) The circles show the experimentally observed peripheral blood leukocyte chimerism in a primary radiation chimera (single values) and in a corresponding cohort of secondary host mice (mean ± SD). The solid lines show average simulations for the chimerism development in the secondary chimeras.

3.2. *Modeling lineage specification*

Although the proposed stem cell model already comprises a large number of phenomena, it is so far not able to account for the process of lineage specification. This process controls the development of progenitor cells into the functionally different cell types (e.g. erythroid, myeloid or lymphoid cells in case of hematopoiesis) during their differentiation.

To incorporate lineage specification into our model, it is assumed to be a competitive process between different interacting and lineage specific factors (e.g. transcription factors). These are denoted by x_i for $i = 1, \ldots, N$ possible lineages. Each of these factors quantitatively represents the commitment status of the cell for one specific lineage. Now, the presumingly complex molecular interactions governing the (transcription) factor control is modeled by a simple stochastic process, leading either to the maintenance of a low level co-expression of lineage factors (characterizing stem cells) or to the dominance of one factor over the others. Depending on cell-extrinsic signals, this process is realized by a gradual suppression or promotion of the particular factors. To functionally relate the process of lineage specification to the model presented in section 3.1, it is particularly assumed that growth enviroment A mediates a repressive and growth environment Ω mediates a progressive lineage control regime. Whereas the repressive control regime forces all factors x_i to fluctuate around a common mean expression level, the progressive regime enhances deviations in the expression of lineage specific factors from the common mean expression in a self-amplifying manner.

Technically, each cell is now characterized by the vector $z(t) = \Big(a(t), c(t), m(t), x(t)\Big)^T$. Additional to the characteristics $a(t), c(t), m(t)$, which had been introduced in section 3.1, another component $x(t) = (x_1(t), x_2(t), ..., x_N(t))^T$ is introduced. It codes for the expression levels of the N lineage specific factors x_i at time t. Because $x(t)$ is normalized (i.e. $\sum_i x_i(t) = 1$), the $x_i(t)$ represent relative expression levels. In a time-discrete stochastic process, one lineage i per time step is chosen randomly with a probability proportional to the size of $x_i(t)$, and x_i is updated according to $x_i(t+1) := x_i(t)(1+m_i)$. A subsequent normalization step guarantees $\sum_i x_i(t) = 1$. The lineage specific rewards m_i are defined as $m_i(x_i) = b_i x_i + n_i$ with $n_i > 0$. Whereas $b_i = 0$ for reward functions under the progressive feedback, we assume $b_i < 0$ for the negative reward function. In the latter case, the root $(\bar{x})$ of the reward function (i.e., $m_i(\bar{x}) = 0$) is assumed to take the value $\bar{x} = 1/N$. In this setting the reward m_i is positive for $x_i < \bar{x}$ and negative for $x_i > \bar{x}$.

Figure 5 outlines the intracellular differentiation dynamics in a simple sketch. As long as the cell is kept under the tight regulation of the repressive regime (mediated by growth environment A) the $N = 4$ lineage specific factors fluctuate around a common mean expression level $\bar{x} = 1/N = 0.25$. Changing to the progressive regime (growth environment Ω), one lineage is favored in a sequence of stochastic decision steps. Ultimately this lineage is up-regulated, while the other lineages are down-regulated. For a mapping onto the phenotypic level, individual cells are classified according to the actual expression levels of their lineage specific factors x_i as undifferentiated cells ($x_i < x_{early}$), early committed cells ($x_{early} < x_i < x_{finally}$) and finally committed cells ($x_{finally} < x_i$). The temporal extension of such a decision process (induced by the gradual shift in the factor ratios) implies the potential for a reversibility of lineage commitment under changing micro-environmental influences.

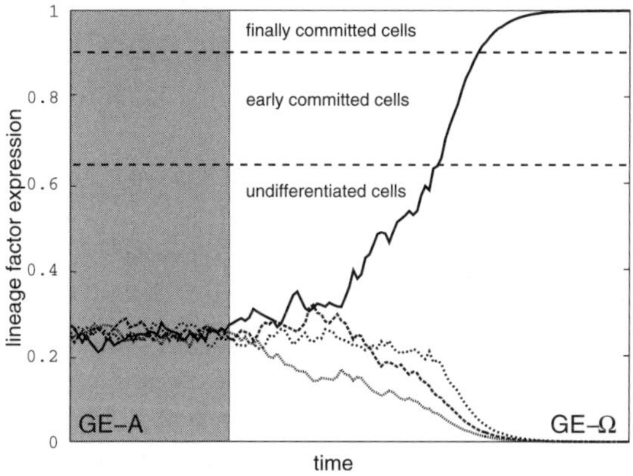

Figure 5. Intra-cellular dynamics of $N = 4$ lineage specific factors under the repressive (GE-A) and the progressive (GE-Ω) feedback regime. Whereas the expression levels fluctuate around the common mean $\bar{x} = 0.25$ under the influence of GE-A (grey background), GE-Ω induces the promotion of one randomly selected lineage. Typical expression levels for a classification of the cells in undifferentiated, early and finally committed cells are given by the dashed horizontal lines.

Comparing simulation results based on the described stem cell model to data describing the multi-lineage potential of individual pre-selected bone marrow cells as well as to cell population growth kinetics of the stem cell-like FDCPmix cell line under different lineage specific culture conditions, we were able to demonstrate the general consistency of the model assumptions with experimental observations (unpublished data).

4. Modeling the dynamics of intra-cellular regulation

Although the processes of stem cell self-renewal, differentiation, and lineage specification are captured by the above described model on the cellular level, there is no link to the underlying molecular regulation. To proceed on the way to achieve a comprehensive systemic understanding of stem cell organization, we are now trying to explain cellular decision processes by explicitly modeling underlying regulatory processes on the molecular level. Due to the high complexity of molecular regulation in eukaryotic systems, consolidated knowledge about the details of regulatory networks is currently very limited. However, particularly with respect to lineage commitment of HSC, at least some important key regulators and there biochemical/physical interactions are known. Therefore, this process is chosen as an example to demonstrate our modeling strategy.

The central objective is to investigate, whether lineage decision processes of HSC (as experimentally observed for erythroid / myeloid fate decisions) can be explained by a simple quantitative model describing the interaction of two regulatory complexes, such as particular transcription factors. The model is motivated by experimental observations on the transcription factors GATA-1 and PU.1, both known to act as key regulators and potential antagonists in the erythroid vs. myeloid differentiation of HSC. Particularly, it will be tested whether such a model is able to account for the observed switching from a state of low expression of both factors (undifferentiated state) to the dominance of one factor (differentiated state). Although our analysis is motivated by experimental observations of specific transcription factor interactions (GATA-1 and PU.1), the model may also be applied in the general context of two interacting transcription factors, which will be denoted as X and Y. In the following, the key idea of the model and some important results are briefly outlined.

The general design of the model structure is based on the following assumptions which are motivated by experimental observations:

- Both transcription factors, X and Y, are able to act as activator molecules:
  - If bound to their own promoter region, X and Y introduce a positive feedback on their own transcription. This process is referred to as *specific* transcription.
  - X and Y are both able to induce an overall transcription which also effects potentially antagonistic transcription factors. Although such an interaction is most likely indirect, for the model we consider a mutual activation of X and Y by the opposing transcription factor, which we refer to as *unspecific* transcription.

  We assume that transcription initiation is only achieved by simultaneous binding of two X and Y molecules, respectively (i.e. binding cooperativity $\nu = 2$).

- There is a mutual inhibition of X and Y. Within this context, two possible mechanisms, based on the formation of two structurally different complexes of X and Y, are considered:
  - Joint binding of X and Y molecules to promoter sites. Here, the DNA-bound XY-complex (Z_1) acts as a transcription re-

pressor, which blocks the binding sites. This represents a mode of competition for free binding sites.

- Formation of another XY-complex, called Z_2, which neither binds to X nor Y DNA binding site. In contrast to Z_1, this represents a competition for free transcription factor molecules.

Both inhibition mechanisms (including combinations of them) are considered for X as well as for Y.

To facilitate the analysis of the mathematical model, the following simplifying assumptions are made: (i) Post-transcriptional regulation is neglected, i.e., the transcription of a gene is considered to ultimately result in the production of the corresponding protein (here, a transcription factor). (ii) Time delays due to transcription and translation processes are neglected. (iii) Simultaneous binding of X/Y monomers together with a Z_1-heterodimer, of two Z_1-heterodimers, as well as of a X and a Y monomer at the same promoter are excluded from the analysis. (iv) Interactions of X, Y as well as the promoter regions of the coding genes with further transcription factors are neglected.

Throughout this section the following notations are used: x, y denote the molecule concentrations of X and Y, respectively. Z_1 denotes the DNA bound XY-complex and Z_2 the structurally different XY- complex, which is not able to bind to promoter DNA. $D_{x/y}$ denotes free DNA binding sites within the promoter region of X and Y, respectively. In contrast, binding sites occupied by X or Y molecules or by the XY-complex Z_1 are denoted as $D_{x/y}^{xx/yy/xy}$.

Based on these assumptions, a set of chemical reaction equations can be set up. The processes of specific and unspecific transcription activation are described by equations (5)-(8):

$$X + X + D_x \overset{K_1}{\rightleftharpoons} D_x^{xx}; \quad D_x^{xx} \overset{s_x}{\rightarrow} D_x^{xx} + X \tag{5}$$

$$Y + Y + D_x \overset{K_2}{\rightleftharpoons} D_x^{yy}; \quad D_x^{yy} \overset{u_x}{\rightarrow} D_x^{yy} + X \tag{6}$$

$$Y + Y + D_y \overset{K_3}{\rightleftharpoons} D_y^{yy}; \quad D_y^{yy} \overset{s_y}{\rightarrow} D_y^{yy} + Y \tag{7}$$

$$X + X + D_y \overset{K_4}{\rightleftharpoons} D_y^{xx}; \quad D_y^{xx} \overset{u_y}{\rightarrow} D_y^{xx} + Y \tag{8}$$

Herein, it is assumed that the DNA binding of X and Y always occurs as the binding of homodimers, i.e. the sequential binding of two monomers is not considered. The process of dimerization, as well as the DNA binding and dissociation, are regarded to be in quasi steady state. The K_i =

$k_i/\bar{k}_i$ $(i = 1, ..., 7)$ denote the dissociation constants of the reactions, with k_i and $\bar{k}_i$ representing the forward and backward reaction rate constants, respectively. Finally, it is assumed that both monomers, X and Y, decay with first order kinetics at rates k_0^x and k_0^y. Dimer-complexes are assumed to be stable.

With regard to the mutual transcription inhibition mechanisms we consider the following two complex formations:
(i) Formation the XY-complex Z_2

$$X + Y \overset{K_5}{\rightleftharpoons} Z_2 \,. \tag{9}$$

Under the quasi steady state assumption Z_2 does not contribute to the mathematical description of the system dynamics.
(ii) Formation of a structurally different heterodimeric complex Z_1, which is able to bind to the promoter regions, acting as a repressor for X and Y transcription, respectively:

$$X + Y + D_x \overset{K_6}{\rightleftharpoons} D_x^{xy}, \tag{10}$$

$$X + Y + D_y \overset{K_7}{\rightleftharpoons} D_y^{xy}. \tag{11}$$

As with the promoter binding of X and Y, we collapse dimerization, which is assumed to be in quasi steady state, and DNA binding into one process, neglecting the sequential binding of monomers.

Under quasi steady state assumptions, equations (5)-(11) lead to the following set of ordinary differential equations:

$$\frac{dx}{dt} = -k_{0_x} x + \frac{s_x K_1 x^2 + u_x K_2 y^2}{1 + K_1 x^2 + K_2 y^2 + K_6 xy} \tag{12}$$

$$\frac{dy}{dt} = -k_{0_y} y + \frac{s_y K_3 y^2 + u_y K_4 x^2}{1 + K_3 y^2 + K_4 x^2 + K_7 xy} \tag{13}$$

To analyze this system with respect to its dynamical behavior (steady state characterization, derivation of bifurcation conditions), we will at this point restrict ourself to the special case of a completely symmetric system, i.e.: $k_{0_x} = k_{0_y} = k_0$, $s_x = s_y = \tilde{s}$, $u_x = u_y = \tilde{u}$, $K_1 = K_3$, $K_2 = K_4$, and $K_6 = K_7$. Using these relations, together with $\mathbf{x} = \sqrt{K_1}x$, $\mathbf{y} = \sqrt{K_1}y$, $k_u = K_2/K_1$, $k_r = K_6/K_1$, $s = \sqrt{K_1}\tilde{s}/k_0$, $u = \sqrt{K_1}\tilde{u}/k_0$, and $\tau = k_0 t$, the system (12), (13) can be written in a dimensionless form as

$$\frac{d\mathbf{x}}{d\tau} = -\mathbf{x} + \frac{s\mathbf{x}^2 + uk_u\mathbf{y}^2}{1 + \mathbf{x}^2 + k_u\mathbf{y}^2 + k_r\mathbf{xy}}, \tag{14}$$

$$\frac{d\mathbf{y}}{d\tau} = -\mathbf{y} + \frac{s\mathbf{y}^2 + uk_u\mathbf{x}^2}{1 + k_u\mathbf{x}^2 + \mathbf{y}^2 + k_r\mathbf{xy}}, \tag{15}$$

Considering the specific transcription rate s as the bifurcation parameter, it can be shown that this system is able to generate two qualitative different bifurcation sequences depending on the unspecific transcription rate u. Whereas the system always facilitates a switch from a stable steady state at the point $x_1^* = y_1^* = 0$ to a stable steady state characterized by one upregulated transcription factor, i.e. $x_{21}^* > y_{21}^*$ or $x_{22}^* < y_{22}^*$, an additional steady state is emerging for sufficiently large values of the unspecific transcription rate u (see figure 6).

These results demonstrate that the presented model is able to generate parameter dependent changes in the system behavior, with alteration in the number of possible stable steady states. Particularly, the model explains a transition from states of stable co-expression to the situation characterized by an over-expression of one factor over the other an vice versa. This is mediated by a quantitative change of transcription rates, which might serve as a possible molecular explanation of the above proposed progressive and repressive control regimes. Moreover, the model predicts two different possibilities to explain the experimentally suggested, stem cell characterizing state of low level transcription factor co-expression. Whereas a sufficiently high unspecific transcription would allow for a stable low-level but non-zero coexpression, such a state is not possible for the situation of no or minor degrees of unspecific transcription.

For technical details, an extended description of bifurcation conditions as well as for an analysis of further scenarios, including asymmetries in the inhibition mechanisms, the reader is referred to Roeder and Glauche, 2006[32].

5. Modeling genesis and treatment of chronic myeloid leukemia

The last section will demonstrate how the proposed mathematical model of HSC organization can be applied to a clinical situation. Summarizing the results of a recent study[33], I will describe model results on the development and the treatment of chronic myeloid leukemia (CML), a hematopoietic disorder that is induced by the mutation of a single cell. Due to a chromosomal

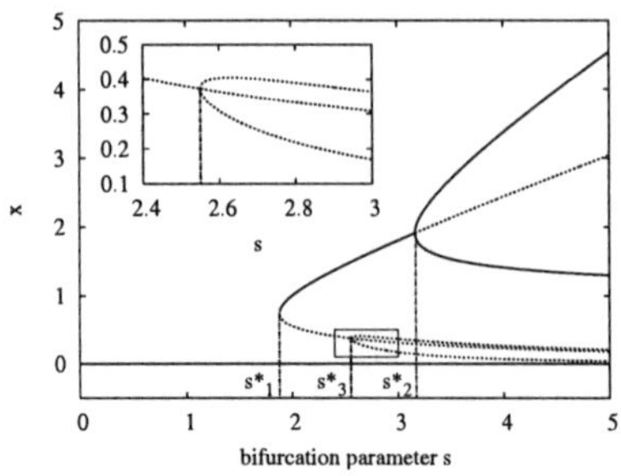

(a)

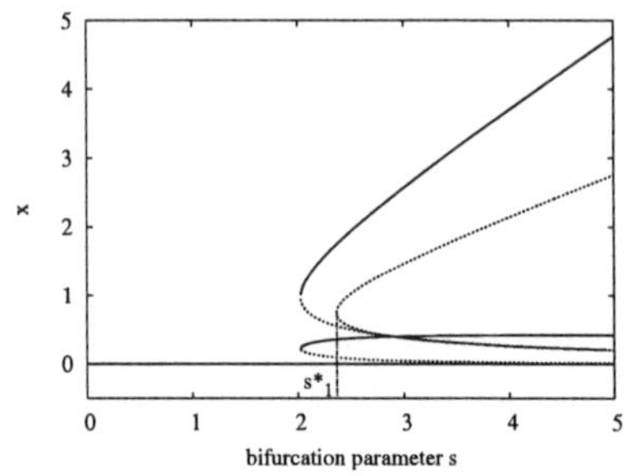

(b)

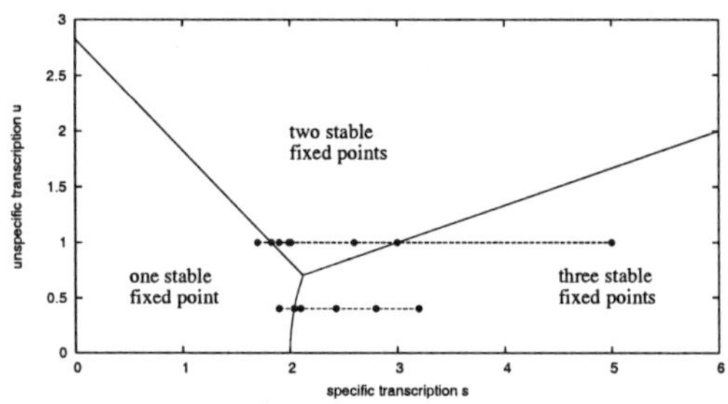

(c)

Figure 6. (a, b) Bifurcation diagrams x vs. s with $k_u = 0.8, k_r = 0, u = 1$ (a) and $u = 0.4$ (b). The stability of the steady states is coded as follows: solid line - stable; dashed line - unstable. (c) Phase space diagram u vs. s with $k_u = 1, k_r = 0$.

translocation (generating the so called Philadelphia(Ph)-Chromosome) in this cell, its whole progeny (referred to as the malignant clone) contains the oncogenic BCR-ABL1 fusion-gene. This gene induces a massive expansion of the malignant clone and results in the ultimate displacement of normal hematopoiesis. The precise mechanisms leading to the competitive advantage of the malignant (BCR-ABL1 - positive) cells are yet unknown.

The current standard treatment of CML patients is the application of the tyrosine kinase inhibitor *imatinib*[34], a drug that selectively acts on BCR-ABL1 positive cells. Experimental results on *imatinib* suggest that it induces a proliferation inhibition as well as an increase of the apoptotic rate of BCR-ABL1 positive cells. However, resulting dynamic effects of *imatinib* on stem cells are currently not sufficiently understood and it is our objective to test potential treatment options with respect to their effect on BCR-ABL1 transcript dynamics.

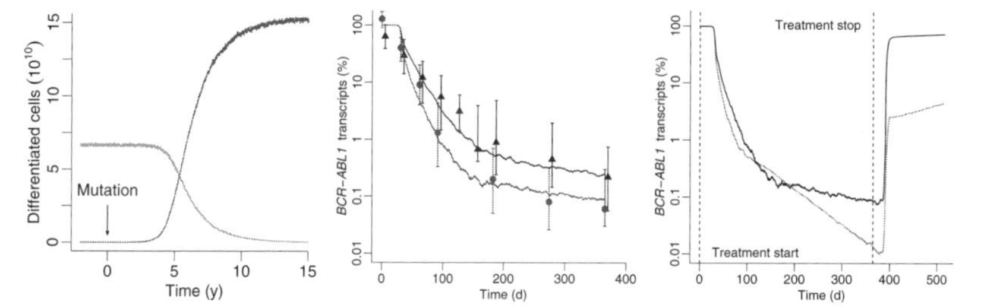

(a) (b) (c)

Figure 7. (a) Simulation of CML genesis: given are the absolute numbers of normal Ph^- (gray) and malignant Ph^+ (black) differentiated cells following a single-cell mutation. (b) BCR-ABL1 transcript dynamics under *imatinib* treatment. Data points: median and interquartile range of BCR-ABL1 transcript levels in peripheral blood; shown are two independent study populations: BCR-ABL1/BCR ratios of 68 CML patients under *imatinib* published in[35] (gray) and BCR-ABL1/ABL1 ratios of 69 *imatinib*-treated CML patients from the German cohort of the IRIS trial (black). Solid lines represent simulation results of BCR-ABL1 levels using slightly different values for the assumed *imatinib* specific killing probability. (c) Predicted effects for the treatment with *imatinib* alone (black line) in comparison with a combination of *imatinib* with an additional hourly activation of 0.1% of all dormant cells into cell cycle (gray line).

In the previously described model context (cf. section 3.1), CML is explained by the assumption of differences of normal and malignant stem cells with respect to their cell-cell / cell-microenvironmental interactions. Particularly, quantitative differences in the transition characteristics f_α and f_ω are assumed (for particular parameter configuration the reader is referred to Roeder et al., 2006[33]). These differences induce an advantage of malignant cells in the competition for the regeneration supporting GE-A. Figure 7a shows a simulation example of CML development based on these assumptions, reproducing the typical CML latency time of 5-7 years. The clinically observable tumor load (determined by the *BCR-ABL1* transcript levels in the peripheral blood) is approximated in the model by the proportion of malignant cells within the compartment of differentiated model cells. Particularly, both clinically used ratios to measure the BCR-ABL1 transcript levels, as there are *BCR-ABL1/BCR* or *BCR-ABL1/ABL1*, are estimated by

$$\frac{\text{number of } \textit{BCR-ABL1} \text{ positive cells}}{\text{number of } \textit{BCR-ABL1} \text{ positive cells} + 2 \cdot \text{number of normal cells}} \cdot 100\%.$$

To investigate the dynamic effects of *imatinib* treatment, the following assumption are made:

- *Imatinib* affects BCR-ABL1 positive cells only.
- *Imatinib* induces a change of cell cycle activation (modeled by altering f_ω). This alteration of f_ω does not occur simultaneously for all cells; each *BCR-ABL1* positive cell has a certain probability to be affected by *imatinib* within one time step (inhibition intensity r_{inh}). Therefore, the proportion of cells with an altered transition characteristic is increasing over time until all *BCR-ABL1* positive cells are affected.
- *Imatinib* induces a degradation of proliferating cells in the stem cell compartment. This is modeled by killing proliferating stem cells with a probability r_{deg} per time step.

Based on these assumption, the simulation results conform with the clinically observed BCR-ABL1 transcript dynamics of CML patients under *imatinib* (Fig. 7b). The clinical data is based on two independent data sets: 68 patients from a study published by Michor et al.[35] as well as data from 69 patients recruited within the German cohort of the IRIS study[36]. The slight difference in the two BCR-ABL1 dynamics is most likely induced by the exclusion of patients with transiently increasing BCR-ABL1 transcript in the cohort published by Michor et al. It is explained within our model by a small quantitative variation in the median *imatinib* effect.

Our assumptions that *imatinib* acts selectively on proliferative (stem) cells suggest the potential of clinical interventions to augment current therapeutic results by influencing the proliferative status of stem cells. Particularly, our simulations of a combination treatment, consisting of *imatinb* + unselective proliferation stimulation, predict a more efficient reduction of tumor load than the treatment by *imatinib* alone (Figure 7c). Herein, the proliferation stimulation had been modeled by an additional activation of GE-A stem cells to GE-Ω at a fixed rate of 0.1% per time step, independent of their cell type (normal, malignant) and independent of the system state (i.e. total cell numbers). It should be noted that the predicted treatment benefit might be diminished by the existence of *imatinib* resistant clones, which frequently arise in CML patients if treated with *imatinib* over a longer time period[37]. For a more detailed description of our modeling results on *imatinib* treated CML as well as for model parameters and technical details the reader is referred to Roeder et al., 2006[33].

Acknowledgment

This work comprises results from different projects and the author would like to acknowledge all collaborators involved in these projects, particularly Markus Loeffler, Ingmar Glauche, Matthias Horn, and Katrin Braesel.

References

1. R. Lanza et al. (eds), *Handbook of Stem Cells* Vol. **1,2**, Elsevier Academic Press (2004).
2. C. Potten and M. Loeffler, *Development* **110**:1001-1020 (1990).
3. I. Weissman, *Cell* **100**, 157-168 (2000).
4. M. Loeffer and I. Roeder, *Cells Tissues Organs* **171**(1), 8-26 (2002).
5. W. Heisenberg, *Zeitschrift fr Physik* **3** 197 (1927).
6. A. Rolink, S. Nutt, F. Melchers et al. *Nature* **401**, 603-606 (1999).
7. T. Sato, J. Laver, M. Ogawa, *Blood* **94**, 2548-2554 (1999).
8. P. Quesenberry, H. Habibian, M. Dooner et al., *Blood Cell. Mol. Dis.* **27**, 934-937 (2001).
9. P. Quesenberry, M. Abedi, J. Aliotta et al., *Blood Cell. Mol. Dis.* **32**, 1-4 (2004).
10. G. Bradford, B. Williams, R. Rossi et al., *Exp. Hematol.* **25**, 445-453 (1997).
11. S. Cheshier, S. Morrison, X. Liao et al. *Proc. Natl. Acad. Sci. USA* **96**, 3120-3125 (1999).
12. M. Goodell, *Blood* **94**, 2545-2547 (1999).
13. G. Van Zant, K. Scott-Micus, B. Thompson, et al. *Exp. Hematol.* **20**, 470-475, (1992).
14. J. Abkowitz, S. Catlin, P. Guttorp, *Nat. Med.* **2**, 190-197 (1996).
15. L. Kamminga, L. Akkerman, E. Weersing et al., *Exp. Hematol.* **28**, 1451-1459 (2000).
16. C. Jordan, I. Lemischka, *Genes Dev.* **4**, 220-232 (1990).
17. N. Drize, Y. Olshanskaya, L. Gerasimova et al., *Exp. Hematol.* **29**, 786-794 (2001).
18. K. Kuramoto, D. Follman, P. Hematti et al., *Blood* **104**, 1273-1280 (2004).
19. Z. McIvor, C. Heyworth, B. Johnson et al., *Br. J. Haematol.* **110**, 674-681 (2000).
20. P. Zhang, G. Behre, J. Pan et al., *Proc. Natl. Acad. Sci. USA* **9**, 8705-8710 (1999).
21. C. Heyworth, S. Pearson, G. May et al., *Embo. J.* **21**, 3770-3781 (2002)
22. M. Cross, T. Enver, *Curr. Opin. Genet. Dev.* **7**, 609-613 (1997).
23. C. Bjornson, R. Rietze, B. Reynolds et al. *Science* ; **283**, 534-537 (1999).

24. J. Grove, E. Bruscia, D. Krause, *Stem Cells* **22**, 487-500 (2004).
25. U. Lakshmipathy, C. Verfaillie, *Blood Rev.* **19**, 29-38 (2005).
26. N. Theise, R. Harris, *Handb Exp Pharmacol.* **174**, 389-408 (2006).
27. C. Bonifer, *Gene* **238**, 277-289 (1999).
28. I. Roeder and M. Loeffer, *Exp. Hematol.* **30**(8), 853 (2002).
29. R. Schofield, *Blood Cells* **4**, 7-25 (1978).
30. L. Calvi, G. Adams, K. Weibrecht et al. *Nature* **425**, 841-846 (2003).
31. I. Roeder, L.M. Kamminga, K. Braesel et al., *Blood* **105**(2) 609-616 (2005).
32. I. Roeder and I. Glauche, *J. Theor. Biol.* **241**, 852 (2006).
33. I. Roeder, M. Horn, I. Glauche et al., *Nat. Med.* **12**(10), 1181-1184 (2006).
34. R. Hehlmann, A. Hochhaus, U. Berger et al., *Ann. Hematol.* **79**, 345-354 (2000)
35. F. Michor, T. Hughes, Y. Iwasa et al., *Nature* **435**, 1267-1270 (2005).
36. S. O'Brien, F. Guilhot, R. Larson et al., *N. Engl. J. Med.* **348**, 994-1004 (2003).
37. A. Hochhaus, P. La Rosee, *Leukemia* **18**, 1321-1331 (2004).

EMERGENCE OF A COLLECTIVE STEADY STATE AND SYMMETRY BREAKING IN SYSTEMS OF TWO IDENTICAL CELLS

RUI DILÃO

Nonlinear Dynamics Group, Instituto Superior Técnico
Av. Rovisco Pais, 1049-001 Lisbon, Portugal
E-mail: rui@sd.ist.utl.pt; ruidilao@gmail.com

We consider a system of two coupled identical cells. The dynamics of the chemical substances in the individual cells are the same, and the coupling is proportional to the differences in the concentrations of its chemical constituents. Without coupling, the cells have a unique and identical stable steady state — the quiescent state. We show that the coupled system of cells can have a new collective stable steady state, not present if the cells were uncoupled. We obtain the conditions for the emergence of this collective steady state. When the collective stable steady state exists, the concentrations of the (two) morphogens assume different values inside the cells, introducing a symmetry breaking in the chemical characterization of the cells. This is a hypothetical mechanism of developmental differentiation in systems with a small number of identical cells.

1. Introduction

In order to describe the emergence of collective effects in aggregates of identical subunits or cells, Turing[1] studied the effect of coupling in finite arrays of identical systems arranged in a periodic geometry. The cells are characterized by its chemical constituents or morphogens, and the coupling between adjacent cells is proportional to the differences in the concentrations of the chemical substances. The mathematical analysis of the dynamics of the coupled (extended) system leads to the conclusion that, depending of the coupling parameters, a stable steady state of the uncoupled identical subsystems can become unstable. According to Turing, this instability, when complemented with the non-linear characteristics of the local dynamics, can eventually be in the origin of stable stationary patterns and of travelling wave type phenomena in spatially extended systems.

In the analysis of Turing, the local systems have identical dynamics and have a stable steady state. In a finite array of cells linearly coupled by diffusion, and near the stable steady state, the time evolution of the concentrations of the morphogens is the sobreposition of time evolving eigenmodes associated with the linearized dynamical system. If the real part of the dominant eigenvalue is positive, an instability in the extended system appears. As the dominant eigenvalue can be real or complex, different types of patterns can eventually emerge in the extended system. If the dominant eigenvalue is real, Turing presented numerical evidence that asymptotically stable pattern can develop. If the dominant eigenvalue is complex, the asymptotic state has a wave type behaviour. In the Turing original approach, these effects are eventually dependent of the type of non-linearity of the local dynamics. If the dominant eigenvalue of the extended dynamical system is real, the extended system is said to have a Turing instability.

The Turing instability has been analyzed by several authors. For example, in systems of reaction-diffusion partial differential equations, Othmer and Scriben[2] have done the linear analysis of the Turing instability for several topologies of the extended system. Their approach essentially, confirms the results of Turing and relates the Turing instability with the spectrum of the Laplacian (diffusion term). If the local dynamics of the system of reaction-diffusion equations has a limit cycle in phase space, Kopell and Howard[3], analyzed the possibility of travelling wave phenomena. In Dilão[4] a necessary and sufficient conditions for the appearance of the Turing instability in two-component systems of reaction-diffusion equations has been obtained.

In general, in reaction-diffusion systems, the asymptotic states can be strongly dependent on the initial conditions of the extended system[5], and it is not known if there is any relation between the attractors of the uncoupled and coupled systems. For the finite dimensional case, Smale[6], analyzed the interaction between two cells and found that, if we have at most four morphogens, a two cells system will oscillate for a choice of the coupling diffusion parameters. In this case, the oscillatory characteristics of the coupled system are already present in the uncoupled one-cell systems, being difficult to speak about emergent phenomena.

Our goal here is to investigate on the possibility of having a steady state of the coupled system that is not an attractor for the dynamics of the individual systems. If this is possible, then it makes sense to speak about emergent phenomena as a result of the aggregation and interaction of identical subsystems.

In order to pursue this program, we construct the simplest possible collective system. This system is formed by two cells in contact. The interaction between the cells is made through a common wall where the flow of mass is made possible. Then, we analyze the dynamics of the system described by a generic vector field obtained from the mass conservation law, and we determine the conditions for the emergence of a collective steady state. This is done in section 2. To characterize chemically the cells, in section 3, we introduce an auto-catalytic kinetic model with two chemical substances, and we analyze the bifurcations of the dynamics of the two-cell system. We show that new stable steady states appear by a pitchfork bifurcation (emergent steady states). Finally, in section 4, we present the main conclusions of the paper, and we discuss the significance of this approach to developmental biology.

2. The dynamics of the two-cell system

We consider two identical cells in contact through a common wall. We suppose that the state of each cell is described by the concentrations of two chemical substances or morphogens X and Y. As the cells are considered equal, in each cell, the chemical processes involving the two morphogens are described by the same vector field. We also assume that the chemical variables in the system other than X and Y are constant inside the cells, and the concentration of the morphogens evolve in time to a stable steady state. We denote by (X^*, Y^*) the steady state value, and to simplify the notation, we can make the choice $(X^*, Y^*) = (0, 0)$. As X and Y evolve in time according to the mass action law[7], the rates of change of X and Y are proportional to sums of products of concentrations of chemical substances. Therefore, the vector field describing the time evolution of the two morphogens has the generic form,

$$\begin{aligned}\frac{dX}{dt} &= f(X,Y) = aX + bY + f^*(X,Y)\\ \frac{dY}{dt} &= g(X,Y) = cX + dY + g^*(X,Y)\end{aligned} \tag{1}$$

where a, b, c and d are constants, and the functions $f^*(X,Y)$ and $g^*(X,Y)$ are homogeneous polynomial of degree greater than one.

Denoting by A the Jacobian matrix of the vector field (1) evaluated at the fixed point $(0, 0)$, if $DetA > 0$ and $TrA < 0$, then this fixed point is asymptotically stable.

The constants a, b, c and d in (1) are the marginal reaction rates, and the role of the morphogens in the chemical reactions is classified according to the signs of these constants. For example, if b is positive, we say that Y activates the production of X, or, simply, Y is an activator. Similarly, if c is negative, X represses the production of Y, or X is a repressor. If the sign of a is positive X is a self-activator, and if the signs of a is negative X is a self-repressor. These definitions have only a local meaning. In a chemical reaction, it can happen that the signs of the marginal reaction rates are different near different fixed points in phase space. In this case, the chemical reaction behaves differently across the phase space.

We consider that the two cells are in contact through a common wall and the morphogens flow from one cell to the other cell through specialized channels on the wall. Assuming that the flow occurs from regions with larger concentrations to regions with lower concentrations, the concentration of morphogens in the two cells evolve in time according to the system of equations,

$$\begin{aligned}\frac{dX_1}{dt} &= f(X_1, Y_1) + \mu(X_2 - X_1)\\ \frac{dY_1}{dt} &= g(X_1, Y_1) + \nu(Y_2 - Y_1)\\ \frac{dX_2}{dt} &= f(X_2, Y_2) + \mu(X_1 - X_2)\\ \frac{dY_2}{dt} &= g(X_2, Y_2) + \nu(Y_1 - Y_2)\end{aligned} \tag{2}$$

where μ and ν are constant flow rates or diffusion coefficients, and the quantities X_i and Y_i refer to the concentration of morphogens inside cell number i.

As, X and Y are different chemical components, and the intercellular communication is done by specialized channels, it is natural to expect that the flow rates or diffusion coefficients are different, $\mu \neq \nu$.

The two-cell dynamical system, described by (2), has a fixed point at $(X_1, Y_1, X_2, Y_2) = (0, 0, 0, 0)$. Now, we want to investigate the change in the stability of the zero fixed point of the two-cell system, when we vary the coupling parameters μ and ν. By construction, if $\mu = \nu = 0$, the fixed point $(0, 0, 0, 0)$ is stable, provided $DetA > 0$ and $TrA < 0$.

To analyze the stability of the zero fixed point of the two-cell system, we linearize (2) around $(0, 0, 0, 0)$. By (1), we obtain the linear system,

$$\frac{d}{dt}\begin{pmatrix} X_1 \\ Y_1 \\ X_2 \\ Y_2 \end{pmatrix} = \begin{pmatrix} a-\mu & b & \mu & 0 \\ c & d-\nu & 0 & \nu \\ \mu & 0 & a-\mu & b \\ 0 & \nu & c & d-\nu \end{pmatrix}\begin{pmatrix} X_1 \\ Y_1 \\ X_2 \\ Y_2 \end{pmatrix} := M\begin{pmatrix} X_1 \\ Y_1 \\ X_2 \\ Y_2 \end{pmatrix} \tag{3}$$

where a, b, c and d are the marginal reaction rates of the local kinetics near the fixed point $(0,0)$. To determine the eigenvalues of the matrix M, we write,

$$M - \lambda I_4 = \begin{pmatrix} A - B - \lambda I_2 & B \\ B & A - B - \lambda I_2 \end{pmatrix} \tag{4}$$

where,

$$A = \begin{pmatrix} a\, b \\ c\, d \end{pmatrix}, \quad B = \begin{pmatrix} \mu\, 0 \\ 0\, \nu \end{pmatrix} \tag{5}$$

and I_n is the $n \times n$ identity matrix. The matrix A is the matrix of the marginal reaction rates near the zero fixed point of system (1). As we show in the Appendix A,

$$Det(M - \lambda I_4) = Det\,(A - \lambda I_2)\,.Det\,(A - 2B - \lambda I_2))\ . \tag{6}$$

Therefore, the characteristic polynomial of the matrix M is the product of two polynomials of degree two, and one of the polynomials is the characteristic polynomial of the matrix A. By (6), the eigenvalues of the matrix M are readily obtained, and we have,

$$\begin{aligned} \lambda_1 &= \tfrac{1}{2}TrA - \tfrac{1}{2}\sqrt{(TrA)^2 - 4DetA} \\ \lambda_2 &= \tfrac{1}{2}TrA + \tfrac{1}{2}\sqrt{(TrA)^2 - 4DetA} \\ \lambda_3 &= \tfrac{1}{2}Tr(A-2B) - \tfrac{1}{2}\sqrt{(Tr(A-2B))^2 - 4Det(A-2B)} \\ \lambda_4 &= \tfrac{1}{2}Tr(A-2B) + \tfrac{1}{2}\sqrt{(Tr(A-2B))^2 - 4Det(A-2B)}\,. \end{aligned} \tag{7}$$

As we are assuming that $DetA > 0$ and $TrA < 0$, by (7), $Real(\lambda_1) < 0$, $Real(\lambda_2) < 0$ and $Real(\lambda_3) < 0$. So, if λ_4 is real and positive, the zero fixed point of the two-cell system is unstable. Hence, by (7), if,

$$Det(A - 2B) = DetA - 2d\mu - 2a\nu + 4\mu\nu < 0 \tag{8}$$

then λ_4 is real and positive.

As $\mu > 0$, $\nu > 0$, $TrA = a + d < 0$, and $DetA = ad - bc > 0$, the condition (8) is verified only if, a and d have opposite signs. On the other hand, as $DetA = ad - bc > 0$, this implies that b and c must have opposite signs.

Therefore, the zero fixed point of the system of equations (2) can be destabilized by diffusion if one of the variables is a repressor and the other variable is an activator, one variable is a self-repressor, and the other variable is a self-activator. In Figure 1, we show the four possible configurations for the signs of the marginal reaction rates near the zero fixed point.

$$\begin{pmatrix} -(sR) & -(R) \\ +(A) & +(sA) \end{pmatrix} \begin{pmatrix} -(sR) & +(A) \\ -(R) & +(sA) \end{pmatrix} \quad \nu > \mu$$

$$\begin{pmatrix} +(sA) & -(R) \\ +(A) & -(sR) \end{pmatrix} \begin{pmatrix} +(sA) & +(A) \\ -(R) & -(sR) \end{pmatrix} \quad \mu > \nu$$

Figure 1. Signs of the marginal reaction rates enabling the destabilization by diffusion of the zero steady state of the two cells system. In the four cases shown, for the zero fixed point of the coupled system to be unstable, the diffusion coefficient of the self activator must be larger than the diffusion coefficient of the self inhibitor.

Inequality (8) gives a simple condition for instability of the zero fixed point of the coupled two-cell system. From (8), it follows that, if $a > 0$, $d < 0$, $DetA > 0$, $0 < \mu < a/2$, and,

$$\nu > \frac{DetA - 2d\mu}{2a - 4\mu} \tag{9}$$

the fixed point $(X_1, Y_1, X_2, Y_2) = (0, 0, 0, 0)$ is unstable. Similarly, if $d > 0$, $a < 0$, $DetA > 0$, $0 < \nu < d/2$, and,

$$\mu > \frac{DetA - 2a\nu}{2d - 4\nu} \tag{10}$$

the $(0, 0, 0, 0)$ fixed point is unstable.

In Figure 2, we show the regions of stability of the $(0, 0, 0, 0)$ fixed point as a function of the diffusion coefficients μ and ν, calculated from (9) and (10).

The main result of this section is summarized in the following theorem:

Theorem 2.1. *We consider systems (1) and (2), with $TrA < 0$ and $DetA > 0$. For the choice, $\mu = \nu = 0$, the zero steady state of system (2) is stable. If the signs of the elements of the diagonal and anti-diagonal entries of the matrix A have opposite signs, $\mu > 0$, $\nu > 0$, and,*

$$DetA - 2d\mu - 2a\nu + 4\mu\nu < 0$$

then, the zero steady state of system (2) is unstable.

If the zero fixed point of the system of equations (2) is unstable, the local structure of the flow in phase space near the zero fixed point splits into a direct sum of a stable and an unstable manifold[8]. By (7), the stable manifold has dimension three and the unstable manifold is one-dimensional.

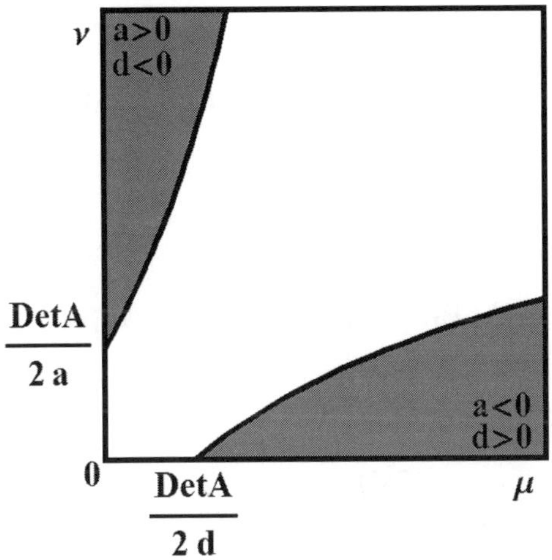

Figure 2. Bifurcation diagram of the zero fixed point of system (2), as a function of μ and ν. For μ and ν in the gray regions, the zero fixed point of system (2) is unstable.

Therefore the unstable fixed point is of saddle-node or saddle-focus type. This suggests that, in the coupled system, by variation of μ and ν a new fixed point appears. To analyze the bifurcations that appear by changing the coupling parameters, we study a specific example.

3. The emergent steady states and symmetry breaking

In order to test the results of the previous section, we consider the Brusselator model[9] of chemical kinetics. This model consists on the kinetic mechanisms,

$$\begin{array}{ll} A & \longrightarrow^{k_1} U \\ B+U & \longrightarrow^{k_2} V+D \\ 2U+V & \longrightarrow^{k_3} 3U \\ U & \longrightarrow^{k_4} E \end{array}$$

where k_1, k_2, k_3 and k_4 are positive rate constants, and A, B, U and V represent different chemical substances, and U is autocatalytic. By the mass action law[7], and assuming that A and B are constants (open system), we obtain the system of equations,

$$\begin{aligned} \frac{dU}{dt} &= Ak_1 - k_4U - Bk_2U + k_3VU^2 \\ \frac{dV}{dt} &= Bk_2U - k_3VU^2 . \end{aligned} \tag{11}$$

This system of equations has a fixed point with coordinates,

$$(U_0, V_0) = (Ak_1/k_4, Bk_2k_4/(Ak_1k_3)) .$$

With the new coordinates, $X = U - U_0$ and $Y = V - V_0$, the system of equations (11) assumes the form,

$$\begin{aligned} \frac{dX}{dt} &= (Bk_2 - k_4)X + \frac{A^2k_1^2k_3}{k_4^2}Y + \frac{Bk_2k_4}{Ak_1}X^2 + \frac{2Ak_1k_3}{k_4}XY + k_3X^2Y \\ \frac{dV}{dt} &= -Bk_2X - \frac{A^2k_1^2k_3}{k_4^2}Y - \frac{Bk_2k_4}{Ak_1}X^2 - \frac{2Ak_1k_3}{k_4}XY - k_3X^2Y . \end{aligned} \tag{12}$$

The system of equations (12) describes the dynamics in each cell in the system of two coupled cells. Assuming that all the constants k_1, k_2, k_3, k_4, A and B are positive, if $B < k_4/k_2 + A^2k_1^2k_3/(k_2k_4^2)$, the system of equations (12) has a stable fixed point for $(X, Y) = (0, 0)$, a stable focus, provided $(A^2k_1^2k_3/k_4^2 - bk_2 + k_4)^2 < 4A^2k_1^2k_3/k_4$. If $B > k_4/k_2 + A^2k_1^2k_3/(k_2k_4^2)$, this fixed point is unstable and system (12) has a stable limit cycle in phase space. The system of equation (12) has a supercritical Hopf bifurcation for $B = k_4/k_2 + A^2k_1^2k_3/(k_2k_4^2)$ and has no other attractors in phase space.

By (5), the matrix of marginal reaction rates of the system of equations (12) is,

$$A = \begin{pmatrix} (Bk_2 - k_4) & \frac{A^2k_1^2k_3}{k_4^2} \\ -Bk_2 & -\frac{A^2k_1^2k_3}{k_4^2} \end{pmatrix} . \tag{13}$$

If $(Bk_2 - k_4) > 0$, we are in the conditions of Theorem 2.1 (Figure 1), and the zero fixed point of the two-cell system is unstable if,

$$\begin{aligned} 0 < \mu &< \frac{Bk_2 - k_4}{2} \\ \nu &> \frac{A^2k_1^2k_3}{k_4^2} \frac{k_4 + 2\mu}{2(Bk_2 - k_4) - 4\mu} . \end{aligned} \tag{14}$$

In this case, the chemical substance X (or U) is a repressor and Y (or V) is an activator. Also, X is a self-activator and Y is a self-repressor.

For the choice of parameters, $A = 2$, $k_1 = k_2 = k_3 = k_4 = 1$, the local system (12) has a supercritical Hopf bifurcations for $B = 5$, and the instability conditions (14) become,

$$\begin{aligned} 0 < \mu < \frac{B-1}{2} \\ \nu > 4\frac{1+2\mu}{2(B-1)-4\mu}\,. \end{aligned} \tag{15}$$

Taking, B, μ and ν as free parameters, the two-cell coupled system (2) with the local vector field (12) becomes,

$$\begin{aligned} \frac{dX_1}{dt} &= (B-1)X_1 + 4Y_1 + \frac{B}{2}X_1^2 + 4X_1Y_1 + X_1^2Y_1 + \mu(X_2 - X_1) \\ \frac{dY_1}{dt} &= -BX_1 - 4Y_1 - \frac{B}{2}X_1^2 - 4X_1Y_1 - X_1^2Y_1 + \nu(Y_2 - Y_1) \\ \frac{dX_2}{dt} &= (B-1)X_2 + 4Y_2 + \frac{B}{2}X_2^2 + 4X_2Y_2 + X_2^2Y_2 + \mu(X_1 - X_2) \\ \frac{dY_2}{dt} &= -BX_2 - 4Y_2 - \frac{B}{2}X_2^2 - 4X_2Y_2 - X_2^2Y_2 + \nu(Y_1 - Y_2)\,. \end{aligned} \tag{16}$$

If conditions (15) are not verified, $(0,0,0,0)$ is the only fixed point of the system of equations (16). Taking $B = 3$ (away from the supercritical Hopf bifurcation), and choosing $\mu = 1/2$, by (15), if $\nu > 4$, the zero fixed point of the system of equations (16) is unstable. For these parameter values, the numerical analysis of the fixed points of the system of equations (16) shows that (16) has a pitchfork bifurcation[8] for $\nu = 4$, and, for $\nu > 4$, two new stable steady states (stable nodes) appear. These stable fixed points are the new steady states of the two-cell system. In Figure 3, we show the X_1 coordinate of the new fixed points as a function of the bifurcation parameter ν.

The two new stable fixed points that appear by the pitchfork bifurcation are the emerging collective steady states associated with the Brusselator model. As we have two distinct stable fixed points, inside the two cells, the steady state of each morphogen assume different values, implying that the emerged collective steady state induces a symmetry breaking in the two-cell system. It can be shown that this symmetry breaking is associated with an equivariant symmetry[8] obtained by changing the roles of the variables of the two-cell system.

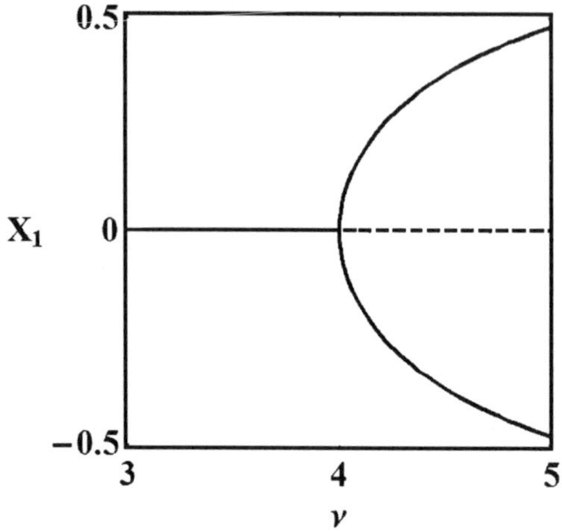

Figure 3. Bifurcation diagram of the zero fixed point of the system of equations (16), as a function of ν, and parameter values: $A = 2$, $B = 3$, $k_1 = k_2 = k_3 = k_4 = 1$ and $\mu = 1/2$. We show the X_1 coordinate of the fixed points. The zero fixed point has a pitchfork bifurcation for $\nu = 4$. For $\nu > 4$, two new stable fixed points are created by a pitchfork bifurcation, and the zero fixed point is unstable (dashed line).

4. Conclusions

We have shown that in a system of two coupled identical cells it is possible to generate a collective stable steady state that does not exists in the dynamical system associated with the individual cells. This collective steady state is a characteristics of the coupled system and suggests that the effect of coupling between identical cell can generate new states — collective states — that are not present in the dynamics of the individual cells.

One of the Roux classical experiments in embryology was to show that at the two-cell stage of the embryo, there is a symmetry breaking in the chemical characterizations of the cells. In this experiments with fertilized frog egg, the destruction of one of the cells results on the development of one-half of the embryo (Gilbert[10], pp. 593-594). On the other hand, similar experiments made by Driesch with sea urchin two-cell and four-cells embryos has shown that cell separation lead to the development of two and four complete larvae. Both experiments can be understood in the framework presented here. In the first experiments, we have a local mechanism that induces a symmetry breaking in the developmental pathway. In the second experiments, the two and four cells remained identical, leading to the development of several organisms.

Acknowledgments

This work has been partially supported by the European Commission project "GENetic NETworks: Emergence and Complexity", grant number FP6-2005-IST-5-FET / 034952 / STREP, and by a Fundação para a Ciência e a Tecnologia (Portugal) pluriannual funding grant to GDNL.

References

1. A. M. Turing, The chemical basis of morphogenesis, *Philo. Trans. Roy. Soc. Lond. Ser. B*, **237** 5-72 (1952).
2. H. G. Othmer and L. E. Scriven, Instability and dynamic patterns in cellular networks, *J. theor. Biol.*, **32**, 507537 (1971).
3. N. Kopell and L. N. Howard, Plane wave solutions to reaction-diffusion equations, *Studies in App. Math.* **52**, 291-328 (1973).
4. R. Dilão, Turing Instabilities and Patterns near a Hopf Bifurcation, *Applied Mathematics and Computation*, **164**, 391-414 (2005)
5. R. Dilão and A. Volford, Excitability in a Model with a Saddle-Node Homoclinic Bifurcation, *Discrete and Continuous Dynamical Systems - series B*, **4**, 419-434 (2004).
6. S. Smale, A mathematical model of the two cells, *in*, The Hopf Bifurcation and its Applications (J. Marsden and M. McCracken, ed.), Springer, New York, 1976.
7. A. I. Volpert, V. A., Volpert and V. A. Volpert, Travelling Wave Solutions of Parabolic Systems, American Math. Soc., Providence, 2000.
8. J. Guckenheimer and P. Holmes, Nonlinear Oscillations, Dynamical Systems, and Bifurcations of Vector Fields, Springer-Verlag, Berlin, 1983.
9. I. Prigogine and R. Lefever, Symmetry breaking instabilities in dissipative systems. II, *J. Chem. Phys.*, **48**, 1695-1700 (1968).
10. S. F. Gilbert, Developmental Biology, Fifth Edition,Sinauer Associates, Sunderland, Massachusetts, 1997.
11. F. R. Gantmacher, The Theory of Matrices, Vol.1, Chelsea Publishing Company, New York, 1960.

Appendix A

Here, we prove the determinant relation (6). The proofs will be based on the iterated application of the generalized Gauss reduction theorem for block matrices, (Gantmacher[11], pp. 45-46).

To prove (6), we consider the block matrix,

$$M' = \begin{pmatrix} A' - B & B \\ B & A' - B \end{pmatrix}$$

where A' and B are square matrices of dimension $n \times n$. Adding the second row to the first one, we obtain,

$$M'' = \begin{pmatrix} A' & A' \\ B & A' - B \end{pmatrix}$$

By the Gauss reduction theorem for block matrices, $DetM' = DetM''$. Multiplying the first row by $-BA'^{-1}$ and adding to the second one, we obtain,

$$DetM'' = \begin{vmatrix} A' & A' \\ 0 & A - 2B \end{vmatrix} = DetA.Det(A - 2B)$$

As, $DetM' = DetM''$, and as $A' = A - \lambda I_2$, we obtain (6).

SURFACE-REACTION MODELS OF PROTOCELLS

ROBERTO SERRA

Università di Modena e Reggio Emilia, Dipartimento di Scienze Sociali via Allegri 9, 42100 Reggio Emilia, Italy

TIMOTEO CARLETTI

Département de mathématique, FUNDP 8 Rempart de la Vierge, Namur B 5000, Belgium E-mail: timoteo.carletti@fundp.ac.be

IRENE POLI

Dipartimento di Statistica, Università Ca' Foscari San Giobbe - Cannaregio 873, 30121 Venezia, Italy

We present a class of models aiming to describe generic protocells hypotheses, improving a model introduced elsewhere[13]. These models, inspired by the "Los Alamos bug" hypothesis, are composed by two coupled subsystems: a self-replicating molecule- SRM- and a lipid container. The latter grows thanks to the replication of the former, which in turn can produce copies of itself thanks to the very existence of the lipid container, as it is assumed that SRMs are preferentially found in the lipid phase. Nevertheless, due to abstraction level of our models, they can be applied to a wider set of detailed protocell hypotheses. It can thus be shown that, under fairly general assumptions of generic non-linear growth law for the container and replication for the SRM, the two growth rates synchronize, so that the lipid container doubles its size when the quantity of self-replicating molecules has also doubled — thus giving rise to exponential growth of the population of protocells. Such synchronization had been postulated a priori in previous models of protocells, while it is here an emergent property. Our technique, combining a continuous time formalism, for the growth between two successive protocell divisions, and a discrete map, relating the quantity of self-replicating molecules in successive generations, allows one to derive several properties in an analytical way.

1. Introduction

The study of primitive cell-like structures, capable of self-replication and endowed with rudimentary metabolism and genetics, is important both for the studies about the origin of life and for possible industrial applications[8,12]. These so-called protocells have not yet been built, although several efforts are under way, hence the study of generic protocell models is particularly relevant.

Different modeling levels address different issues related to protocell behavior; here we concentrate on a class of models that allows us to study the *evolution* problem of a *populations* of *protocells*. Indeed, the evolvability of such populations is a key issue both in the origin of life problem and for application purposes, where by applying a suitable selection pressure one tries to develop populations specialized in a desirable task (e.g. drug design).

In order to be manageable, such models need to abstract from many details providing the further advantage that their conclusions can be relevant for many specific protocells that can be developed in the future.

We analyze and improve here a protocell model, previously introduced and studied in[13], loosely inspired by the so-called "Los Alamos bug" (briefly Labug in the following) hypothesis[6,7], which however abstracts from many details of Labug and can therefore be compatible also with other specific protocell models. The level of details can be compared to that of a model by Kaneko[3], who however considered the interaction of two molecular types, which catalyze each other's formation, in a way similar to that of nucleic acids and proteins. In the Labug hypothesis and also in our model, on the contrary, one deals with a single kind of Self-Replicating Molecule[a] — briefly SRM in the following — and a lipid micelle container, which in our model can be either a micelle or a vesicle. On the one hand, the presence of the SRM affects the growth rate of the container, e.g. by favoring the formation of amphiphiles from precursors, which exist in the neighborhood of the protocell outer surface (amphiphiles are supposed to be quickly incorporated in the lipid membrane). On the other hand, the very existence of the lipid container is a necessary condition for the working of the protocell, as it is assumed that SRM are preferentially found in the lipid phase.

So SRM catalytic activity favors the growth of the lipid container, which provides in turn the physical conditions appropriate for the replication of SRM, without being however a catalyst. The relationship between container and SRM is different from the one considered by Kaneko and therefore requires a different analysis. One of the main assumptions of our

[a]Actually, PNA: but here we will not make any specific hypothesis about the chemical identity of the self-replicating molecules, and we will only suppose that they can be found in the lipid phase.

models is that all the reactions occur close to the surface of the protocell, that's why we called them *surface reaction models.*

In our model the SRM replication rate can be linear or sublinear, as suggested by the Labug papers[6,7] and others[5,11], coupled with the container growth, which also can be non-linear. The model is continuous in time, and the dynamics is smooth during the growth of a protocell, but it is assumed that once the membrane size reaches a critical threshold, the protocell splits into two daughters units, as in the Chemoton model[2].

We will then consider the evolution of a population of protocells, ignoring for the time being mutations in the SRM. In particular the concentration of SRM affects the growth rate of the protocell itself, and therefore the doubling time of the population. Starting from the first protocell, which is born with a certain amount of SRM, the rate of replication of SRM will in general be different from that of the growth of the container. A consequence is that the amount of SRM at the protocell division time may be different from twice its initial value, so each daughter protocell could start with a quantity of SRM different from that of the parent protocell. Therefore the duplication time of the second generation will also be different from that of the first one. A natural question is how will these two quantities change in time, under the combined action of continuous growth and sudden division, hence the occurrence of a possible synchronization mechanism.

The synchronization phenomenon is a key ingredient to ensure a possible Darwinian evolution[1,10]. In fact if the two subsystems do synchronize then death by dilution[4] is avoided and moreover the population size grows exponentially, independently of the actual replication rate of the SRMs and/or the container, if no exogenous events arise. But exponential growth is necessary condition to have survival of the fittest in a competitive environment, hence selection among protocells.

Our main result is that, under very general assumptions, the container growth and the duplication of genetic material do synchronize in successive generations. Note that the problem of assuring consistency between the replication rates of the different protocell components is present in every Chemoton-like model, where protocell division is assumed to take place at a certain critical size. In the original Chemoton model[2] the issue is handled by assuming a priori a stoichiometric coupling between the different processes, while here *synchronization* is an *emergent property* of the model, derived without assuming ad hoc stoichiometric ratios.

In order to prove this result we introduce a mathematical technique which is well suited for this kind of problems: the continuous growth between two successive divisions allows conserved quantities, which are used to derive an iteration map for the value of the SRM quantity in successive generations. The map tends to a fixed point (thus proving synchronization) and provides quantitative information about the kinetics of protocell replication.

2. The basic model

Let us start by recalling the main model introduced in[13] which will be the starting point for the successive investigations.

Let C be the total quantity of "container", e.g. lipid membrane in vesicles or bulk of the micelle (since we assume constant density, it does not matter whether we measure C as mass or volume). Let us denote by S the surface area, which is a function of C: typically, S is approximately proportional to C for a large vesicle with a very thin surface (a condition which will be referred to as the "thin vesicle case"), and to $C^{2/3}$ for a micelle. Let X to denote the total quantity of genetic material in the protocell lipid phase.

We assume that only the fraction of the total X, which is near the external surface, is effective in catalyzing amphiphiles formation, that's because precursors are found outside the protocell. For the same reason this applies also to the replication of X itself, here the precursors are nucleotides. Let us denote volume concentrations with square brackets, therefore the total fraction of active X is proportional to $\delta S[X]_S$, where $[X]_S$ is the volume concentration of X in a layer of width δ below the external surface.

Let $[P]$ be the concentration, in the external solution near the protocell surface, of precursors of amphiphiles: assuming it to be buffered, then it is just a constant. If the growth of the lipid membrane and the replication of SRM take both places near the surface, we have:

$$\begin{cases} \dfrac{dC}{dt} = \alpha' S[X]_S[P] + \chi S[P] - \gamma\varphi(C) \\ \frac{dX}{dt} = \eta' S[X]_S^{\nu} - \lambda\psi(X)\,, \end{cases} \qquad (1)$$

for some positive constants denoted by Greek letters.

The first term of equation (1) is the growth due to the transformation of precursors into amphiphiles, $P \to A$, catalyzed by the X-SRM, assuming the amphiphile A to be quickly incorporated in the membrane once produced. The second term is a spontaneous growth, due to spontaneous formation of amphiphiles, while the third term accounts for possible release of amphiphiles previously incorporated in the membrane (note that the exact form for the decay term has not been specified).

The second equation in (1) describes autocatalytic growth of the X-SRM (with a possible non first order kinetics described by the exponent $\nu > 0$) with degradation, because of the last term $\lambda\psi(X)$.

We now neglect the term of spontaneous amphiphile formation, which is assumed to be smaller than the catalyzed term, we assume $[P] =$ constant, and we suppose that S is proportional to C^β (β ranging between 2/3 and 1). By slightly redefining the constants we obtain:

$$\begin{cases} \dfrac{dC}{dt} = \alpha\text{"} C^\beta [X]_S - \gamma\varphi(C) \\ \frac{dX}{dt} = \eta\text{"} C^\beta [X]_S^\nu - \lambda\psi(X) \,. \end{cases}$$

But $[X]_s$ is proportional to the concentration of X in the whole lipid phase, which is[b] X/C. Therefore, again incorporating constant terms in the rate constants:

$$\begin{cases} \dfrac{dC}{dt} = \alpha X C^{\beta-1} - \gamma\varphi(C) \\ \frac{dX}{dt} = \eta X^\nu C^{\beta-\nu} - \lambda\psi(X) \,. \end{cases} \tag{2}$$

To get a feeling for the behavior of equations (2), let us consider the growth of a vesicle container with a very thin membrane ($\beta \simeq 1$) in the case where X is constant and $\phi(C) = C$. Then the first equation rewrites:

$$\frac{dC}{dt} = k - \gamma C \,,$$

where $k = \eta X_0$ is a constant, X_0 being the initial concentration of SRM. This equation can be explicitly solved and thus we can describe the growth of the lipid container up to its asymptotic value k/γ (provided that the initial value C_0 is smaller than k/γ).

We will assume that the protocell breaks into two identical daughters units when its size reaches a certain threshold θ. Moreover, we will assume that the growth is essentially exponential, i.e. that the rate limiting steps in Eq. (2) above do not play a significant role when $C < \theta$. Therefore the growth of a protocell up to its critical size is approximately ruled by the following equations (coming back to a generic container and non constant

[b]We assume here that transport in the lipid phase is extremely fast, leading to homogeneous concentrations of SRM in the whole vesicle membrane or micelle.

X):

$$\begin{cases} \dfrac{dC}{dt} = \alpha X C^{\beta-1} \\ \frac{dX}{dt} = \eta X^{\nu} C^{\beta-\nu} \,. \end{cases} \tag{3}$$

This system of equations (3) will be the starting point for our further analysis in the next sections.

Let just observe that in the case two different, non-interacting SRMs were present in the same protocell, the model (3) can be generalized into:

$$\begin{cases} \dfrac{dC}{dt} = \alpha' C^{\beta-1} X + \alpha" C^{\beta-1} Y \\ \frac{dX}{dt} = \eta' C^{\beta-\nu} X^{\nu} \\ \frac{dY}{dt} = \eta" C^{\beta-\nu} Y^{\nu} \,, \end{cases} \tag{4}$$

assuming the general growth rate of the container depending on both the SRMs.

These two models, (3) and (4) have been introduced in[13] and extensively analyzed there, we thus invite an interested reader to refer to them for further details. Nevertheless, for the sake of completeness, we will report in the next paragraph some relevant results obtained in[13].

2.1. *Summary of some known facts concerning the basic model*

One main feature of our model is that is it able to provide a unified treatment of both micelle and vesicle cases. More precisely it has been proved that, up to an appropriate non-linear rescaling of time, the behavior of the micelle model and the thin vesicle case are asymptotically qualitatively the same. Thus all our computations we will be explicitly ruled out in the thin vesicle case, $\beta = 1$, because computationally simpler.

Let us sketch this proof here for sake of completeness. Indeed let us observe that $C(t)$ is positive for any finite t, so one can define a new time by:

$$\tau = \tau_0 + \int_0^t C(s)^{\beta-1} ds \,,$$

note that $d\tau/dt = C(t)^{\beta-1}$. Let now $\omega(t)$ be a quantity which satisfies a differential equation of the form,

$$\frac{d\omega(t)}{dt} = C(t)^{\beta-1} F(\omega(t)) \,,$$

for an arbitrary function F. Define $\psi(\tau) \equiv \omega(t(\tau))$, i.e. the same quantity ω but read in the new time variable, then its evolution is given by:

$$\frac{d\psi(\tau)}{d\tau} = \frac{dt}{d\tau}\frac{d\omega(t)}{dt} = F(\psi(\tau)) .$$

Let us now apply this idea to equation (3) by defining $c(\tau) = C(t(\tau))$, $x(\tau) = X(t(\tau))$: in terms of the new variables, using the previous remarks, these equations become:

$$\begin{cases} \dfrac{dc}{d\tau} = \alpha x \\ \dfrac{dx}{d\tau} = \eta x , \end{cases}$$

which can be considered as the equations describing the container growth and SRM replication in a vesicle protocell.

To summarize the known results let us consider separately the case where the growth of the self-replicating molecules is linear, from the one where it is sublinear. In the former case it has been proved that:

i) When only one kind of SRM is present, then the doubling time depends only upon the rate constant for self-replication (so if there are two kinds of protocells, one with higher α and lower η than the other, the former will eventually be outperformed by the latter);
ii) If there are two different SRMs in the same protocell, the one, which is slower in replicating itself, vanishes in the long time limit, even if it can provide a faster growth rate for the container (and in the case of fast parasites these will dominate and lead to halting the growth).

In the case where the growth of the self-replicating molecules is sublinear, it has been shown that:

iii) The synchronization still occurs in the case of only one kind of SRM;
iv) When two different SRMs exist in the same protocell, the ultimate fate of the system is coexistence of both SRMs, reaching fixed ratios.

3. Non-linear growths for the container

The main hypothesis used to derive the model (3) is that the container growth is linear in the concentration of X-SRM. This can be considered true in first approximation if the involved concentrations are small; on the other hand some non-linear phenomena can occur when concentrations increase. The main goal of this section is to prove that our analysis can be extended as to consider generic growth rates for the container, in particular also an auto inhibitory effect can be taken into account: large concentration of SRM can stop the container growth.

Let the container growth be described by some *positive* function $\psi(s)$ of a real positive variable s, and assume the following:

a) $\psi(0) = 0$, namely the container doesn't grow if there are no SRM at all;
b) There exists a positive constant L, such that for all $s > 0$, we have: $\psi(s) \leq L$, namely the instantaneous growth rate of the container is always finite.

Assuming a linear reproduction law for the SRM, model (3) can thus be generalized into:

$$\begin{cases} \dfrac{dC}{dt} = C^{\beta}\psi(\dfrac{X}{C}) \\ \frac{dX}{dt} = \eta X C^{\beta-1} . \end{cases} \tag{5}$$

The main result of this section is that synchronization is a emergent property of our model once non-linear growth for the lipid container is take into account, as stated by the following:

Theorem 3.1. *Let us denote by X_k the initial amount of X-SRM, inside the container, at the kth division and let $\eta > 0$ and $2/3 \leq \beta \leq 1$ be assigned positive constants. Then under the previous assumptions of the function ψ, we have the following mutually exclusive results:*

(1) If $\eta > \psi(\xi)$ for all ξ, then $X_k \to \infty$ when k increases.
(2) There exist $N \geq 1$ positive values ξ_i such that $\psi(\xi_i) = \eta$ assume moreover such roots to be transversal [c] and ordered in increasing magnitude, $\xi_1 < \ldots < \xi_i < \ldots < \xi_N$. Then there are $N' = [(N+1)/2]$ possible asymptotic values for X_k: $\Xi_{2l-1} = \theta\xi_{2l-1}/2$ for $l = 1, \ldots, N'$. More precisely the actual value is fixed by the initial condition: if X_0 belongs to (Ξ_{2l-2}, Ξ_{2l}), for some $l = 1, \ldots, N'$, $X_k \to \Xi_{2l-1}$.

The proof of this result will be given in the next paragraph.

[c]This means that the function $\eta - \psi(\xi)$ changes signs at $\xi = \xi_i$.

3.1. *Synchronization for generic non-linear growth of lipid containers*

To analyze model (5) let us introduce an auxiliary variable $\xi(t) = X(t)/C(t)$, which allows to rewrite it as follows:

$$\begin{cases} \dfrac{dC}{dt} = C^{\beta}\psi(\xi) \\ \frac{dX}{dt} = C^{\beta-1}\xi(\eta - \psi(\xi))\,. \end{cases} \tag{6}$$

Let us distinguish to cases, according to η is larger than $\psi(\xi)$ for all ξ or there exists at least one positive values ξ_1 such that $\psi(\xi_1) = \eta$.

In the first case, the second equation in (6) implies that $\xi(t)$ is an increasing function, except the trivial case: $\xi(0) = 0$, which means complete absence of SRM at the beginning, and thus it can be discarded. Hence for all positive t, we have $\xi(t) > \xi(0)$. On the other hand the first equation in (6) implies that also $C(t)$ is an increasing function, and thus starting from $C(0) = \theta/2$, there always exists a positive time T, such that $C(T) = \theta$. An estimate for T can be obtained using assumption b) on ψ, in fact the first equation in (6) gives:

$$\frac{dC}{dt} = C^{\beta}\psi(\xi) \leq LC^{\beta}$$

thus by simple integration we obtain:

$$\begin{cases} \beta < 1 \Rightarrow T \geq \dfrac{\theta^{1-\beta}}{L(1-\beta)}(1 - \dfrac{1}{2^{1-\beta}}) \\ \beta = 1 \Rightarrow T \geq \frac{1}{L}\log 2\,. \end{cases}$$

Back in the original variables, X and C, we get:

$$X(0) = C(0)\xi(0) = \frac{\theta}{2}\xi(0) \quad \text{and} \quad X(T) = C(T)\xi(T) = \theta\xi(T)$$

and recalling the halving hypothesis at the division: each offspring will start with an initial concentration of SRM equal to $X_1 = X(T)/2$, we thus obtain:

$$X_0 = \frac{\theta}{2}\xi(0) \quad \text{and} \quad X_1 = \frac{\theta}{2}\xi(T)$$

Hence we can conclude that if $\eta > \psi(\xi)$ for all ξ, then for any initial concentration of SRM X_0 the successive generation will start with a larger amount of SRM:

$$X_1 = \frac{\theta}{2}\xi(T) > \frac{\theta}{2}\xi(0) = X_0$$

This of course holds true for any division and thus we conclude that in this case the number of SRM grows unbounded.

Still using the new auxiliary variables, ξ and C, let us now consider the remaining case: there exist N positive values ξ_i such that $\psi(\xi_i) = \eta$. This means that the function $f(\xi) = \xi(\eta - \psi(\xi))$ has $N+1$ roots, the N ones of ψ and ξ_0=0. Each root of f corresponds to a steady solution of the second equation in (6) while discarding the division mechanism. By assumption each root is transversal and they are ordered by increasing magnitude, thus performing a local stability analysis in neighborhoods of each root, we can prove that *even indexed* roots are *unstable equilibria*, while *odd indexed* are *stable*[d].

A simple analysis of the one-dimensional system, still discarding the division mechanism, tell us that for any $\xi(0)$ in $I_{2l} = (\xi_{2l-2}, \xi_{2l})$, for some $l = 1, \dots, N'$, then $\xi(t)$ will asymptotically converge to ξ_{2l-1}, observe the minus sign in front of $\psi(\xi)$ which reverses the usual stability condition related to the sign of the first derivative. Moreover $\xi(t) - \xi_{2l-1}$ has the same sign than $\xi(0) - \xi_{2l-1}$ and $|\xi(t) - \xi_{2l-1}| < |\xi(0) - \xi_{2l-1}|$, namely the orbit never leaves the interval I_{2l}.

Let us now introduce the division mechanism and go back to the original variables. The first equation in (6) still implies again that $C(t)$ is an increasing function of time and thus there always exists a division time T, $C(T) = \theta$ if $C(0) = \theta/2$. To each roots ξ_l we associate the values $\Xi_l = \theta\xi_l/2$ and the intervals $J_{2l} = (\Xi_{2l-1}, \Xi_{2l+1})$.

The behavior in the ξ variables can be translated for the X variables, as follows. For X_0 belonging to some J_{2l}, then, calling T the division time, we have:

$$X(0) = C(0)\xi(0) = \frac{\theta}{2}\xi(0) \quad \text{and} \quad X_1 = \frac{X(T)}{2} = \frac{C(T)\xi(T)}{2} = \frac{\theta}{2}\xi(T)$$

moreover:

$$|X_1 - \Xi_{2l}| = |\frac{\theta}{2}\xi(T) - \frac{\theta}{2}\xi_{2l}| < |\frac{\theta}{2}\xi(0) - \frac{\theta}{2}\xi_{2l}| = |X_0 - \Xi_{2l}|$$

and $X_1 - \Xi_{2l}$ has the same sign that $X_0 - \Xi_{2l}$. Namely X_1 still belongs to J_{2l} and it is closer to Ξ_{2l} than X_0. The same considerations hold for the second generation, which start with an initial amount of X-SRM of X_1; we can thus construct in this way a sequence of initial amount of X-SRM at each generation X_k, converging to Ξ_{2l}. Once X_k converges to a fixed

[d]Setting $u = \xi - \xi_i$ we get from the second equation in (6):$\dot{u} = -C^{\beta-1}(u+\xi_i)\psi'(\xi_i)u + \dots$, and the stability claim is thus obtained.

value, the same holds true for the division time and thus synchronization is obtained as *emergent property* of the model, moreover the asymptotic value for the amount of X-SRM depends on η but also on the growth rate of the container.

Let us observe that the transversality assumption, even if it is generically verified - if for some value of η we have a non-transversal root, just by slightly changing η the root is preserved and it becomes a transversal one - can be relaxed without changing the analysis of the model, except the rate at which now the synchronization is reached, which will be slower in this case.

4. Non-linear replication rate for the SRM

In this section we relax the second hypothesis of the basic model (3) by allowing general non-linear rates for the replication of SRM. The linear (or the sublinear) rates analyzed in[13] representing particular cases, while the analysis described here is definitely more general and largely independent from specific hypotheses.

Let the replication rate be described by some positive function $\phi(s)$ of a real positive variable s, and assume the following:

c) $\phi(0) = 0$, namely the SRM doesn't replicate if there are no SRM at all. More precisely there exists $0 < p < 2$ such that for s sufficiently small we have $\phi \sim s^p$.

d) There exists a positive constant M, such that for all $s > 0$, we have: $\phi(s) \leq \mathrm{M}$, namely the instantaneous replication rate of SRM is always finite.

Observe that if we introduce the hypothesis that $\phi(s)$ goes to 0 when s becomes unbounded, this model can describe an auto inhibitory mechanism for the template replication: too many SRMs prevent the container growth.

Assuming a linear container growth with respect to the amount of X and the previous assumptions on the replication rate of SRMs, then model (3) can be rewritten as follows:

$$\begin{cases} \dfrac{dC}{dt} = \alpha C^{\beta-1} \\ \dfrac{dX}{dt} = C^{\beta}\phi(\dfrac{X}{C})\,. \end{cases} \tag{7}$$

The main result of this section is that synchronization is a emergent property of our model once self-replication has a generic non-linear, as stated by the following theorem which will be proved in the next paragraph.

Theorem 4.1. *Let us denote by X_k the initial amount of X-SRM, inside the container, at the kth division and let $\alpha > 0$, $\eta > 0$ and $2/3 \leq \beta \leq 1$ be assigned positive constants. Under the previous assumptions of the function ϕ, let us call Λ the set of roots of the equation $g(\zeta) = \phi(\zeta) - \alpha\zeta^2$ and assume they are all transversal. Then $X_k \to \theta\zeta_i/2$ where ζ_i is an element of Λ such that $\phi'(\zeta_i) - 2\alpha\zeta_i < 0$; toward which element actually X_k does converge depend on the initial concentration X_0.*

4.1. *Synchronization for generic non-linear growth of self-replicating molecules*

Once again to analyze this model let us introduce an auxiliary variable $\zeta(t) = X(t)/C(t)$, which allows to rewrite (7) as follows:

$$\begin{cases} \dfrac{dC}{dt} = \alpha C^{\beta}\zeta \\ \dfrac{d\zeta}{dt} = C^{\beta-1}(\phi(\zeta) - \alpha\zeta^2)\,. \end{cases} \tag{8}$$

Let us analyze this model by starting with the trivial fixed point $\zeta = 0$. Assumption c) implies that if ζ is sufficiently small, but positive, then $\phi(\zeta) - \alpha\zeta^2 \sim \zeta^p$, hence positive, namely the origin is an unstable equilibrium point.

Assumptions c) and d) imply that the functions $\phi(\zeta)$ and $\alpha\zeta^2$ have at least a non-zero intersection point; moreover all the intersection points are contained in the bounded interval $[0, \Delta]$, where $\Delta = \sqrt{M/\alpha}$. We can also assume these intersection points to be transversal ones.

Let us call ζ_i the Q distinct ($Q = 1$ is allowed) positive intersections points of the functions $\phi(\zeta)$ and $\alpha g\zeta^2$, namely the roots of the function $g(\zeta) = \phi(\zeta) - \alpha\zeta^2$. Once again let us divide these points into two groups: a first group denoted by $\Lambda^<$ is formed by those for which $\phi'(\zeta_i) - 2\alpha\zeta_i < 0$, and a second group, denoted by $\Lambda^>$ for which $\phi'(\zeta_i) - 2\alpha\zeta_i > 0$ together with $\zeta = 0$. Moreover to each ζ_i in $\Lambda^<$ we can uniquely associate two elements $\zeta_1^> < \zeta_i < \zeta_2^>$ in $\Lambda^>$ and an interval[e] $I_i = [\zeta_1^>, \zeta_2^>]$.

Discarding for the time being the division mechanism, a simple analysis of the one-dimensional system in the second equation in (8), tell us that for any $\zeta(0)$ in I_i, then $\zeta(t)$ will converge to ζ_i as t increases. Moreover $\zeta(t) - \zeta_i$

[e]In the case there exists only one root, we set $I_i = (0, \infty)$.

has the same sign than $\zeta(0) - \zeta_i$ and $|\zeta(t) - \zeta_i| < |\zeta(0) - \zeta_i|$, namely the orbit never leaves the interval I_i.

Let us now consider the division mechanism and go back to the original variables. The first equation in (8) implies again that $C(t)$ is an increasing function of time and thus there always exists a division time T, $C(T) = \theta$ if $C(0) = \theta/2$. To each root ζ_i we can associate a value $Z_i = \theta\zeta_i/2$, hence an interval $J_i = (Z_1^>, Z_2^>)$, where for $l = 1, 2$: $Z_l^> = \theta\zeta_l^>/2$, and $\zeta_l^>$ hasbeen defined previously.

The behavior of the ζ variables can be straightforwardly translated into the following one for the X variables. For X_0 belonging to some J_i then, calling T the division time, we have:

$$X(0) = C(0)\zeta(0) = \frac{\theta}{2}\zeta(0) \quad \text{and} \quad X_1 = \frac{X(T)}{2} = \frac{C(T)\zeta(T)}{2} = \frac{\theta}{2}\zeta(T)\,,$$

moreover:

$$|X_1 - Z_i| = |\frac{\theta}{2}\zeta(T) - \frac{\theta}{2}\zeta_i| < |\frac{\theta}{2}\zeta(0) - \frac{\theta}{2}\zeta_{2l}| = |X_0 - Z_i|$$

and $X_1 - Z_i$ has the same sign of $X_0 - Z_i$. Namely X_1 still belongs to J_i and it is closer to Z_i than X_0. Thus can repeat the same consideration for the second generation which start with an initial amount of X-SRM of X_1, thus we construct in this way a sequence of initial amount of X-SRM at each generation X_k, converging to Z_i.

Hence we can conclude that synchronization is achieved and it is an emergent property of the model; moreover the asymptotic value for the amount of X-SRM depends, of course, on the function ϕ but also on α, namely the "speed" of growth rate of the container.

Let us observe that the transversality assumption can be relaxed without changing the analysis of the model, the only change is in the rate at which now the synchronization is reached, which will be slower in this second case.

A "simple" function which verify our assumptions is

$$\phi(s) = \frac{s^p}{a^2 + b^2 s^{p+q}}$$

for some positive $p < 2$ and $q > 0$.

5. Conclusions

In this paper we have improved a basic model introduced in[13] to describe a class of abstract protocells hypotheses, called *surface reaction model* because the mechanisms responsible for the growth of the lipid container and the self-replicating molecules are assumed to take place near the cell surface.

Although the inspiration was drawn from the Los Alamos bug hypothesis, the high abstraction level of our models may allow their application to a broader set of detailed models.

We also introduced a powerful analytical technique to study the behavior of this class of protocell models, which combines continuum methods, used to describe the growth between two successive protocell duplications, and discrete maps which relate the initial value of the relevant quantities of two successive generations. This technique also allows us to draw conclusions on the asymptotic properties of a micelle or a thin vesicle, by analyzing the "thin vesicle" case only, i.e. $\beta = 1$.

It has been shown that, under general non-linear growth for the container or the replication rate of SRM, the replication rate becomes constant in the long time limit, which in turn implies exponential growth of the population of protocells, unless there are other limitations to growth. Synchronization of container and SRM duplication is here an emergent property, while in earlier models, like the Chemoton, it was imposed a priori through a stoichiometric coupling. This phenomenon of exponential growth for the population size could eventually produce a Darwinian selection in the group.

We recently became aware of the fact that a similar synchronization has been demonstrated by Rasmussen and co-workers in another protocell model, using a different approach[9] assuming linear growth for the container and a sublinear one for the SRM replication. This suggests that such synchronization phenomena may be "generic", i.e. common to several protocell models

In the case where the growth of the self-replicating molecules is fully non-linear, it has been shown here that several possible asymptotic values for the amount of SRM in the protocell, can be present, the one which will be chosen depend on the initial amount of SRM and on a peculiar relation between the function describing the growth rate of the SRM and the "speed" of the container growth. A similar result is obtained when considering a non-linear growth for the container.

The models we studied are quite general and they can be applied to describe several specific systems.

Acknowledgment

This work was funded by PACE (Programmable Artificial Cell Evolution), a European Integrated Project in the EU FP6-IST-FET Complex Systems Initiative. T.C. would also thank the *BIOMAT consortium* and the Belgian "*Fonds National de la Recherche Scientifique*, F.N.R.S." to provide the travel and stay funding. We also thank the participants to the PACE workshop on "Evolution and Self-assembly" held at the European Center for Living Technology on March 16-19, 2006, with which we had stimulating and useful discussions, in particular John McCaskill, Norman Packard, Steen Rasmussen and Marco Villani for their helpful comments. We also gratefully acknowledge the help of Michele Forlin in the preparation of this paper.

References

1. M. Eigen and P. Schuster, Naturwiss., **64**, 541, (1977).
2. T. Gánti, New York, Kluwer Academic/Plenum Publishers, (2003).
3. K. Kaneko, Advances in Chemical Physics **130B**, 543, (2005).
4. P.L. Luisi, F. Ferri and P. Stano, Naturwiss., **93**, 1, (2006).
5. A. Munteanu, C.S. Attolini, S. Rasmussen, H. Ziock, and R.V. Solé, submitted, (2006).
6. S. Rasmussen, L. Chen, M. Nilsson and S. Abe, Artificial Life **9,** 269, (2003).
7. S. Rasmussen, L. Chen, B. Stadler and P. Stadler, Origins Life & Evol. Biosph. **34,** 171, (2004).
8. S. Rasmussen et al, Science **303**, 963-965, (2004).
9. T. Rochelau et al, submitted, (2006).
10. E. Szathmáry and J. Maynard Smith, J. Theor. Biol., **187**, 555, (1997).
11. B. Stadler and P. Stadler, Adv. Comp. Syst. **6,** 47, (2003).
12. D. Szostak, P.B. Bartel, and P.L. Luisi, Nature **409**, 387, (2001).
13. R. Serra, T. Carletti and I. Poli, Artificial Life, **13:2**, pp 1-16, (2007)

STABILITY OF THE PERIODIC SOLUTIONS OF THE SCHNAKENBERG MODEL UNDER DIFFUSION

MARIANO R. RICARD
Faculty of Mathematics & Computer Science
Havana University, C. Habana 10400, Cuba
E-mail: rricard@matcom.uh.cu

YADIRA H. SOLANO
Faculty of Mechanical Engineering
Instituto Superior Politécnico José A. Echevarría
Calle 114, #11901, Marianao, La Habana, Cuba
E-mail: yhernandez@mecanica.cujae.edu.cu

In this paper we study the stability under diffusion of the lymit cycle solution to the Schnakenberg system. It is shown that diffusive instabilites for this periodic spatially homogeneous solution may lead to pattern formation. We explore conditions for such instabilities.

1. Introduction

A great attention from many different points of view has been given to pattern formation via Turing instabilities for the Schnakenberg system, in correspondence to many aspects to this phenomena. After Turing's paper[1] in 1952 several studies have been done about conditions for diffusive instability and its interpretation in the frame of morphogenesis. Nowadays it is well-known that reaction-diffusion models have been proposed to account for pattern formation in a wide variety of biological situations. Turing showed that it was possible for diffusion to cause an instability under conditions in which the reaction kinetics admit a linearly stable spatially uniform steady state. Such instabilities lead to spatially varying profiles in reactant concentration. It is well known that two chemical reactants have the attribute of diffusive instability only if the interaction represents an activator-inhibitor or a positive-feedback interaction[2]. Further, the diffusion coefficients of reactants must be dissimilar for diffusive instability occur. The mathematical basis for these assertions is the stability analysis of spatially homogeneous steady states so, leading to spectral analysis of operators. The Turing's standard procedure for the stability analysis is done considering only perturbation normal modes of the type

$$\overrightarrow{A} \exp i\,(kx - i\sigma t) \tag{1}$$

$\overrightarrow{A} \neq \overrightarrow{0}$, $\overrightarrow{A} \in \mathbb{R}^2$. which are sufficient to study the stability on bounded domains due to the fact that any perturbation can be written as a linear combination of basic functions in the form of Eq. (1), that is, in Fourier series. If, additionally, is required that perturbations satisfy specific boundary conditions, this represent a restriction to the set of basic functions of the form in Eq. (1). Note that such functions in Eq. (1) does not belong to any function space $\mathbf{L}_p(\Omega)$ for any $p < \infty$ if Ω be an unbounded domain, and in general cannot be considered to be small perturbations of the steady state in spite of the smallness of its non-zero amplitude. In this analysis, it is well known that rather than to study the general shape of the neutral stability curves on the plane of diffusion parameters, it is most convenient to identify the ratio $\frac{D_2}{D_1}$ as a parameter causing instabilities or, as usually says, Turing bifurcation.

Turing bifurcation, i.e. the evolution to a spatially patterned state as a certain parameter was varied, is the basic one generating spatial pattern in biology and chemistry[3], and the Schnakenberg's system[4] has a simple structure but its solutions are very consistent with the observed behavior in experiments, and has had strong influence on experimental design[2,3,4,5]. Unfortunately, only few of the reaction-diffusion models in morphogenesis exhibit patterns consistent with those in experiments. Turing model is based on the fact that spatially homogeneous equilibrium can be unstable if reactants diffuse with sufficiently distinct diffusion coefficients. As the ratio of the diffusion coefficient of the inhibitor over that of the activator increases from unity, there is a critical value at which the uniform equilibrium becomes locally unstable. Further complications on the mathematical study of the Turing bifurcation appear due to the degeneracy[6] when are considered bounded domains with particular boundaries. It has been shown that boundary conditions can have a profound effect on mode selection and robustness of patterning, at least in the 1-dimensional case.

Scant presence in literature has received the instability of the periodic solution due to Hopf's bifurcation in reaction–diffusion systems. This situation takes place when both Hopf and Turing instabilities appears simultaneously in biological or chemical reactions[7]. For this reason this kind of simultaneous instabilities are called Turing-Hopf bifurcations, which results in periodic spatial patterns with temporal oscillations. We find Turing-Hopf bifurcations in many different situations, even in semiconductor heterostructures[8]. Our main intention in this paper is to analyze, in connection with pattern formation, the orbital stability of the non-trivial spatially homogeneous periodic solution to the PDE system representing the Schnakenberg's model of chemical reaction but including now the fact that reactants may diffuse:

$$\begin{aligned} u_t &= D_1\ \Delta u + u^2 v - u + b \\ v_t &= D_2\ \Delta v + a - u^2 v\ . \end{aligned} \tag{2}$$

Here u, v are concentrations of reactants under diffusion and a, b represent the concentrations of reactants in such abundance that they can be assumed approximately constant. We denote by Δ the laplacian operator. Such periodic solutions are associated to the limit cycle which appears due to a Hopf Bifurcation as the parameters a and b varies but in the absence of diffusion. So we want to explore the influence in pattern formation of differences between the diffusion coefficients of the reactants with concentrations u and v, when they vary periodically over the limit cycle which appears due to Hopf bifurcation. From the general paper of Leiva[9] is known that for bounded domains and natural boundary conditions can be expected the existence of an open neighborhood of the ray $D_1 = D_2$ in the set of positive diffusion coefficients of the reaction diffusion system also for the case

$$\frac{\partial V}{\partial t} = \begin{pmatrix} D_1 & 0 \\ 0 & D_2 \end{pmatrix} \Delta V + G(V) \tag{3}$$

in which is proved that non-constant spatially homogeneous periodic solutions to Eq. (3) are orbitally asymptotically stable. In the mentioned paper of Leiva there is of course no mention to the shape of the open neighborhood which obviously depends on the form of the reaction terms $G(V)$. So, the periodic and homogeneous solution to system Eq. (2) is orbitally asymptotically stable if the diffusion coefficients D_1 and D_2 are sufficiently close together. More exactly, if (D_1, D_2) belongs to certain open neighborhood of the main diagonal in $\mathbb{R}_+ \times \mathbb{R}_+$, then stability is guaranteed. Here we study the question of stability of such solution on a bounded domain, considering the nonflow condition across the boundary. In the Turing analysis near the steady state is considered stable the reaction mixture in the absence of diffusion, to analyze solely the destabilizing influence of diffusion. But, if we are considering the appearance of a stable limit cycle for the reaction without diffusion then we implicitly are considering that the steady state is unstable. Nevertheless, in spite of the unstable character of the steady state, for any initial condition close to the punctual orbit the corresponding solution is bounded for all t and so, the instability of the steady state do not contribute to the formation of infinitely growing instabilities in the sense of Turing bifurcation.

In this paper we give elements to the study linear ordinary differential systems with periodic coefficients, the Flocquet theory, and also about the formation of a limit cycle due to Hopf bifurcation in the case of the Schnakenberg's model of chemical reaction. We construct explicitly an asymptotic expansion of the periodic solution to this model which is associated to the limit cycle in formation. Then, we obtain a hierarchy of systems corresponding to a formal asymptotic development of the solution in the small parameter, which is also the parameter associated to the Hopf bifurcation. The proposed here methodology, which follows the Turing procedure, may serve as a justification of Turing-Hopf bifurcations for the Schnakenberg model.

2. Preliminaries

2.1. *Flocquet theory*

Flocquet Theory is devoted to the stability analysis of solutions to ODE systems with periodic coefficients. More exactly, to systems of the form

$$\dot{X} = A\left(t\right)X \tag{4}$$

where $A\left(t\right)$ is a $n \times n$ continuous and T-periodic matrix. There it is proved that the fundamental matrix of Eq. (4) has the form

$$X\left(t\right) = P\left(t\right)C\left(t\right) \tag{5}$$

where $P\left(t\right)$ is non-singular T-periodic matrix, and $C\left(t\right) = \exp\left(Rt\right)$ and R is a constant matrix. The eigenvalues ν_j of the monodromy matrix $C\left(T\right)$ are called *multipliers* of the system Eq. (4). Of course, the asymptotic behavior of the fundamental matrix will depend on the eigenvalues of the matrix R. The eigenvalues ρ_j of the matrix R are called *Flocquet exponents.* Note that Flocquet exponents are not determined uniquely, but its real parts can be so. The connections between multipliers and the real part of Flocquet exponents are given by

$$\mathrm{Re}\rho_j = \frac{1}{T}\ln\left|\nu_j\right| \tag{6}$$

for each $j = 1, .., n$.

The following theorems are well-known and give us the connections between Floquet exponents and the solutions of the system Eq. (4):

Theorem 2.1. *For each multiplier ν there is a non-trivial solution $X(t)$ of Eq. (4) such that*

$$X(t+T) = \nu\ X(t)\,. \tag{7}$$

Further, if for any nontrivial solution $Y(t)$ takes place the relation Eq. (7), then ν should be a multiplier.

Theorem 2.2. *The linear T-periodic system has a T-periodic solution if and only if at least one of its multipliers is* 1.

Observe that, from Eq. (6), the system Eq. (4) has a T-periodic solution if at least one of its Floquet exponents is 0. To each Floquet exponent are associated one or more Jordan cells of the matrix R. Further, to assure the stability of the system Eq. (4) is necessary that Jordan cells corresponding to a Floquet exponent with zero real part have order 1. More exactly,

Theorem 2.3. *(Flocquet) The T-periodic system Eq. (4) is stable if and only if all of its Flocquet exponents have negative real part or been purely imaginary they have simple elemental divisors.*

2.2. *Limit cycles*

For a dynamical system on the plane

$$\dot{Y} = F(Y) \tag{8}$$

an isolated non-punctual closed orbit is called a *limit cycle.* Limit cycles correspond to periodic solutions and they cannot appear in linear systems. The problem of the (linear) stability of such limit cycles can be reduced by the standard procedure to the study of the stability of linear system with periodic coefficients like Eq. (4). Let $\Theta(t)$ be a nontrivial T-periodic vector solution to Eq.??dyn system) then, doing the substitution $Y = \Theta(t) + Z(t)$ we obtain the following linear equation with periodic matrix for the vector perturbation

$$\dot{Z} = J_{\Theta}(t)\,Z \tag{9}$$

where $J_{\Theta}(t)$ is the Jacobian matrix of the function F evaluated on the periodic solution $\Theta(t)$. Obviously, the periodic system Eq. (9) have the nontrivial periodic solution $\dot{\Theta}(t)$ so, one of the Flocquet exponents of the

Eq.(9) is zero. For the stability of the system we need to have also another Flocquet exponent with negative real part.

Definition 2.1. Let $\theta(t)$ be a non-constant periodic solution to Eq.??dyn system) and let Γ be the corresponding orbit, say $\Gamma = \{\theta(t) \, |t \in \mathbb{R}\}$. Then, $\theta(t)$ is orbitally asymptotically stable if there exist positive constants ρ, δ, μ such that, for any solution $\psi(t)$ to Eq. (8) such that

$$d(\psi(t_0), \Gamma) \leq \rho$$

for some t_0, then there is $h \in \mathbb{R}$ such that

$$\|\psi(t+h) - \theta(t)\| \leq \delta \exp(-\mu t)$$

for all $t \geq t_0$.

The negative parameter $(-\mu)$ can be selected in general as an upper bound of the nonzero Flocquet exponents for the system Eq.??periodic system Z). In our case, it can be taken itself as the non-zero Flocquet exponent of the system Eq. (9) if the periodic solution $\Theta(t)$ is orbitally asymptotically stable. It is known an expression of the nonzero Flocquet exponent in terms of the mean value of the trace of the Jacobian matrix

$$\rho_2 = \frac{1}{T} \int_0^T \left(\frac{\partial F_1(\Theta(s))}{\partial y_1} + \frac{\partial F_2(\Theta(s))}{\partial y_2} \right) ds \, .$$

2.3. *Hopf bifurcation*

The appearance of limit cycles for dynamical systems depending on a real parameter, say a,

$$\dot{Y} = F(Y, a) \tag{10}$$

may occur due to variation of the parameter through some critical value a_0, called *bifurcation value*[10]. Shortly, the situation may be sketched as follows: for $a < a_0$ the vector steady state Y_0 defined by $F(Y_0, a) = 0$ is unstable, but for $a > a_0$ this steady state is stable. So, the behavior of the solutions change abruptly at the bifurcation value.

Theorem 2.4. *(Hopf bifurcation, case $n = 2$) Consider the system Eq. (10) where F is a C^1-function, $Y \in \mathbb{R}^2$, $a \in \mathbb{R}$. Suppose that for each value of a the equations admit a steady state whose value may depende on a,say $\overline{Y}$. Let $J(a)$ the jacobian matrix of F at the point $\overline{Y}$. Let the eigenvalues of $J(a)$ be given by $\alpha(a) \pm i\beta(a)$, and suppose that there is a value a_0, the* *bifurcation value, such that* $\alpha(a) = 0$, $\beta(a) \neq 0$ *and as* a *varies through* a_0 *the real parts of the eigenvalues change signs. Then, one of the following posibilities takes place: A) at the value* $a = a_0$ *the steady state is a center, and thus infinitely many neutrally stable concentric closed orbits surround the point* $\overline{Y}$*. B) For values of* a *in some open set* $]a_0, c[$ *there is an isolated closed orbit surrounding* $\overline{Y}$*. The diameter of the limit cycle changes in proportion to* $|a - a_0|^{\frac{1}{2}}$*. C) For values of* a *in some open set* $]c, a_0[$ *there is an isolated closed orbit surrounding* $\overline{Y}$*. The diameter of the limit cycle changes in proportion to* $|a - a_0|^{\frac{1}{2}}$*.*

Moreover, in[10] or[2] the reader can see a criterion for stability of the limit cycles produced by Hopf bifurcation.

2.4. *The Schnakenberg kinetics*

Let us consider the Schnakenberg's modified system

$$\begin{cases} \dot{u} = u^2 v - u + b \\ \dot{v} = a - u^2 v \end{cases} \tag{11}$$

in which is included a reversible reaction. This system has a single stationary point

$$(u_0, v_0) = \left(a + b, \frac{a}{(a+b)^2}\right) . \tag{12}$$

Here the parameters a and b are both positive. The bifurcation occurs when takes place the following relation

$$\frac{a - b}{a + b} = (a + b)^2 \tag{13}$$

and, the eigenvalues at this situation will be

$$\lambda_{1,2} = \pm (a + b)\, i \,.$$

It is usual to consider $b \ll a$ in order that the bifurcation parameter still remains $a_0 \approx 1^2$. Note that for $a > a_0$ the steady state (u_0, v_0) is stable, and turns to be an unstable focus for $a < a_0$.

The Jacobian matrix of the right hand part of Eq. (11) at the steady state (u_0, v_0) is

$$J_0 = \begin{pmatrix} 2u_0 v_0 - 1 & u_0^2 \\ -2u_0 v_0 & -u_0^2 \end{pmatrix} = \begin{pmatrix} \frac{a-b}{a+b} & (a+b)^2 \\ -\frac{2a}{a+b} & -(a+b)^2 \end{pmatrix} . \tag{14}$$

Considering the new variables (x, y) defined by the relations

$$x = u - u_0$$
$$y = v - v_0$$

we reflect perturbations near the stationary point, and get the following system

$$\left\{ \begin{pmatrix} \dot{x} \\ \dot{y} \end{pmatrix} = J_0 \begin{pmatrix} x \\ y \end{pmatrix} + \begin{pmatrix} v_0 x^2 + 2u_0 xy + x^2 y \\ -v_0 x^2 - 2u_0 xy - x^2 y \end{pmatrix} \right. \tag{15}$$

having the steady state at $(0, 0)$.

3. Stability analysis to the periodic solution

3.1. *Limit cycle to the Schnakenberg system*

Let us now sketch the construction of the limit cycle. First note that the features of the nonlinear part in Eq. (15) suggest to consider the new variable

$$z = x + y \; . \tag{16}$$

Consequently, direct calculations conduce us to the relation

$$\dot{z} = -x \; . \tag{17}$$

So, from the pair of functions (x, y) can be determined $\left(z, \dot{z}\right)$ and reciprocally. Let us now consider the second order equation for the unknown z being equivalent to the system Eq. (15):

$$\ddot{z} + (a+b)^2 z = \delta \dot{z} + \gamma \left(\dot{z}\right)^2 + 2(a+b) z\dot{z} - z\left(\dot{z}\right)^2 - \left(\dot{z}\right)^3 \tag{18}$$

where

$$\delta = \frac{a-b}{a+b} - (a+b)^2$$
$$\gamma = 2(a+b) - \frac{a}{(a+b)^2}$$

and δ can be also considered as the bifurcation parameter. More exactly, bifurcation occurs when $\delta = 0$, as shown in Eq. (13). For shortly, in the following we will say that the parameters a, b correspond to a state after bifurcation if $\delta > 0$ and, we assume further that δ be a small parameter.

Let us assume that the solution of Eq. (18) represents an oscillation with small, but finite and positive, amplitude ε. Then, we do a change of variables

$$z(t) = \varepsilon\varsigma(t) \tag{19}$$

and, from Eq. (18) follows for ς:

$$\ddot{\varsigma} + (a+b)^2\varsigma = G\left(\varsigma, \dot{\varsigma}\right) \tag{20}$$

where

$$G\left(\varsigma, \dot{\varsigma}\right) = \delta\dot{\varsigma} + \varepsilon\left(\gamma\left(\dot{\varsigma}\right)^2 + 2(a+b)\varsigma\dot{\varsigma} - \varepsilon\varsigma\left(\dot{\varsigma}\right)^2 - \varepsilon\left(\dot{\varsigma}\right)^3\right).$$

In the above Eq. (20) two small parameters appear, but δ can get also negative small values. As the Eq. (20) represents the equation of a weakly nonlinear oscillator we shall apply the Krylov-Bogoliubov averaging method[11] in order to find an asymptotic approximation to the solution. So, let us consider the new variables $r = r(t)$ and $\theta = \theta(t)$ defined as follows

$$\begin{cases} \varsigma = r\cos(t+\theta) \\ \dot{\varsigma} = -r\sin(t+\theta) \end{cases}$$

then,

$$\begin{cases} \dot{r} = -\frac{1}{2\pi}\int_0^{2\pi}\sin\phi\, G(r\cos\phi, -r\sin\phi)\, d\phi \\ \dot{\theta} = -\frac{1}{2\pi r}\int_0^{2\pi}\cos\phi\, G(r\cos\phi, -r\sin\phi)\, d\phi \end{cases}$$

finally,

$$\begin{cases} \dot{r} = -\frac{r}{2}\left(\delta - \frac{3}{4}r^2\varepsilon^2\right) \\ \dot{\theta} = \frac{1}{8}r^2\varepsilon^2 \end{cases}. \tag{21}$$

From the first equation in the above Eq. (21) follows the existence of an orbitally asymptotically stable limit cycle if

$$\delta > 0 \text{ and } r^2 = \frac{4\delta}{3\varepsilon^2}. \tag{22}$$

In that case,

$$\dot{\theta} = \frac{\delta}{6}$$

from which we will obtain the angular speed of the oscillation.

We finally obtain the following uniform asymptotic expansion of the solution to Eq. (18)

$$z(t) = 2\sqrt{\frac{\delta}{3}}\cos\left(1+\frac{\delta}{6}\right)t + O(\delta)$$

and, from Eqs. (16-17) follows

$$\overline{u}(t) = u_0 + 2\sqrt{\frac{\delta}{3}}\sin\left(1+\frac{\delta}{6}\right)t + O(\delta) \tag{23}$$

$$\overline{v}(t) = v_0 + 2\sqrt{\frac{\delta}{3}}\left(\cos\left(1+\frac{\delta}{6}\right)t - \sin\left(1+\frac{\delta}{6}\right)t\right) + O(\delta) \tag{24}$$

as the components of the periodic solution $\Theta(t) = (\overline{u}(t), \overline{v}(t))$ to Eq. (11) leading to the limit cycle. Here (u_0, v_0) are the given in Eq. (12). We remark that the bifurcation parameter should remain of order $\delta = O(\varepsilon^2)$, where ε is a small parameter proportional to the diameter of the limit cycle. Furthermore, the period of the limit cycle is

$$T = \frac{2\pi}{\left(1+\frac{\delta}{6}\right)}$$

so it be a bit shorter than 2π.

Note that the stability analysis via normal modes as in the Turing's paper is well enough if we are considering bounded domains. More precisely, considering the problem defined in a bounded domain we do guarantee that any perturbation can be expanded via superposition of normal modes of the type in Eq. (1). On the other hand, the stability analysis via normal modes gives enough information about the relations between growth rate, wavelength and amplitude under which instabilities occur.

3.2. *Orbital stability in parabolic PDE*

Let us now consider the parabolic system

$$\frac{\partial Y}{\partial t} = \begin{pmatrix} D_1 & 0 \\ 0 & D_2 \end{pmatrix} \Delta Y + F(Y, a) \tag{25}$$

resulting from a dynamical system like in Eq. (10) when it is incorporated a diffusion process. Assuming that Eq. (10) have a limit cycle solution due to Hopf bifurcation as the parameter a varies we might consider the corresponding periodic nontrivial solution to Eq. (10). This solution is also a spatially homogeneous solution to Eq. (25) also satisfying Neumann boundary conditions. For a system with the form described in Eq. (25) but considering that F represents a predator-prey interaction was proved in[12] the orbital stability of the homogeneous periodic solution if $|D_1 - D_2|$ be small enough and the nonzero Flocquet exponent of the linearization around the periodic orbit is negative. Following the proof in that situation we can prove the orbital stability of the periodic solution which result from the Hopf bifurcation in our case.

Further, the concept of orbitally stable solution to the Eq. (25) is extended in a natural way considering now instead of Eq. (10) the dynamical system defined in an appropriate infinite dimensional Sobolev space, obtained when we look for the solution to Eq. (2) with Neumann conditions. More exactly, this dynamical system is defined by

$$Y_t = -AY + F(Y) \tag{26}$$

where A is the positive sectorial unbounded operator in $L_2(\Omega)$ associated to $-\Delta$ and defined on the dense set $\mathcal{D} = \left\{\phi \in H^2(\Omega) \,|\, \frac{\partial \phi}{\partial n} = 0 \text{ on } \partial\Omega\right\}$ and, the definition of orbital stability is as follows:

Definition 3.1. The spatially homogeneous periodic solution $\Theta(t)$ to Eq.(26) is orbitally asymptotically stable if there exist positive constants ρ, δ, μ such that, if $\Psi(t)(x)$ is a solution to Eq. (26) such that

$$\min_t \|\Psi(t_0) - \Theta(t)\| \leq \rho$$

for some t_0, then there is $h \in \mathbb{R}$ such that

$$\|\Psi(t+h) - \Theta(t)\| \leq \delta \exp(-\mu t)$$

for all $t \geq t_0$.

Then, it is also proved in[9] the following assertion

Theorem 3.1. *Let ρ_1, ρ_2 be the Flocquet exponents for the two-dimensional periodic system Eq. (9), with $\rho_1 = 0$ being simple and $Re\rho_2 < 0$. Then, if $|D_1 - D_2|$ be small enough follows $\Theta(t)$ is orbitally asymptotically stable for the boundary value problem Eq. (2) with Neumann boundary conditions.*

3.3. *The stability of the periodic but spatially homogeneous solution*

Let us now consider the boundary value problem Eq. (2) with Neumann conditions at the boundary:

$$\frac{\partial u}{\partial n} = \frac{\partial v}{\partial n} = 0 \text{ on } \partial\Omega \,. \tag{27}$$

We shall denote here by the same letters the unknowns in the Schnakenberg system Eq. (11) as in the system with diffusion Eq. (2). More exactly, here we are considering $u = u\,(t,x)$ and $v = v\,(t,x)$, $x \in \Omega$. Suppose we are considering small perturbations near the periodic solution $\Theta\,(t)$ to Eq. (2) due to Hopf bifurcation, which clearly satisfy the conditions Eq. (27). Denoting the corresponding perturbation by capital letters we get,

$$u\,(t,x) = \overline{u} + U\,(t,x)$$
$$v\,(t,x) = \overline{v} + V\,(t,x)$$

and linearizing we get the linear system with periodic coefficients for the perturbations

$$\begin{aligned} \frac{\partial U}{\partial t} &= D_1 \Delta U + (2\overline{u}\,\overline{v} - 1)\,U + (\overline{u})^2\,V \\ \frac{\partial V}{\partial t} &= D_2 \Delta V - (2\overline{u}\,\overline{v})\,U - (\overline{u})^2\,V \end{aligned} \tag{28}$$

where $\overline{u}\,(t)$ and $\overline{v}\,(t)$ denote the components of $\Theta\,(t)$ given in Eq. (23) and Eq. (24) respectively. We already know that the periodic solution $\Theta\,(t)$ to Eq. (2) is orbitally asymptotically stable if the nonzero Flocquet coefficient of the linear system

$$\dot{Z} = J_\Theta\,(t)\,Z$$

where $Z = (U,V)^T$ and

$$J_\Theta\,(t) = \begin{pmatrix} (2\overline{u}\,\overline{v} - 1) & (\overline{u})^2 \\ -(2\overline{u}\,\overline{v}) & -(\overline{u})^2 \end{pmatrix} .$$

We remark that from Eq. (21) we show that $\Theta\,(t)$ is a solution orbitally asymptotically stable for Eq. (11).

Remembering that $\delta > 0$ after bifurcation, let us assume that the solutions to Eq. (28) can be asymptotically developed in the small parameter δ as follows,

$$\begin{aligned} U &= U_0 + \delta^{\frac{1}{2}} U_1 + O\,(\delta) \\ V &= V_0 + \delta^{\frac{1}{2}} V_1 + O\,(\delta) \end{aligned} \tag{29}$$

then, substituting Eqs. (23-24-29) into the system Eq. (28) we get the following hierarchy of equations:

$$\begin{aligned}\frac{\partial U_0}{\partial t} &= D_1 \Delta U_0 + \left(2u_0 v_0 - 1\right) U_0 + \left(u_0\right)^2 V_0 \\ \frac{\partial V_0}{\partial t} &= D_2 \Delta V_0 - \left(2u_0 v_0\right) U_0 - \left(u_0\right)^2 V_0\end{aligned} \tag{30}$$

which corresponds to terms $O\left(1\right)$, and

$$\begin{aligned}\frac{\partial U_1}{\partial t} &= D_1 \Delta U_1 + \left(2u_0 v_0 - 1\right) U_1 \\ &\quad + \left(u_0\right)^2 V_1 + 2\left(u_0 v_1 + u_1 v_0\right) U_0 + 2u_0 u_1 V_0 \\ \frac{\partial V_1}{\partial t} &= D_2 \Delta V_1 - \left(2u_0 v_0\right) U_1 \\ &\quad - \left(u_0\right)^2 V_1 - 2\left(u_0 v_1 + u_1 v_0\right) U_0 - 2u_0 u_1 V_0\end{aligned} \tag{31}$$

corresponding to the $O\left(\delta^{\frac{1}{2}}\right)$ terms. We remark the expressions of u_1 and v_1 participating in the coefficients of the above system, which follows from Eqs. (23-24)

$$\begin{aligned}u_1 &= \frac{2}{3}\sqrt{3}\sin\left(1 + \frac{\delta}{6}\right)t \\ v_1 &= \frac{2}{3}\sqrt{3}\left(\cos\left(1 + \frac{\delta}{6}\right)t \quad \sin\left(1 + \frac{\delta}{6}\right)t\right).\end{aligned}$$

From Eqs. (30) for the leading term in the expansion of the solution to Eq. (28) follows that perturbations are governed by the same equations than those governing the perturbations to the steady state so, we expect similar conditions to the appearance of instabilities and also for Turing bifurcations. Instabilities now will have the same structure as the corresponding in the analysis near the steady state, and be only different in that the parameters a and b take different values too. More precisely, if we take perturbations of the form

$$\begin{aligned}U_0 &= A \exp i\left(kx - i\sigma t\right) \\ V_0 &= B \exp i\left(kx - i\sigma t\right)\end{aligned}$$

and substitute them in the system Eq. (30), we get an homogeneous linear system for the amplitudes. Then, imposing the determinant of the linear system to be zero this leads to the characteristic equation in the unknown σ, having coefficients depending on the values of the wavenumber k, and on the parameters in the reaction a, b. Turing instabilities will occur for the set of parameters k, a, b for which the corresponding characteristic value σ

have positive real part, $\text{Re}\sigma > 0$. At the beginning of the Hopf bifurcation the behavior of U_1, V_1 are less transcendent to the whole solutions. But, from Eq. (31) we may expect resonance or space-time chaos, so U_1, V_1 might turns to be relevant as a consequence either of growing values of the bifurcation parameter δ or of the secularity.

4. Conclusions

Note that, in spite of the unstable character of the steady state a after Hopf bifurcation, the system for the leading term in the expansion Eq. (30) shows that Turing instabilities behaves as usual before bifurcation, only taking into account the continuous change of the values u_0 and v_0 in terms of the parameters D_1, D_2, a and b in Eq. (2). So, both Hopf and Turing bifurcations superpose and produce spatial patterning and temporal oscillations. Further, we may expect circumstances in which a growing influence of the $O\left(\delta^{\frac{1}{2}}\right)$ terms in Eq. (31) occurs as time pass. The analysis of the leading perturbation term should take into account boundary conditions in the selection of normal modes, or more precisely, in the selection of appropriate basic functions to the corresponding Fourier development respect to the spatial coordinates.

We show that, assuming $0 < \delta \ll 1$, sufficiently distinct diffusion coefficients may produce patterning even in the case of unstable steady states. The appearance of these instabilities is a consequence of diffusive instabilities of the periodic spatially homogeneous solution associated to the Hopf's bifurcation limit cycle. We propose here a methodology which may serve as a justification for the formation of Turing-Hopf bifurcations for the Schnakenberg model.

References

1. A.M. Turing, Philos. Trans. R. Soc. London, Sect. B, 327, 37 (1952)
2. L. Edelstein-Keshet, "*Mathematical Models in Biology*", Birkhauser, NY (1988)
3. P.K. Maini, K.J. Painter and H.N.P. Chau, J.Chem.Soc., Faraday Trans., 93(20), 3601 (1997)
4. J. Schnakenberg, J. Theor. Biol., 81, 389 (1979)
5. J.D. Murray, *Mathematical Biology*, Springer, Berlin (1993)
6. R.A. Satnoianu, P.K. Maini and M. Menzinger, Physica D, 160, 79 (2001)
7. L.Yang, I.Berenstein, I.R. Epstein, Phys.Rev.Letters, 95,3,38303-1 (2005)
8. W. Just, M. Bose, S. Bose, H. Engel, E. Schöll, Phys.Rev. E, 64, 026219(1) (2001)

9. H. Leiva, Applicable Analysis, 60, 277 (1996)
10. J.E. Marsden, M. McKraken, *The Hopf Bifurcation and its Applications*, Springer-Verlag, New York (1976)
11. F. Verhulst, *Nonlinear Differential Equations and Dynamical Systems*, Springer-Verlag, Berlin (1990)
12. D. Henry, *Geometric Theory of Semilinear Parabolic Equations*, Springer-Verlag, New York (1981)
13. M. Golubitsky, E. Knobloch, I. Stewart, J. Nonlinear Sci., 10, 333 (2000)

HIV EPIDEMIOLOGY AND THE IMPACT OF NONSTERILIZING VACCINES*

RUY M. RIBEIRO, ALAN S. PERELSON

Theoretical Biology & Biophysics Group, Los Alamos National Laboratory
Los Alamos NM 87545, USA
E-mail: ruy@lanl.gov

DENNIS L. CHAO

Fred Hutchinson Cancer Research Center, Seattle, WA, USA

MILES P. DAVENPORT

Department of Hematology, Prince of Wales Hospital
and Centre for Vascular Research, University of New South Wales, Kensington, NSW, Australia

Human immunodeficiency virus (HIV) is the cause of the most severe pandemic that the world has ever seen. In 2005, there were 40 million people living with this infection and 2.8 million people died, the vast majority in the 15-49 age group. Altogether, acquired immunodeficiency syndrome (AIDS), a condition that follows from HIV infection and leaves the host unable to fight infectious challenges, has resulted in over 25 million deaths worldwide. Unfortunately, the spread of this disease continues at a fast pace, and the best hope for any successful intervention is the development of a vaccine against this virus. However, studies and trials of HIV vaccines in animal models suggest that it is difficult to induce complete protection from infection ('sterilizing immunity'), and it may only be possible to reduce viral load and to slow or prevent disease progression following infection. What would be the effect of such vaccine on the spread of the epidemic? We have developed an age-structured epidemiological model of the effects of a disease modifying HIV vaccine that incorporates intra host dynamics of infection (transmission rate and host mortality that depend on viral load), the possible evolution and transmission of vaccine escape mutant virus, a finite duration of vaccine protection, and possible changes in sexual behavior. Using this model we investigate the long-term outcome of a disease modifying vaccine and utilize uncertainty analysis to quantify the effects of our lack of precise knowledge of various parameters.

*This work was made possible by Grant Number P20 RR18754 from the National Center for Research Resources (NCRR), a component of the National Institutes of Health (to RMR).

1. Introduction

1.1. *The Global Spread of Human Immunodeficiency Virus*

The United Nations AIDS organization (UNAIDS) reported that, at the end of 2005, 40.3 million people were living with human immunodeficiency virus (HIV). Of these, 25.8 million (64%) were located in the developing countries of sub-Saharan Africa (Figure 1a). Since the beginning of the epidemic, detected in the early eighties, over 25 million people have died of the complications occurring in late stage HIV [1], when acquired immunodeficiency syndrome (AIDS) manifests itself. In 2005 alone, 3.1 million people died, but at the same time 4.9 million new infections were registered, which indicates that this pandemic is still growing. The most affected region is sub-Saharan Africa, where the proportion of infected adults has reached more than 25% in countries like Botswana and South Africa (Figure 1b) [1], however, the incidence of this infection is growing in other large countries like China, India and Russia [1]. Even in developed countries there is worry that the trend of stable or declining prevalence may be changing with increases in high risk behavior, including injectable drug use and risky sexual behavior.

The population wide effects of HIV are devastating. By crippling the main productive age group in the population (young adults), HIV has reduced the life expectancy in several African countries, reverting a trend for increased quality of life that had begun before the explosion of the epidemic. For example, in the last 15 years, the life expectancy in Botswana has been reduced from 65 to 33 almost exclusively due to the AIDS epidemic. Indeed, this pandemic has completely changed the profile of age at death in southern Africa (Figure 1c). It has also made millions of children orphan, many of them also infected, who have very little prospect of escaping the vicious circle of disease, poverty, and hunger. The 2006 UNAIDS report mentioned this problem explicitly: "What for example is the likely long-term damage — social, economic, psychological — wrought by the orphaning of millions of children? What we do know is that [AIDS] impacts will continue to be felt for years to come and the situation will get significantly worse before it gets better" [1].

1.2. *Biology of HIV Infection*

HIV is a retrovirus and encodes its genetic material in two RNA positive-strands, that are about 9800 nucleotides long and include three main coding regions: *gag*, *pol* and *env* [2, 3]. These three regions are common to all

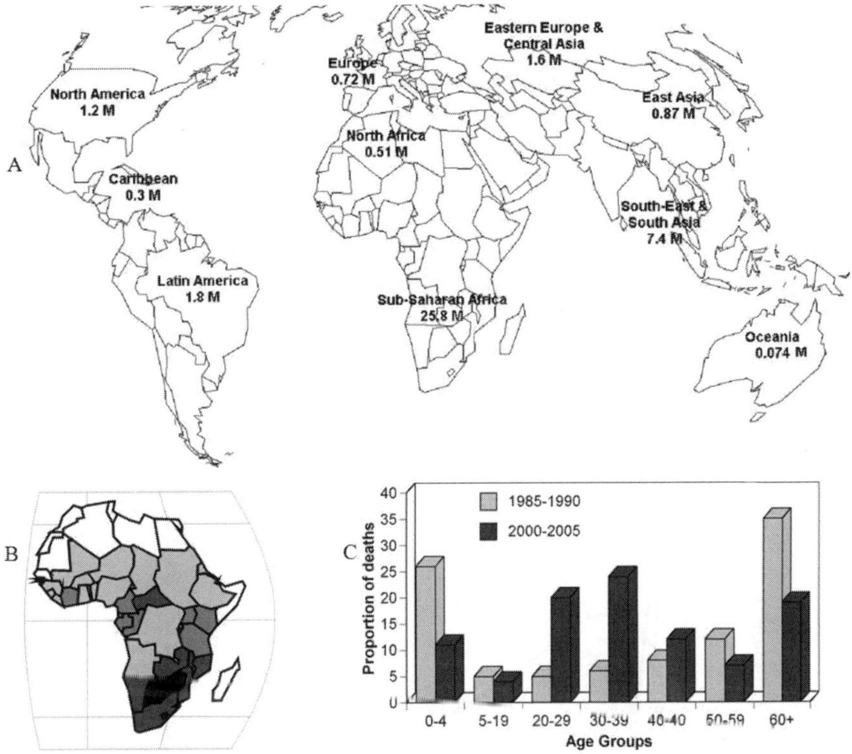

Figure 1. The burden of HIV in the world. (a) Number of infected people at the end of 2005. (b) Prevalence of HIV infection among the adult population in Africa, white is less than 1% and darker grays represent countries with prevalence between 1%-5%, 5%-10%, 10%-20%, and 20%-34%, respectively. (c) Distribution of the proportion of deaths by age group in Southern Africa, before and after the effects of the AIDS epidemic (light gray and dark gray, respectively) [1].

retroviruses.

The main target cell populations for HIV infection are those expressing the membrane surface protein CD4 and an appropriate co-receptor [2]. Two of the targets for infection are T-helper lymphocytes and macrophages, which are essential components of the immune system. Briefly, HIV infection cycle occurs by a succession of steps [2] including: 1) Binding and fusion of the viral envelope proteins with the plasma membrane; 2) Uncoating of the virus and reverse transcription of the viral RNA into DNA, with the help of the viral reverse transcriptase enzyme; 3) Integration of the pro-viral DNA into a random locus of the host-cell genome, via the HIV-1 integrase enzyme; 4) Copying by the cell machinery of the viral DNA into viral RNA for new virions and mRNA for synthesis of the viral proteins; 5) Assembly of the structural proteins near the plasma membrane and release of new virions from the infected cell. The newly created virions then mature, through the action of the viral protease, which splices the long polyprotein formed earlier by transcription of the viral mRNA.

As the infection progresses, the number of CD4+ T-helper cells declines steadily [3]. Why this happens on a time scales of a decade or so is still one of the biggest mysteries of HIV infection. During this long period, the immune system vigorously attempts to fight off the virus. However, HIV-1's rapid replication and mutation rates [4] allow the virus to continuously escape the immune response mounted against it. Thus, viral evolution away from the control of the immune response has been detected [5-7].

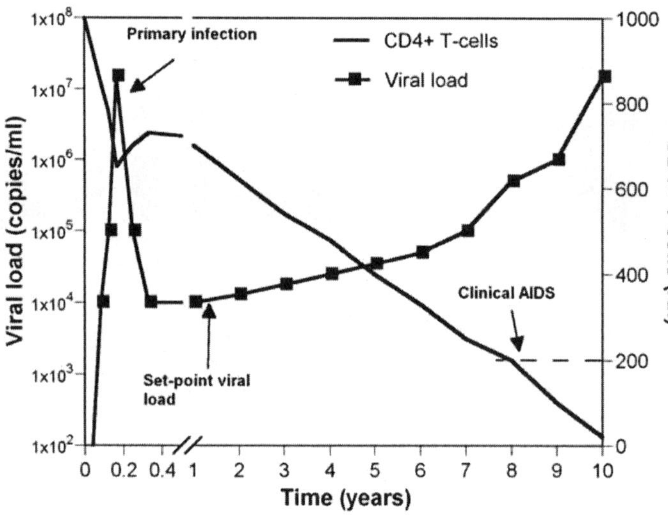

Figure 2. Schematic profile of viral load and CD4+ T-cell count in the peripheral blood of a HIV infected individual. Notice the break in the x-axis. The onset of clinical AIDS occurs when the CD4+ T-cell count drops below 200 cells $\mu\ell^{-1}$ (figure adapted from [4]).

When the CD4+ T-cell count drops below 200 cells $\mu\ell^{-1}$, from the normal steady state of 1000 cells $\mu\ell^{-1}$, the patient enters the stage of clinical AIDS (Figure 2). From this point on, and as the CD4+ T-cells decline even further, the patient's immune system has greater and greater difficulty fighting common opportunistic infections, which at higher CD4 counts may have been innocuous. Eventually the person dies from these infections. After primary infection, the viral load stays more or less constant at the so-called viral set-point until the late stages of infection when it increases dramatically, in association with disruption of the lymphatic system [8].

Once infected, there are no confirmed cases of an individual being capable of eradicating the virus. There is a small proportion of infected individuals that are able to control the virus at low levels without significant loss of CD4+ T-cells over very long periods of time. These individuals are known as long-term non-progressors. However, for the majority of infected people the only way to control the virus, maintain immune function, and lead a normal life is through medical treatment with HIV-specific medications.

1.3. *Treatment of HIV Infection*

In principle, any of the steps in the viral lifecycle could serve as targets for therapy. In practice, most of the drugs used today against HIV can be classified in two classes: reverse transcriptase (RT) inhibiting (further subdivided into nucleoside and non-nucleoside analogues) and protease inhibiting drugs [9, 10]. As their names indicate, the two classes of drugs are directed at distinct steps of the infection cycle, and lead to different outcomes. The RT-inhibitors (RTI) prevent new infections of uninfected cells, by blocking the reverse transcription of the incoming viral RNA into DNA, thus preventing viral integration into the cell's genome. Protease inhibitors (PI) prevent the formation of new infectious viruses from infected cells, by abrogating the final maturation step in the viral lifecycle, which is mediated by the viral protease. In addition to these drugs, there is one approved fusion inhibitor in the clinic, which prevents the fusion of the virus with the cell membrane. There are also new types of drugs being developed targeted at other steps of the lifecycle [11-15], such as proviral DNA integration and viral gene regulation.

All the drugs available at the moment select for drug resistant mutants when used in monotherapy [10]. Many of the primary resistant mutants, those appearing first during therapy, differ from the consensus sequence by only a few point mutations [16] (see also http://resdb.lanl.gov/Resist_DB/default.htm). Hence, it is not surprising that they are selected quickly, since it is expected that all viable one-point mutants are present in the viral quasi-species [17-19]. For these reasons, monotherapy with any of these drugs is not recommended, unless there are no alternatives [9]. Therefore, combination therapy is the preferred treatment against HIV, in principle, with at least one drug of each category (PI and RTI) [20, 21]. The hope is that in these situations simultaneous multiple drug resistance will be much harder to develop.

Unfortunately, these multiple drug regimes with HIV-specific drugs are expensive to administer, and even after significant price reductions due to the introduction of generic brand medications, these treatments cost hundreds of dollars per year for each infected person. For countries where the total health service budget is of the order of tens of dollars per year per person or less, HIV treatment is unaffordable [22]. Thus, medical treatment is mostly only available to people living in richer countries, unless the government has a national strategy to fight the disease, such as is the case in Brazil. In poorer countries, even with the help of international organizations (such as the United Nation's Global Fund for HIV, Malaria and Tuberculosis), treatment is available to only a very small proportion of the people that need it.

With the continued spread of HIV, it is generally believed that only an inexpensive and efficient vaccine can curb this pandemic. Unfortunately, so far, all efforts to develop such a vaccine have faced difficult scientific problems that have yet to be resolved [23].

2. Vaccines for HIV-1 Infection

One of the most important questions in HIV research is what kind of immune response can prevent infection. Since there are no documented cases of people able to eradicate the virus after being infected, it is not known what the necessary ingredients are for a successful immune response. Moreover, there are cases of people who were infected by one strain of HIV, and later were super-infected by a different strain, suggesting that ongoing infection and concomitant immune response does not protect against acquisition of new infection. However, HIV infection, by definition, cripples the immune system, so that these super-infection cases may not be representative of what happens during initial infection. In Table 1, we summarize these and other scientific obstacles for the development of a HIV vaccine [23]. Nonetheless, much effort is being dedicated in pursuit of a safe and effective HIV vaccine, with more than 35 vaccine candidates in clinical trials (from Phase I – safety – to phase III – large efficacy trials) [24], and many dozens other under experimental development.

In general, vaccines work by stimulating in a controlled way either the humoral immune response, mediated by antibodies; and/or the cell-mediated immune response, including the expansion of specific cytolytic T-lymphocytes (CD8+ T-cells called CTL) [25]. In HIV infection, it is difficult to find functional antibody responses, in part because the virus can mutate very fast and escape that response [26-29]. However, specific CD8+ T-cells are readily detected. Indeed, several lines of evidence seem to indicate that these cells are important in HIV control: (i) the decline in virus in acute infection coincides with the peak expansion in CTL numbers [30, 31], (ii) experiments in monkeys in which CD8+ T-cells are depleted leads to increases in virus [32, 33], (iii) infusion of CTL leads to decrease in viral loads [34, 35], (iv) viral mutations that lead to 'escape' from CTL recognition leads to increases in viral loads [5, 34, 36], and (v) there appears to be an association between CTL numbers and disease outcome in HIV infection [37]. Thus, much of the work to develop a vaccine against HIV infection has been centered on stimulating the cell-mediated immune response, although research into antibody vaccines is becoming more common.

Table 1. Scientific challenges for the development of a HIV vaccine (adapted from [23]).

Scientific issue	*Challenges for the development of HIV vaccine*
Virus	•HIV is variable around the world, and even within countries
	•HIV infects and destroys the immune system
	•Poor animal models for HIV infection
Host	•No cases of immune response eradicating HIV infection
	•The needs for a successful immune response not known
	•Roles of different forms of immune response are unclear
	•Super-infection is possible
Pathogenesis	•Generation of latently infected cells, with integrated provirus
	•HIV quickly spreads within the host, destroying gut-associated lymphoid tissue within a few weeks of infection
	•Pathogenesis has both short and long time scales (weeks to peak of infection/years until AIDS)
Transmission	•Multiple forms and routes of transmission
	•Ascertaining effects of "non-sterilizing" vaccine includes understanding reductions in secondary transmission

2.1. *Vaccines to Prevent Infection or to Attenuate Infection?*

The ideal vaccine would prevent infection of a vaccinated person exposed to HIV infection. Indeed this is the concept that applies to other vaccines such as those against measles, hepatitis B or *Hemophilus influenzae*. However, so far such an HIV vaccine has proved elusive, and none of the hundreds of research projects dedicated to HIV vaccine development has been able to convincingly demonstrate a vaccine protocol that truly prevents infection [38].

On the other hand many vaccine experiments have shown that it is possible to attenuate infection after exposure, at least temporarily [39-43]. That is, vaccinated animals (macaques) have much lower viral loads after infection and progress much more slowly to AIDS than control animals that were not vaccinated. These vaccines that do not prevent infection are called "non-sterilizing" or "disease-modifying" vaccines [44]. The mechanisms of disease attenuation in these vaccines are not well known. One possibility is that the vaccine-elicited immune response leads after primary infection to a lower set-point viral load (Figure 3) [44].

Although these vaccines are clearly beneficial for the individual, who may be able to live longer, with a better quality of life, and without expensive drug treatment, it is not clear what the epidemiological consequences of such "disease attenuating" vaccines are. Two factors may contribute to a worse outcome at the population level of vaccination campaigns with such vaccines. First, the individual is infectious for longer and hence could spread the disease longer. Second, there is the possibility that the population's risky behavior increases. In this regards, increases in risky sexual behavior have been observed with the introduction of effective antiretroviral therapy [45, 46], and have been modeled before [47, 48]; the same possibility has been raised concerning the introduction of a vaccine [49, 50].

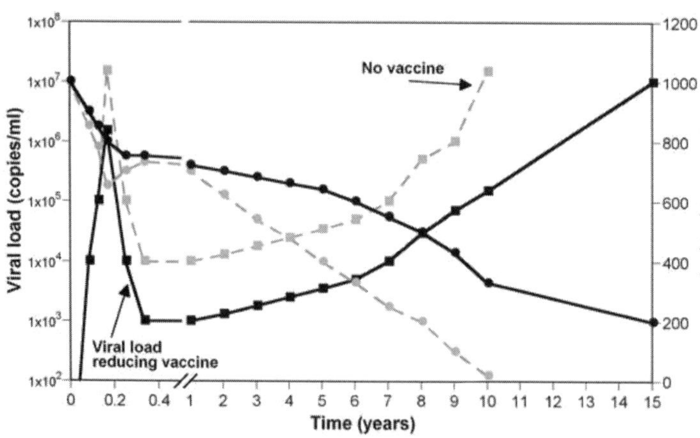

Figure 3. Schematic of possible "non-sterilizing" vaccine effect. The dashed gray lines represent the baseline without vaccine (see Figure 2). In black, we show the effect of a viral "load reducing vaccine", with a lower set-point viral load but increases similar to the no vaccine case (the two viral load lines are approximately parallel). In this case, the loss of CD4+ T-cells is obviously slower than in the no vaccine case.

3. Epidemiology of Non-Sterilizing Vaccines

We developed a mathematical and computational model to study the epidemiological effects of non-sterilizing vaccines against HIV infection [44]. In essence this is a susceptible-infectious (SI) epidemiological model, structured by sex and age of the individuals. But, because vaccine does not prevent infection, we also need to take into account the within-individual dynamics of HIV viral loads — these determine the infectious state of the individual, as well as the effect of the vaccine. Thus, the model populations are also structured by duration of infection and vaccination status (Figure 4). The model discussed here is a simplified version of the one we studied earlier, because it does not include the possibility of the virus escaping the protection conferred by the vaccine [51].

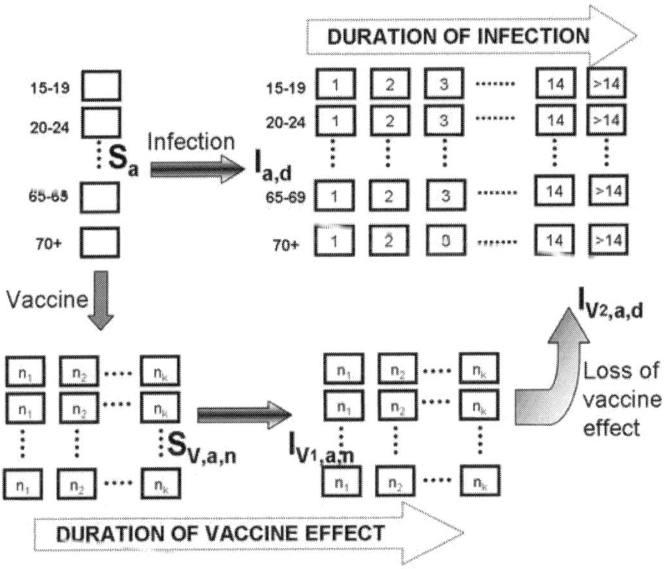

Figure 4. Schematic of the model structured by age (a), duration of infection (d) and duration of vaccine protection (n) used to assess the impact of a "non-sterilizing" vaccine.

3.1. *Modeling "Non-sterilizing" Vaccines for HIV Infection*

The details of the model are presented in Davenport *et al.* [44]. Briefly, uninfected individuals may become vaccinated (at rate V) and enter the population, S_V, of susceptible-vaccinated individuals, may become infected with HIV (at rate λ) and enter the infected population, I, or may die at their sex- and age-specific mortality rate ($\mu_{s,a}$) [52]. Infected individuals (I) are also followed according to the time since they were infected (in yearly increments) (Figure 4) and progress through one year categories of duration of infection (at rate p) with empirically derived increases in viral loads, and rates of HIV associated death ($D_{s,a,d}$), determined from field epidemiology studies [53]. Vaccinated susceptible individuals (S_V) have different levels of protection from vaccination. The level of protection in the vaccinated population is defined by the reduction in viral load (compared to unvaccinated individuals) these individuals would experience following infection. It is assumed that vaccine induced protection wanes with time (at rate ω), as is the case for example with tetanus vaccine, which needs to be re-taken every 5-10 years. Vaccinated individuals pass through progressively lower levels of protection until all protection is lost and they return to the unvaccinated population (S). Vaccinated individuals may become infected at a higher rate than unvaccinated individuals, if vaccination increases high-risk activities (R). Vaccinated individuals who become infected move into different stages of the vaccinated infected population (I_V) at disease progression rate p_V. The vaccinated infected population is divided into those with low viral loads and low mortality [54, 55] due to vaccination (I_{V1} in Figure 4) and those who have lost vaccine protection, and therefore have viral loads and mortality equivalent to those seen in natural infection (I_{V2}). Thus, vaccine protection wanes with time.

We can write the equations for this simplified model as follows [44]. For the unvaccinated susceptible population (S),

$$\frac{dS_a}{dt} = \pi_0 + \omega S_{Va,20} + \alpha_{(a-1)}S_{(a-1)} - (\lambda_a + V + \alpha_a + \mu_a)\, S_a\,, \quad (1)$$

where the subscript a represents the age group. Thus, π_0 represents the influx of susceptibles (example into the youngest age group), and α is the aging rate. Note that for simplification purposes, here we only show the equations for one of the sexes (no subscript s). The vaccinated and uninfected population (S_V) evolve as

$$\frac{dS_{Va,n}}{dt} = VS_a + \omega S_{Va,(n-1)} + \alpha_{(a-1)}S_{V(a-1),n} - (R\lambda_a + \omega + \alpha_a + \mu_a)S_{Va,n}\,, \quad (2)$$

with the subscript $n = 1, 2, \ldots$ representing progressively lower levels of vaccine protection. The vaccination term (VS_a) only applies to the first stage with $n = 1$, and correspondingly the waning vaccination at rate ω

only applies for $n > 1$. Infected unvaccinated individuals are described by,

$$\frac{dI_{a,d}}{dt} = \lambda_a S_a + p_{(d-1)} I_{a,(d-1)} + \alpha_{(a-1)} I_{(a-1),d} - (\alpha_a + p_d + \mu_a + D_{a,d}) I_{a,d}\,, \tag{3}$$

where progression through time since infection is indicated by the subscript d. The infection term $(\lambda_a S_a)$ only applies for newly infected individuals $(d = 1)$, whereas the progression term (p_{d-1}) affects those people that have been infected with $d > 1$, and p_d is zero for the most advanced disease stage. For infected vaccinated individuals, the corresponding equation is

$$\begin{aligned}\frac{dI_{V1a,n}}{dt} &= R\lambda_a S_{Va,n} + p_{V(d-1)} I_{V1a,(n-1)} + \alpha_{(a-1)} I_{V1(a-1),n} \\ &\quad -(\alpha_a + p_{Vd} + \mu_a + D_{V1a,n}) I_{V1a,n}\end{aligned} \tag{4}$$

where vaccinated individuals with protection level n $(S_{Va,n})$ enter the infected, vaccinated population at the equivalent level and then progress through stages of infection at rate p_V. Once viral loads rise to values equal to that seen following natural infection individuals enter category I_{V2},

$$\begin{aligned}\frac{dI_{V2a,d}}{dt} &= p_{Vd} I_{V1a,20} + p_{V(d-1)} I_{V2a,(d-1)} + \alpha_{(a-1)} I_{V2(a-1),d} \\ &\quad -(\alpha_a + p_{Vd} + \mu_a + D_{a,d}) I_{V2a,d}\end{aligned} \tag{5}$$

The per capita force of infection (λ) was calculated in two steps [44, 56]. First, the 'infectiousness' of the population of age a and sex s was calculated, taking into consideration the number of infected individuals in each category, their 'virological infectiousness' and age specific transmission rate [57]. We then calculated the force of infection for a given sex and age group $(\lambda_{s,a})$ using the empirical effective average rate of partner change [58, 59], and the matrix of sexual mixing between age groups [60, 61] (Figure 5). For details, see Davenport *et al.* [44].

Using this model based on population data, we can analyze the effects that a vaccine that reduces viral load has on mortality and transmission at the population level. Such viral load reducing vaccines should lead to lower virological transmission early after infection (Figure 3). Using the full model in [44], we explored factors such as increases in risky behavior, duration of vaccination protection, phase of the epidemic, the emergence of virus escaping the control of the vaccine, *etc.*.

We solved the model using standard numerical integration techniques to simulate the differential equations with parameters based on the literature. For the analysis, we compared the prevalence and mortality of a baseline population, without vaccination, with those quantities in different

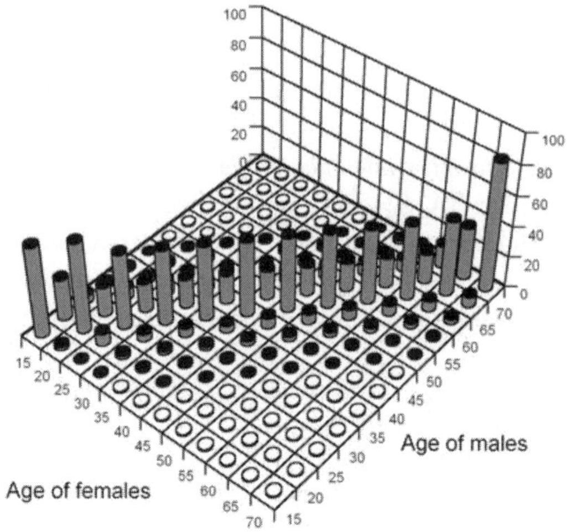

Figure 5. Mixing of sexual contacts among different age groups. Most people have sex with people in their own age group, but there is also some mixing with other younger and older people. This empirically derived mixing matrix is used in the model to determine who can potentially infect whom.

vaccinated populations at 10 and 25 years post-introduction of a vaccine. We performed 1000 simulations of the effect of vaccination using different values for the vaccine parameters. That is, we built 1000 sets of parameters by randomly selecting each from a pre-defined range based on the literature [44]. For example, we simulated the impact of potential increases in risky sexual behavior by increasing the partner exchange rate above that reported in field studies between 0 and 30% [62-64]. We then performed risk and sensitivity analysis [65] to interpret the results. That is, we analyzed which input factors most significantly affected the outcome in terms of prevalence of HIV infection and mortality due to it.

3.2. *Epidemiological Impact of a "Non-sterilizing" Vaccine*

Looking at the results of our full model, which is an extension of the model presented above including the possibility of virus escaping vaccine control, (see [44]), we can make a number of observations. The mean number of deaths averted by vaccination reached 10.7% by 10 years and 35.2% by 25 years (Figure 6). Moreover, the number of deaths was not increased in any of the vaccinations scenarios. In terms of new infections, vaccination averted a mean of 8.4% by 10 years and 26.7% by 25 years. However, mean

HIV incidence rose in the first few years compared to no vaccination (by 1% at 2 years), before commencing a decline in most scenarios. The fraction of scenarios in which the cumulative HIV incidence increased compared to using no vaccine was 11.2% at 10 years and only 3.5% by 25 years (Figure 6). If no cure for HIV/AIDS is found in the interim, these small probabilities of increased incidence will also lead to increases in mortality, albeit beyond the 25 year horizon discussed here. However, in the majority of those cases it is likely that death would occur from other causes before AIDS sets in. Thus, a viral load reducing vaccine has a clear potential to reduce HIV deaths and increase quality of life at the population level, however with a small risk of a long-term increase in HIV incidence [44, 66].

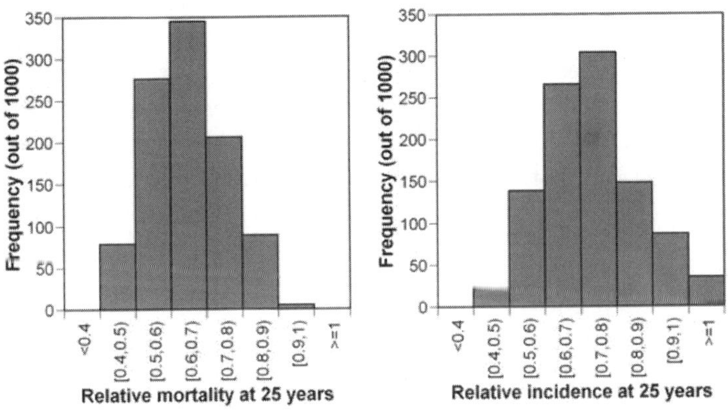

Figure 6. Percentage of deaths and infections averted over the 1000 simulations performed with the full model. The distribution of mortality and incidence at 25 years relative to baseline (that is, no vaccine) is shown. The number of deaths and infections in the scenario without vaccination is defined as 1, and the improvement in those measurements corresponds to simulations with mortality/incidence values less than one.

The rise seen in new infections in the first few years after the introduction of vaccination clearly is a consequence of increases in risky sexual behavior accompanying vaccination. The vaccine neither prevents infection, nor does it affect individuals already infected. Therefore, individuals infected before the vaccination programs continue to be just as "infectious" as before, but vaccinated individuals may become more "susceptible", if they increase their sexual risk behavior. As the vaccination campaign continues through the years, many of the infected individuals would be vaccinated prior to infection, and thus will have low viral loads and a lower rate of transmission. Our simulations showed that if the reduction in viral load set point is at least $1.0 \log_{10}$ copies $m\ell^{-1}$, the reduced transmission due to vaccination would be sufficient to counterbalance rises in risk behavior of up to 30% (data not shown).

To analyze the influence of different input parameters in our results, we performed a sensitivity analysis, based on the partial rank correlation between the HIV mortality and incidence at 25 years and each input variable [65]. The input parameters most correlated with reductions in both incidence and mortality were: *i)* higher levels of vaccine induced viral load reduction, *ii)* lower levels of increased risky sexual behavior, and *iii)* higher proportion of population vaccinated, which is a composite of both the rate of vaccination and rate of waning of vaccine effect. By contrast, the rate of disease progression had little influence on either incidence or mortality at 25 years [44].

We then extended these results by studying other biological and epidemiologically relevant situations, such as the possibility of emergence of mutant viruses that escaped the effect of the vaccine; and the consequences of introducing vaccination programs in populations with different epidemic parameters (such as prevalence and growth rate). These results are detailed in Davenport *et al.* [44].

4. Clinical Trials of "Non-sterilizing" Vaccines for HIV Infection

Obviously, the adoption of a given vaccine will only be possible after extensive clinical trials prove its efficacy, and perhaps indicate its method of action. There is a vast literature on how to conduct clinical trials, from how to design them, how to allocate individuals, how to follow-up during the trial, how to study outcome measures, to how to perform statistical analysis [67, 68]. However, it is not at all clear that existing techniques can directly be applied to clinical trials of a "non-sterilizing vaccine". For instance, traditional double-blind clinical trials follow two groups of participants, one taking the experimental treatment and the other usually taking a placebo. But in the case of a "disease modifying vaccine" it is necessary to also follow HIV infections that may occur or be prevented outside those two groups of trial participants. Indeed, our model and others clearly show that much of the benefit of this type of vaccine would be in preventing or reducing transmission from vaccinated infected individuals to other people [44, 66, 69]. The extent of this reduction in transmission depends on the efficacy of the vaccine in reducing the viral load in infected patients, so that monitoring of viral load may be important, albeit difficult in a potentially resource poor setting. Moreover, because this is a secondary effect, it may take longer to pick up its signal. For instance our model showed that if risky behavior increases there is a possibility of early increases in the number of infected people, even though in the long term the vaccine is efficacious.

Because early increases may be due to changes in high-risk behavior in vaccinated individuals or their partners, monitoring this parameter will also be crucial. Indeed, this is another important issue in "non-sterilizing vaccines". Since the participants are "healthy" and could become infected during the trial, it is an ethical imperative that other measures like counseling and condom distribution also be undertaken during the trial. In turn, these measures could confound the effect of the vaccine, and we could reject vaccines that have small efficacy in the trial setting (with the extra measures), but that would have higher efficacy in a "real-world" setting.

Another question highlighted by our model is the possible impact of the phase of the epidemic on the clinical trial outcome. Since the effect of the vaccine depends on the initial HIV prevalence, as well as growth rate of the epidemic in the particular population where the vaccine is tested [44], results from a trial in a geographic region may not be valid in another region. In this context, an additional issue to take into account is the structure of the population itself. How does the contact structure (e.g., possible existence of super-spreaders) affect the results and interpretation of a clinical trial of these vaccines?

Given the vast investment needed to conduct these clinical trials, it is fundamental that the problems described above and others be thoroughly analyzed before embarking on the trials. Some of these issues will benefit enormously from clear, quantitative thinking and modeling.

5. Conclusions

Vaccines are undoubtedly one of humankind's greatest medical achievements, and they have been instrumental in drastically reducing mortality from infectious diseases, including the global eradication of smallpox, a scourge of the Middle Ages. Indeed, they continue to play a fundamental role in our fight against disease. In the case of the ever-growing pandemic of HIV/AIDS, a safe and efficacious vaccine is seen by most as the best (if not the only) way to curb its spread and ameliorate its effects.

There is a huge effort, measured in hundred of millions of dollars and thousands of man-hours of work, invested in developing HIV vaccines. There is also significant political and social pressure to develop such vaccines and proceed rapidly towards clinical trials. However it is becoming clear that sterilizing vaccines are unlikely to be developed in the near future, and only disease modifying vaccines may be possible in the near term. What the impact of these vaccines will be and how this impact should be measured is not obvious. Thus, efforts must continue to be dedicated to a better theoretical and practical understanding of the epidemiology of these "non-sterilizing" vaccines.

Acknowledgment

ASP was supported by NIH grants RR06555 and AI28433. MPD is supported by a James S. McDonnell Foundation 21st Century Award and the NHMRC. MPD is a Sylvia and Charles Viertel Senior Medical Research Fellow. Part of this work was done under the auspices of the U.S. Department of Energy under contract DE-AC52-06NA25396.

References

1. UNAIDS/WHO AIDS Epidemic Update, UNAIDS/WHO, Geneva, 2005.
2. J.A. Levy HIV and the Pathogenesis of AIDS, American Society of Microbiology, Washington, 1998.
3. N. Nathanson, R. Ahmed, F. Gonzlez-Scarano, et al., Viral Pathogenesis, Lippincot-Raven Publishers, Philadelphia, 1997, pp. 940
4. R.M. Ribeiro, Modeling the in vivo dynamics of viral infections, in: Mondaini R.P., Dilao R. (Eds.), BIOMAT 2005, World Scientific, Singapore, 2006, p.
5. R.E. Phillips, S. Rowland-Jones, D.F. Nixon, et al., Human immunodeficiency virus genetic variation that can escape cytotoxic T cell recognition, Nature 354 (1991) 453-459.
6. P. Borrow, H. Lewicki, X. Wei, et al., Antiviral pressure exerted by HIV-1 specific cytotoxic T lymphocytes (CTLs) during primary infection demonstrated by rapid selection of CTL escape virus, Nat. Med. 3 (1997) 205-211.
7. P. Borrow, G.M. Shaw, Cytotoxic T-lymphocyte escape viral variants: how important are they in viral evasion of immune clearance in vivo?, Immunol. Rev. 164 (1998) 37-51.
8. G. Pantaleo, A.S. Fauci, Immunopathogenesis of HIV infection, Annu. Rev. Microbiol. 50 (1996) 825-854.
9. G. Barbaro, A. Scozzafava, A. Mastrolorenzo, C.T. Supuran, Highly active antiretroviral therapy: current state of the art, new agents and their pharmacological interactions useful for improving therapeutic outcome, Curr. Pharm. Des. 11 (2005) 1805-1843.
10. A.M. Vandamme, K. VanLaethem, E. DeClercq, Managing resistance to anti-

HIV drugs - An important consideration for effective disease management, Drugs 57 (1999) 337-361.
11. D. Daelemans, E. Afonina, J. Nilsson, et al., A synthetic HIV-1 Rev inhibitor interfering with the CRM1-mediated nuclear export, Proc. Natl. Acad. Sci. U. S. A. 99 (2002) 14440-14445.
12. D. Daelemans, A.M. Vandamme, E. De Clercq, Human immunodeficiency virus gene regulation as a target for antiviral chemotherapy, Antivir. Chem. Chemother. 10 (1999) 1-14.
13. G.A. Donzella, D. Schols, S.W. Lin, et al., AMD310, a small molecule inhibitor of HIV-1 entry via the CXCR4 co- receptor, Nat. Med. 4 (1998) 72-77.
14. B.N. Fields, Fundamental Virology, Lippincont-Raven Publishers, Philadelphia, 1996, pp. 1340
15. Y. Pommier, A.A. Pilon, K. Bajaj, et al., HIV-1 integrase as a target for antiviral drugs, Antivir. Chem. Chemother. 8 (1997) 463-483.
16. K. Van Vaerenbergh, K. Van Laethem, J. Albert, et al., Prevalence and characteristics of multinucleoside-resistant human immunodeficiency virus type 1 among European patients receiving combinations of nucleoside analogues, Antimicrob. Agents Chemother. 44 (2000) 2109-2117.
17. A.S. Perelson, P. Essunger, D.D. Ho, Dynamics of HIV-1 and CD4+ lymphocytes in vivo, AIDS 11 (1997) S17-24.
18. J.M. Coffin, HIV population dynamics in vivo: implications for genetic variation, pathogenesis, and therapy, Science 267 (1995) 483-489.
19. R.M. Ribeiro, S. Bonhoeffer, M.A. Nowak, The frequency of resistant mutant virus before antiviral therapy, AIDS 12 (1998) 461-465.
20. R.F. Schinazi, B.A. Larder, J.W. Mellors, Mutations in retroviral genes associated with drug resistance, International Antiviral News 5 (1997) 129-142.
21. P.G. Yeni, S.M. Hammer, M.S. Hirsch, et al., Treatment for adult HIV infection: 2004 recommendations of the International AIDS Society-USA Panel, JAMA 292 (2004) 251-265.
22. S. Chao, K. Kostermans Improving Health for the Poor in Mozambique: The Fight Continues, The World Bank, Washington, D.C., 2002.
23. IAVI AIDS Vaccine Blueprint 2006, New York, 2006.
24. IAVI Database of AIDS vaccines in human trials, accessed Sep 2006.
25. C.A. Janeway, Jr., P. Travers, M. Walport, M.J. Shlomchik Immunobiology, Garland Science, New York, 2005.
26. D.D. Richman, T. Wrin, S.J. Little, C.J. Petropoulos, Rapid evolution of the neutralizing antibody response to HIV type 1 infection, Proc. Natl. Acad. Sci. U. S. A. 100 (2003) 4144-4149.
27. X. Wei, J.M. Decker, S. Wang, et al., Antibody neutralization and escape by HIV-1, Nature 422 (2003) 307-312.
28. P.D. Kwong, M.L. Doyle, D.J. Casper, et al., HIV-1 evades antibody-mediated neutralization through conformational masking of receptor-binding sites, Nature 420 (2002) 678-682.
29. D.R. Burton, R.C. Desrosiers, R.W. Doms, et al., HIV vaccine design and the neutralizing antibody problem, Nat Immunol 5 (2004) 233-236.
30. P. Borrow, H. Lewicki, B.H. Hahn, et al., Virus-specific CD8+ cytotoxic

T-lymphocyte activity associated with control of viremia in primary human immunodeficiency virus type 1 infection, J. Virol. 68 (1994) 6103-6110.
31. R.A. Koup, Virus escape from CTL recognition, J. Exp. Med. 180 (1994) 779-782.
32. X. Jin, D.E. Bauer, S.E. Tuttleton, et al., Dramatic rise in plasma viremia after CD8(+) T cell depletion in simian immunodeficiency virus-infected macaques, J. Exp. Med. 189 (1999) 991-998.
33. J.E. Schmitz, M.J. Kuroda, S. Santra, et al., Control of viremia in simian immunodeficiency virus infection by CD8+ lymphocytes, Science 283 (1999) 857-860.
34. S. Koenig, A.J. Conley, Y.A. Brewah, et al., Transfer of HIV-1-specific cytotoxic T lymphocytes to an AIDS patient leads to selection for mutant HIV variants and subsequent disease progression, Nat. Med. 1 (1995) 330-336.
35. J. Lieberman, P.R. Skolnik, G.R. Parkerson, 3rd, et al., Safety of autologous, ex vivo-expanded human immunodeficiency virus (HIV)-specific cytotoxic T-lymphocyte infusion in HIV-infected patients, Blood 90 (1997) 2196-2206.
36. D.H. Barouch, J. Kunstman, M.J. Kuroda, et al., Eventual AIDS vaccine failure in a rhesus monkey by viral escape from cytotoxic T lymphocytes, Nature 415 (2002) 335-339.
37. L. Musey, J. Hughes, T. Schacker, et al., Cytotoxic-T-cell responses, viral load, and disease progression in early human immunodeficiency virus type 1 infection [see comments], N. Engl. J. Med. 337 (1997) 1267-1274.
38. X.F. Shen, R.F. Siliciano, AIDS - Preventing AIDS but not HIV-1 infection with a DNA vaccine, Science 290 (2000) 20.
39. R.R. Amara, F. Villinger, J.D. Altman, et al., Control of a mucosal challenge and prevention of AIDS by a multiprotein DNA/MVA vaccine, Science 292 (2001) 69-74.
40. D.H. Barouch, S. Santra, J.E. Schmitz, et al., Control of viremia and prevention of clinical AIDS in rhesus monkeys by cytokine-augmented DNA vaccination, Science 290 (2000) 486-492.
41. X. Chen, G. Scala, I. Quinto, et al., Protection of rhesus macaques against disease progression from pathogenic SHIV-89.6PD by vaccination with phage-displayed HIV-1 epitopes, Nat. Med. 7 (2001) 1225-1231.
42. M.A. Egan, W.A. Charini, M.J. Kuroda, et al., Simian immunodeficiency virus (SIV) gag DNA-vaccinated rhesus monkeys develop secondary cytotoxic T-lymphocyte responses and control viral replication after pathogenic SIV infection, J. Virol. 74 (2000) 7485-7495.
43. J.W. Shiver, T.M. Fu, L. Chen, et al., Replication-incompetent adenoviral vaccine vector elicits effective anti-immunodeficiency-virus immunity, Nature 415 (2002) 331-335.
44. M.P. Davenport, R.M. Ribeiro, D.L. Chao, A.S. Perelson, Predicting the impact of a nonsterilizing vaccine against human immunodeficiency virus, J. Virol. 78 (2004) 11340-11351.
45. N. Dukers, J. Goudsmit, J.B.F. de Wit, et al., Sexual risk behaviour relates to the virological and immunological improvements during highly active antiretroviral therapy in HIV-1 infection, AIDS 15 (2001) 369-378.

46. T. Kellogg, W. McFarland, M. Katz, Recent increases in HIV seroconversion among repeat anonymous testers in San Francisco [letter], AIDS 13 (1999) 2303-2304.
47. S.M. Blower, A.N. Aschenbach, H.B. Gershengorn, J.O. Kahn, Predicting the unpredictable: transmission of drug-resistant HIV, Nat. Med. 7 (2001) 1016-1020.
48. S.M. Blower, H.B. Gershengorn, R.M. Grant, A tale of two futures: HIV and antiretroviral therapy in San Francisco, Science 287 (2000) 650-654.
49. S.M. Blower, A.R. McLean, Prophylactic vaccines, risk behavior change, and the probability of eradicating HIV in San Francisco [see comments], Science 265 (1994) 1451-1454.
50. S.M. Blower, K. Koelle, D.E. Kirschner, J. Mills, Live attenuated HIV vaccines: predicting the tradeoff between efficacy and safety, Proc. Natl. Acad. Sci. U. S. A. 98 (2001) 3618-3623.
51. D.H. Barouch, J. Kunstman, J. Glowczwskie, et al., Viral Escape from Dominant Simian Immunodeficiency Virus Epitope-Specific Cytotoxic T Lymphocytes in DNA-Vaccinated Rhesus Monkeys, J. Virol. 77 (2003) 7367-7375.
52. Vital statistics of the United States, Vol II: Mortality, Part A, Government Printing Office, Washington, DC, various.
53. Anonymous, Time from HIV-1 seroconversion to AIDS and death before widespread use of highly-active antiretroviral therapy: a collaborative reanalysis., Lancet 355 (2000) 1131-1137.
54. K. Anastos, L.A. Kalish, N. Hessol, et al., The relative value of CD4 cell count and quantitative HIV-1 RNA in predicting survival in HIV-1-infected women: results of the women's interagency HIV study, AIDS 13 (1999) 1717-1726.
55. J.W. Mellors, C.R. Rinaldo, Jr., P. Gupta, et al., Prognosis in HIV-1 infection predicted by the quantity of virus in plasma [published erratum appears in Science 1997 Jan 3;275(5296):14], Science 272 (1996) 1167-1170.
56. M.P. Davenport, J.J. Post, Rapid disease progression and the rate of spread of the HIV epidemic, AIDS 15 (2001) 2055-2057.
57. T.C. Quinn, M.J. Wawer, N. Sewankambo, et al., Viral load and heterosexual transmission of human immunodeficiency virus type 1. Rakai Project Study Group [see comments], N. Engl. J. Med. 342 (2000) 921-929.
58. S.M. Blower, Exploratory data analysis of three sexual behaviour surveys: implications for HIV-1 transmission in the U.K, Philosophical Transactions of the Royal Society of London - Series B: Biological Sciences 339 (1993) 33-51.
59. B.C. Leigh, M.T. Temple, K.F. Trocki, The sexual behavior of US adults: results from a national survey, Am. J. Public Health 83 (1993) 1400-1408.
60. S.O. Aral, J.P. Hughes, B. Stoner, et al., Sexual mixing patterns in the spread of gonococcal and chlamydial infections, Am. J. Public Health 89 (1999) 825-833.
61. J.E. Darroch, D.J. Landry, S. Oslak, Age differences between sexual partners in the United States, Fam. Plann. Perspect. 31 (1999) 160-167.
62. T. Smith American Sexual Behaviour: Trends, sociodemographic differences, and risk behaviour, National Opinion Research Centre, Chicago, 1994.
63. A. van der Straten, C.A. Gomez, J. Saul, et al., Sexual risk behaviors among

heterosexual HIV serodiscordant couples in the era of post-exposure prevention and viral suppressive therapy, AIDS 14 (2000) F47-F54.
64. S. Blower, E.J. Schwartz, J. Mills, Forecasting the future of HIV epidemics: the impact of antiretroviral therapies & imperfect vaccines, AIDS Rev 5 (2003) 113-125.
65. S. Blower, H. Dowlatabadi, Sensitivity and uncertainty analysis of complex models of disease transmission: and HIV model, as an example, International Statistical review 62 (1994) 229-243.
66. M. van Ballegooijen, J.A. Bogaards, G.J. Weverling, et al., AIDS vaccines that allow HIV-1 to infect and escape immunologic control: a mathematic analysis of mass vaccination, J. Acquir. Immune Defic. Syndr. 34 (2003) 214-220.
67. D.G. Altman Practical Statistics for Medical Research, Chapman & Hall/CRC, Boca Raton, 1999.
68. S.J. Pocock Clinical Trials: A Practical Approach, Wiley, Chichester, 1983.
69. R.J. Smith, S.M. Blower, Could disease-modifying HIV vaccines cause population-level perversity?, Lancet Infect Dis 4 (2004) 636-639.

MATHEMATICAL AND COMPUTER MODELING OF MOLECULAR-GENETIC MECHANISMS OF LIVER CELLS INFECTION BY HEPATITIS B VIRUS*

BAHROM R. ALIEV
Uzbek Institute of Virology
700194, Murodov, 7, Tashkent, Uzbekistan
E-mail: regulatorika@yahoo.com

BAHROM N. HIDIROV, MAHRUY SAIDALIEVA, MOHINISO B. HIDIROVA
Institute of Informatics
700125, F. Khodjaev, 34, Tashkent, Uzbekistan
E-mail: regulatorika@yahoo.com

This work deals with some results of the analysis of molecular-genetic mechanisms of interconnected activity between hepatocytes and hepatitis B virus (HBV) based on mathematical modeling. The functional-differential equations of functioning regulatory mechanisms (regulatorika) of molecular-genetic systems of cells in mul ticellular organisms are used as equations class for the quantitative analysis of activity regularities of "hepatocytes-HBV" genetic system. Results obtained during qualitative studies show that there are different scenarioes for fulfilling the infectious process on cellular level at HBV, including symbiotical coexistence between hepatocyte and virus, also periodical excitement of the virus infection.

1. Introduction

At viruses penetration into an organism there are the complex processes leading to (i) acute infections diseases development with expressed clinical picture or (ii) to asymptomatic disease or (iii) to lingering chronic disease, depending on viruses type and cell response. Hepatitis B virus takes the first place among human hepatitis viruses due to diseases prevalence and the frequency of infection persistence which can subsequently lead to a cirrhosis and liver cancer. About one third of all population in the world is infected by hepatitis B virus and over 350 million (5-6 percents) individuals

*This work is supported by grants F-4.3.5 and A-14-010 of CST RUZ

are positive for HBsAg. Annually from one up to two millions patients die due to complications after hepatitis B.

The features of mutual relations between virus and hepatocytes are the basis of complex infection process formation at each type of virus hepatitis. At present, main mechanisms of interconnected activity between hepatocytes and hepatitis viruses are not clear.

Results of the analysis of articles reviews show that quantitative research in virus diseases has begun half century ago[1]. In 1975 G.I. Marchuk have developed and investigated elementary mathematical model for virus diseases in an organism[2]. Further the model has been modified and investigated in detail by his disciples in[3,4]. In these models a liver (or other modelled organ) is considered as the cells population, which at norm (at viruses absence) capable to keep own stable and constant number. In a number of the works, devoted to research the processes at virus disease by quantitative methods, the elementary models for the dynamic theory of populations[5,6,7] are applied. Unlike the previous approach, in this simulation method the basic attention is given to disease dynamics in an organ. Mathematical principle for the models class (in works this models are called "predator-prey" type models) construction have been developed by V. Volterra, V.A. Kostitzin and A.N. Kolmogorov in the decade 1930-1940. These models are based on the ordinary differential equations and take into account the values, expressing the number of "predators", "preys", and sometimes some medium feature for considered population[7,8].

In this work we quantitatively investigate regulatory mechanisms (regulatorika) for interconnected activity between molecular-genetic systems of hepatocyte and hepatitis B virus. Understanding interconnected activity between hepatocyte and hepatitis B virus on molecular-genetic level assumes the analysis of genes activity mechanisms in a functioned cell. Regulation of a gene activity is realized by means of a genetic regulatory systems based on the interactions between cell DNA, virus DNA, RNA and protein-enzymes. Intuitive understanding activity of a genetic regulatory system during genes regulation in multi-component system connected by means of complex interactions between positive and negative feedback mechanisms is very difficult. Here we need formal mathematical methods and computer tools for imitation and modeling the regulatory mechanisms of cells molecular-genetic system, for carrying out computing experiments.

The prevalence of using the ordinary differential equations (ODE) for modeling dynamic systems activity (this class of equations is widely used for the analysis of genetic regulatory systems too) is indisputable. The

ODE formalism allows to simulate dynamics of RNA concentration, protein-enzymes and other molecules using variables and values changing over the time t and belonging to the set of real, non-negative numbers. Regulatory interactions between elements of considered systems take the form of functional and differential equations for concentration variables.

Usually, at the ODE formalism for molecular-genetic systems, the regulatory mechanisms are modeled by "velocity equations", expressing productivity speed for systems element as a function of other elements concentration.

Temporal parameters in a genetic system for cell regulation consisting of times for transcription, translation, genetic products action and feedback can be taken into account using the equations:

$$\frac{dx_i}{dt} = a_i r_i(x_1(t-h_{i1}), x_2(t-h_{i2}), ..., x_n(t-h_{in})) - \gamma_i x_i, \qquad (1)$$

where $h_{i1}, ..., h_{in} > 0$ are the discrete times necessary for realization of genetic regulation loop as a whole; $\{a\} > 0$ are the synthesis constant and $\{\gamma\} > 0$ are the decay constants. "Velocity equations" (1) express balance between quantity of the molecules, synthesizing and dissociating for a time unit. In this case the integral equation can be used for modeling genetic processes taking into account temporal mutual relations in system for cell regulation[9,10,11,12].

In the XX century a powerful mathematical methods have been developed for modeling an intracellular processes using the "velocity equations"[13,14,15]. Based on these methods the kinetic models for a genetic regulatory mechanisms can be developed by determination a concrete type for functions r_i $(i = 1, 2, ..., n)$. The following section is devoted to one of possible methods for quantitative researching the interconnected activity between hepatocyte and HBV molecular-genetic systems based on the functional-differential equations for living system regulatorika.

2. The basic equations

If for (1) we consecutively apply methods for modeling living systems regulatorika, using approaches by B, Goodwin[16], M. Eigen[17], V.A. Ratner[18], J. Smith[19], J. Murray[20] and cooperative character of the biological processes, the time relations, the end-product inhibition in a cell, then we have the following functional-differential equations for cells molecular-genetic system regulatorika[21,22,23]:

$$\frac{dx_i(t)}{dt} = \Lambda_i^n(X(t-h))\exp\left(-\sum_{k=1}^{n}\delta_{ik}x_k(t-h_{ik})\right) - \frac{1}{\tau_i}x_i(t) \qquad (2)$$

with

$$\Lambda_i^n(X(t-h)) = \sum_{j=1}^{n}\left(\sum_{k_1,\ldots,k_j=1}^{n} a_{ik_1,\ldots,k_j}\prod_{m=l}^{j} x_{k_m}(t-h_{ik_m})\right),$$

where $x_i(t)$ are the values describing products quantity, developed by i-th genetic system at the time moment t; h_{ik} are the time intervals necessary for activity change of i-th genetic system under influence upon k-th genetic system; $a_{ik_1,\ldots,k_j}, \tau_i$ are parameters of i-th product formation velocity and "life duration" for i-th product, accordingly; δ_{ik} are parameters of i-th genetic system repression by activity products of k-th genetic system; $ik_1,..,k_j,i,j,k_j = 1,2,...,n$.

The system (2) is concerned to class of the delay functional-differential equations and if we have continuous functions on the initial time interval with length h $(h = \max_{i,j} h_{ij}, \quad (i,j = 1,2,...,n))$, then we get its continuous solutions using the method of consequent integration[21,24,25].

Since HBV molecular-genetic system can function only under "assistance" from a hepatocyte molecular-genetic system which can act independently, then the functional-differential equations for the regulatorika of interconnected activity between hepatocyte and HBV molecular-genetic systems have the following form

$$\begin{aligned}
\frac{dX_i(t)}{dt} &= \alpha_i\left(\prod_{l=1}^{n} X_l(t-h)\right)e^{-A} - \frac{1}{\tau_{X_i}}X_i(t);\\
\frac{dY_i(t)}{dt} &= \beta_j\left(\prod_{l=1}^{m} Y_l(t-h)\right)\left(\prod_{k=1}^{n} X_k(t-h)\right)e^{-B} - \frac{1}{\tau_{Y_j}}Y_j(t),
\end{aligned} \qquad (3)$$

where

$$A = \sum_{l=1}^{n} c_{1il}X_l(t-h) + \sum_{l=1}^{m} c_{2il}Y_l(t-h);$$

$$B = \sum_{p=1}^{n} d_{1jp}X_p(t-h) + \sum_{p=1}^{m} d_{2jp}Y_p(t-h),$$

$X_i(t), Y_j(t)$ are the values, expressing an activity of hepatocyte and HBV molecular-genetic systems at time moment t; h is the time radius for cell (the time necessary for feedback fulfillment in molecular-genetic systems); $\{\alpha, \beta\}$ and $\{c, d\}$ are non-negative parameters for system (3) expressing levels of maintenance and inhibition for considered gene systems, accordingly; $\{r\}$ are parameters of "life duration" for activity products in genes systems; n, m are the quantity of considered hepatocyte and HBV genetic systems accordingly; $i = 1, 2, ..., n$; $j = 1, 2, ..., m$.

The equations system (3) has a complex character and its application for the qualitative and quantitative analysis of mechanisms for interconnected activity between hepatocyte and HBV molecular-genetic systems requires development of its model systems. The model systems, preserving the basic properties of the initial equations are more simple equations system with minimally possible number of relations and parameters[26,27,28]. The model system in many cases allows to carry out the successful analitical analysis of characteristic solutions and to define the basic behaviour regimes for considered mathematical models. Usually qualitative, quantitative analyses and computing experiments realized during modeling, allow choosing admissible model system for the initial equations, which is basis in mathematical model for considered process.

At the equations (1) application for a quantitative research regulatory mechanisms of interconnected activity between hepatocyte and HBV, it is necessary to consider some features for the given process. The infections process in hepatocyte is realized at close interaction between HBV genome and hepatocyte genome, what leads to "self-conjugacy" (it is equal to two or is greater than two) acceptance by a hepatocyte genetic systems[22,23]. On the other hand at modeling HBV genetic system activity we must take into account required participation of both genomes (corresponding "inter-conjugacy"[22,23] is equal to two). This means that the model system for regulatorika equations of interconnected activity between HBV and hepatocyte genetic systems has the form

$$\frac{dX(t)}{dt} = \alpha X^2(t-h)e^{-c_1 X(t-h) - c_2 Y(t-h)} - \frac{1}{\tau_x} X(t);$$
$$\frac{dY(t)}{dt} = \beta X(t-h)Y(t-h)e^{-d_1 X(t-h) - d_2 Y(t-h)} - \frac{1}{\tau_y} Y(t), \tag{4}$$

where $X(t), Y(t)$ are the values, expressing activity of HBV and hepatocyte molecular-genetic systems, accordingly; α, β are the velocities constants of products formation for considered genetic systems; c_1, c_2, d_1, d_2 are the parameters expressing repression degree in HBV and hepatocyte molecular-genetic systems; h is the time necessary for feedback realization in considered system; τ_x, τ_y are the parameters describing "life duration" for products in HBV and hepatocyte molecular-genetic systems; all parameters are positive.

3. Conditions for the equations applicability

The minimal claims to the mathematical equations for modeling an intracellular processes are:

- activity non-negativity for HBV and hepatocyte genetic elements;
- resources limitation in a cell.

These demands lead to following conditions for the equations (4) applicability for investigations the interconnected activity between HBV and hepatocyte genetic systems:

- solutions are in the first quadrant of phase space;
- solutions are limit.

The following theorems define conditions then the equations (4) have these properties.

Theorem 3.1. *If parameters and initial functions are non-negative then solutions of the equations (4) are non-negative.*

Proof. Let us have the following initial conditions:

$$X(t) = \varphi_1(t); \qquad Y(t) = \varphi_2(t) \quad at \quad t_0 - h \le t \le t_0,$$

where $\varphi_1(t), \varphi_2(t)$ are integrate, non-negative function over $[t_0 - h, t_0]$; t_0 is the initial time of considered regulatorika process (usually, t_0 greater than h).

Using the construction method for solutions of functional-differential equations (4) over $[t_0, t_0 + h]$ we have

$$X(t) = \varphi_1(t_0)e^{-\frac{t-t_0}{\tau_x}} + \int_{t_0}^{t} e^{-\frac{t-\theta}{\tau_x}}(\alpha\varphi_1^2(\theta - h)e^{-c_1\varphi_1(\theta-h)-c_2\varphi_2(\theta-h)})d\theta;$$

$$Y(t) = \varphi_2(t_0)e^{-\frac{t-t_0}{\tau_y}} + \int_{t_0}^{t} e^{-\frac{t-\theta}{\tau_y}}(\beta\varphi_1(\theta-h)\varphi_2(\theta-h)e^{-d_1\varphi_1(\theta-h)-d_2\varphi_2(\theta-h)})d\theta;$$

It is clear that if conditions of the theorem are true then constructed solutions are non-negative over $[t_0, t_0+h]$. If we take this solution as initial functions for $[t_0, t_0+h]$ we get non-trivial solution for $t > t_0$. Obviously, the solution is unique and continuous at $t > t_0$.

Theorem 3.2. *Infinite points of the system (4) in the first quadrant of phase space are unstable, i.e. its non-negative solutions are limited.*

Proof. Since

$$\lim_{X,Y\to\infty} \alpha X^2 e^{-c_1X-c_2Y} = 0, \qquad \lim_{X,Y\to\infty} \beta XY e^{-d_1X-d_2Y} = 0$$

and if the variables values are very great, then system (4) can be approximated by the following equations

$$\frac{dX(t)}{dt} = -\frac{1}{\tau_x}X(t) < 0; \qquad \frac{dY(t)}{dt} = -\frac{1}{\tau_y}Y(t) < 0.$$

Solutions have the form

$$X(t) = X(t_0)e^{-\frac{t-t_0}{\tau_x}}; \qquad Y(t) = Y(t_0)e^{-\frac{t-t_0}{\tau_y}}, \quad t > t_0.$$

We see that this solutions decreases. Consequently, the solutions of system (4) are limited in the first quadrant of phase space.

4. The qualitative research

Carrying out the elementary transformations for (4) we have

$$\begin{aligned}\epsilon_1\frac{dX(t)}{dt} &= aX^2(t-1)e^{-X(t-1)-cY(t-1)} - X(t);\\ \epsilon_2\frac{dY(t)}{dt} &= bX(t-1)Y(t-1)e^{-dX(t-1)-Y(t-1)} - Y(t),\end{aligned} \tag{5}$$

where

$$\epsilon_1 = \frac{\tau_x}{h}; \quad \epsilon_2 = \frac{\tau_y}{h}; \quad a = \frac{\alpha\tau_x}{c_1}; \quad b = \frac{\beta\tau_y}{c_1}; \quad c = \frac{c_2}{d_2}; \quad d = \frac{d_1}{c_1}.$$

Here ϵ_1, ϵ_2 are the regulatorika parameters; a, b are the velocity constants of products formation; c, d are the parameters of inter-repression degree for HBV and hepatocyte molecular-genetic systems; all parameters are positive.

We assume that

- hepatocyte activity is normal at HBV absence;

- HBV can function in hepatocyte.

It follows that we need $A(X_0, 0)$ type equilibrium point (X_0 is positive) and the area in phase space of (5) where $\frac{dY(t)}{dt} > 0$. Results of qualitative research show, that these conditions are fulfilled if

$$a > e, \qquad b > ec. \tag{6}$$

We accept (due to independent self-regulation of both genomes), that parameters values of inter-repression are lesser than parameters values of self-repression:

$$c < 1, \qquad d < 1. \tag{7}$$

For getting equilibrium points for the functional-differential equations (5) we take[23,24,25,26,27] that $X(t) = X_0 = const$; $\quad Y(t) = Y_0 = const$. Therefore, for an equilibrium points for (5) we have

$$aX_0^2 e^{-X_0 - cY_0} - X_0 = 0;$$

$$bX_0 Y_0 e^{-dX_0 - Y_0} - Y_0 = 0.$$

It is obvious that there is the trivial equilibria. Location character for other equilibrium points can be investigated based on analysis of main isoclinic lines behaviour with taking conditions (6) into consideration:

$$Y_1 = \frac{ln(ax) - x}{c}; \qquad Y_2 = ln(bx) - dx.$$

Analyzing possible variations for main isoclinic lines locations we find that there are $A(X_0, 0)$ type equilibrium points with positive X_0 and $B(X_0, Y_0)$ type non-trivial equilibrium points, where X_0, Y_0 are positive.

From the conditions for non-trivial equilibrium points

$$aX_0 e^{-X_0 - cY_0} - 1 = 0;$$

$$bX_0 e^{-dX_0 - Y_0} - 1 = 0,$$

based on the qualitative research we have the following conditions for presence of non-trivial equilibrium points

$$a > b^c \left(\frac{1 + cd}{1 - c} e \right)^{1-c}. \tag{8}$$

The trivial equilibria is attractor. $A(X_0, 0)$ and $B(X_0, Y_0)$ type equilibrium points (X_0, Y_0 are positive) can be attractors under some parameters values.

Let us consider possibility of stability failure for the equilibrium points of (5) based on Lyapunov method. We replace $X(t)$ by $X_0 + x(t)$, $Y(t)$ by $Y_0 + y(t)$, where $x(t), y(t)$ are small. Linearizing (5) neighborhood of the equilibrium (X_0, Y_0) we get

$$\epsilon_1 \frac{dx(t)}{dt} = (2 - X_0)x(t-1) - X_0 c y(t-1) - x(t);$$
$$\epsilon_2 \frac{dy(t)}{dt} = Y_0(1/X_0 - d)x(t-1) - (1 - Y_0)y(t-1) - y(t). \tag{9}$$

The characteristic equation for (9) has the form

$$(\xi_1\lambda + (X_0 - 2)e^{-\lambda} + 1)(\xi_2\lambda + (1 - Y_0)e^{-\lambda} + 1) + Y_0 c(1 - X_0 d)e^{-2\lambda} = 0.$$

Analyzing the characteristic equation for (9) we find that the equilibria of (5) can be unstable.

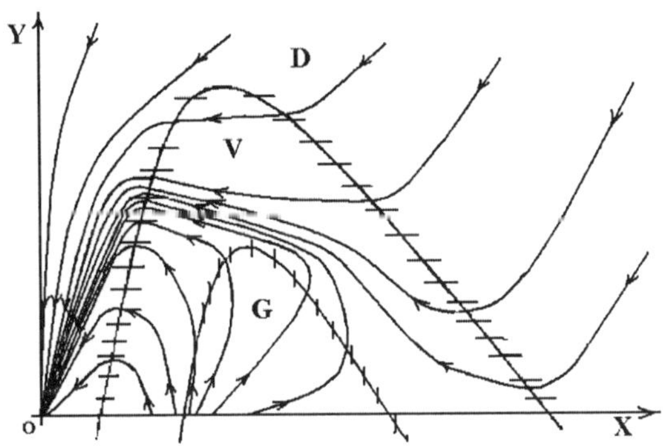

Figure 1. "Hepatocyte-HBV" system interaction with lethal outcome, according to (5)

The qualitative research of (5) taking into account (6), (7) and (9) shows that there are different regimes for "hepatocyte-HBV" system activity. At the certain values of (5) parameters (when the main isoclinals have not points in common and (8) is not true) there is activity suppression in molecular-genetic systems of hepatocyte and HBV (Figure 1, area G) or activation only HBV molecular-genetic system under hepatocyte activity suppression (Figure 1, area V). The infection is completed by getting "hepatocyte-HBV" system into attraction area of the trivial attractor (Figure 1, area D). It corresponds to fulminant lethal end for considered infection.

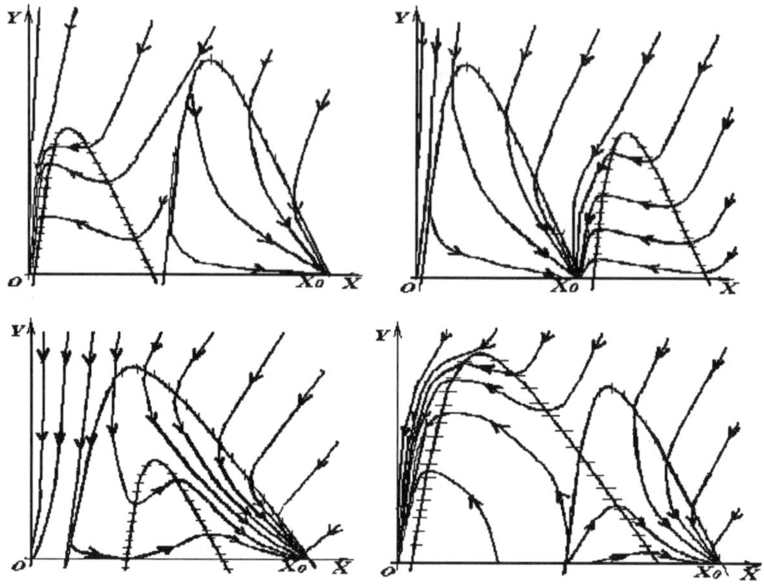

Figure 2. Different variants of hepatocyte domination in "hepatocyte-HBV" system, according to (5)

Analysis of characteristic solutions for (5) shows that there exists set of parameters values under which the hepatocyte genome is dominant (Figure 2). Depending on the initial state of "hepatocyte-HBV" system the solutions can be attracted by attractor x_0 or trivial attractor.

The regime when hepatocyte and HBV genomes work in concert (the chronic hepatitis B) is more interesting (Figure 3). Here the main isoclinic lines for equation (5) have points in common and the condition (8) is true. It is necessary to note that in this case the main isoclinic lines can have two common points in the first quadrant of phase space. The analysis shows that one of them is attractor, another is anti-attractor.

Under the certain values of parameters in the system of functional-differential equations (5) the considered attractor is stable and corresponding solutions have stationary character (Figure 3). Stability loss for considered critical point is realized by Hopf bifurcation by appearance of stable oscillations.

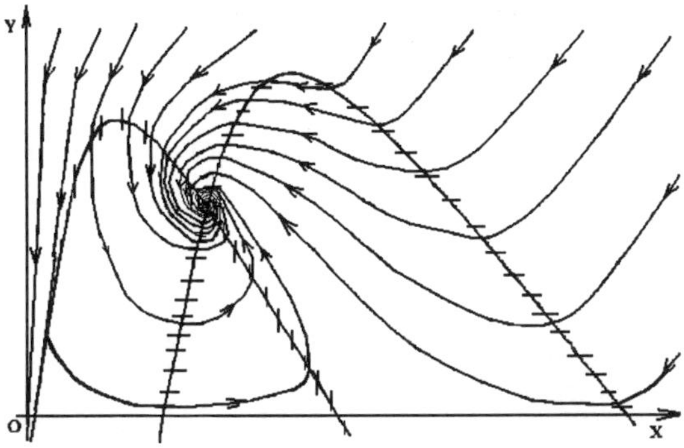

Figure 3. The symbiotic regime in "hepatocyte-HBV" system, according to (5)

5. The computer investigations

Qualitative research of the functional-differential equations (5) was accompanied by computer simulation on PC (Figure 4). Main attention was given to the symbiotic co-existence (chronic hepatitis B).

At quantitative research on PC using vectors with delayed identifiers the equations (5) are realized by the method of consecutive integration (integration step ($\triangle t$) is equal to $0,001$) and

$$\xi_1 = 1.8; \qquad \xi_2 = 0.8; \qquad c = 0.7; \qquad d = 0.5.$$

Results of the computer researches have shown presence of an irregular oscillations and "black hole" effect during interconnected symbiotic activity between HBV and hepatocyte (besides regimes observed at quantitative research). At "black hole" effect the solutions of (5) are broken down into the trivial attractor (Figure 4*c*, 4*d*).

The area of irregular oscillations is characterized by disturbance in hepatocyte regulation system with consecutive worsening its functional activity. For qualitative representation of the given process in general we assume that self-control realization in the hepatocyte during HBV infection is difficult ($h \gg \tau_i$, therefore $\xi_i \sim 0$, $\quad i = 1,2$). These assumptions allow to approximate equations (5) by discrete equations

$$X_{k+1} = aX_k^2 exp(-X_k - cY_k);$$
$$(10)$$
$$Y_{k+1} = bX_kY_kexp(-dX_k - Y_k).$$

Using (10) and computer we investigate quantitative characteristics for irregular oscillatory solutions of (5). In more detail the method for reduction of functional-differential equations for living systems regulatorika is given in[28]. Quantitative research on PC of the structural organization in irregular oscillations area (area of dynamic chaos) shows its strong heterogeneity with sharp uneven chaotic changes of Lyapunov number (Figure 5).

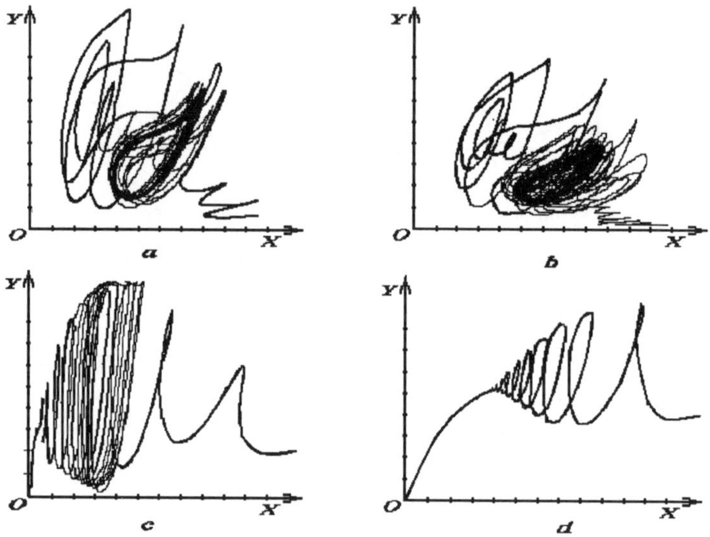

Figure 4. The characteristic phase trajectories for (5) (a is auto-oscillatory regime; b is irregular oscillations (chaos) ; c, d are "black hole" variants (trajectories go from right to left))

Computer investigations have shown that in the area of dynamic chaos there are a small regions (r-windows) with the regular solutions (Figure 5, arrows specify r-windows). It follows that there exists the provisional improvement in hepatocyte state during HBV infection. However such conditions improvement has temporary character and under small disturbance the hepatocyte molecular-genetic system turns into dynamic chaos, again. The input in the area of irregular oscillations can be predicted: it is preceded with splashes series in Lyapunov number by Hopf bifurcation under Feigenbaum scenario. The splashes can be fixed by the solutions analysis using PC. It follows to predict coming destructive changes in the hepatocyte under HBV influence.

Thus, developed functional-differential equations for regulatorika of interconnected activity between hepatocyte and HBV molecular-genetic systems allow to quantitatively investigate the basic regularities for infections

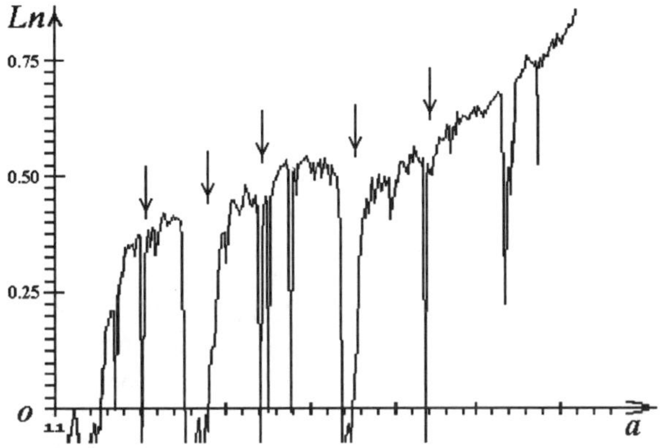

Figure 5. Irregular oscillations area of hepatocyte molecular-genetic system activity (by reduced discrete equation (10))

process in the hepatocyte during hepatitis B based on the qualitative analysis and computer researches. During quantitative researches the following modes for considered process: cleaning, symbiosis, regular and irregular oscillations, sharp destructive changes which define various clinical forms of disease have been observed. The forecasting opportunity for beginning above mentioned regimes and their basic characteristics allows to establish molecular-genetic bases of pathogenesis, to carry out diagnostics and prognosis of characteristic stages for disease current using computer simulation parallel with laboratory and clinical investigations of infections process during hepatitis B.

References

1. J.G.Hege, G.Cole, *J. Immunol.* **97** (1966).
2. G.I.Marchuk, *VTs SO ANSSSR* (1975).
3. N.I. Nisevich, G.I.Marchuk, I.I.Zubikova, I.B.Pogojev, *Moscow, Nauka* (1981).
4. A.A. Romanyukha, *Rus.J. Numer. Anal. Math. Modelling*, **15** (2000).
5. M.A. Nowak, C.R.M. Bangham *Science*, **272** (1996).
6. M.G.M. Gomes, G.F. Medley *Springer Verlag, New York* (2002).
7. A.S. Perelson, *Cambridge University Press, Cambridge* (1980).
8. R.V. Culshaw, S. Ruan, *Math Biosci.*, *165(1)* (2000).
9. H. D. Landahl, *Math. Biophys*, **31** (1969).
10. J. M. Mahaffy, *J. Math. Biol.*, **106** (1984).
11. N. MacDonald, *Cambridge University Press, Cambridge* (1989).
12. P. Smolen, D. A. Baxter, J.H. Byrne, *Bull. Math. Biol.*, **62** (2000).
13. A. Cornish-Bowden, *Portland Press, London*(1995).
14. R. Heinrich, S. Schuster, *Chapman and Hall, New York* (1996).

15. E.O. Voit, *Cambridge University Press, Cambridge* (2000).
16. B. Goodwin, *Academic Press, London and New York* (1963).
17. M. Eigen, P.Shuster, *J. Mol. Evol.*, **19**: 47-61 (1982).
18. B.V. Ratner, V.V. Shamin, *Mathematical models of evolutionary genetics*, 60-82 (1980) (in russian).
19. J. Smith, *Cambridge, Cambridge Univ. Press* (1968).
20. J. Murray, *Clarendon Press, Oxford* (1977).
21. B.N. Hidirov, *Voprosy kibernetiki*, **128**. 41-46 (1984)(in russian).
22. B.N. Hidirov, *Mathematical modeling*, **16**, (2004) (in russian).
23. B.N. Hidirov, *Scientiae Mathematicae Japonicae*, **58**, **2**, 407-413 (2003).
24. R. Bellman, K.L. Cooke, *Academic Press, London* (1963).
25. L. Glass, M. Mackey, *Princeton University Press* (1988).
26. B.I. Arnold, *Ijevsk*, (2000) (in russian).
27. A.D. Bazykin, Yu.A. Kuznetsov, A.I. Hibnik, *Pushino, NCBI, AN USSR* (1985) (in russian).
28. M.B. Hidirova, *Differential equations*, **39 (6)** (2003) (in russian).

MODELING THE GEOGRAPHIC SPREAD OF INFECTIOUS DISEASES USING POPULATION- AND INDIVIDUAL-BASED APPROACHES

LISA SATTENSPIEL

Department of Anthropology, University of Missouri-Columbia
107 Swallow Hall, Columbia, MO, 65211, USA
E-mail: SattenspielL@missouri.edu

Mathematical models are a useful tool to help understand patterns of global spread of infectious diseases and to help prepare for these risks and develop and implement appropriate control strategies. Examples are presented of how two modeling approaches, a population-based ODE model and an individual-based computer model, are used to study the geographic spread of the 1918-19 influenza epidemic in central Canada. The basic structure and major results of each of the models is presented and the insights derived from each approach are compared. Results from both models show that movement between communities serves to introduce epidemic diseases into the communities, but that within-community social factors have a stronger influence on disease severity. However, results of the two models are sometimes significantly different. Discussion of these differences highlights the advantages of using multiple approaches to address similar questions.

1. Introduction

In February, 2003 the World Health Organization began to receive reports of an unusual respiratory illness striking individuals in China. Over the next five months the world watched as this new disease, given the name Severe Acute Respiratory Syndrome, or SARS, spread to all parts of the globe, eventually resulting in 774 reported deaths. The epidemic spread from China to Hong Kong, then to Vietnam, Singapore, Canada, and elsewhere, eventually reaching 26 countries on 5 continents. In spite of this dramatic and rapid worldwide spread, only a few countries experienced outbreaks with significant numbers of deaths. Nonetheless, the rapid diffusion across the globe was enough to convince the world's health authorities that the increasing globalization of the world and associated rates of travel now provide the necessary foundation for a truly global pandemic at some time in the (perhaps near) future. As a consequence, an increasing number of mathematical models have been developed to try to improve our ability to determine when, where, and why diseases spread across geographic space (see [1-6] for recent examples).

The development of mathematical models to describe the transmission of infectious diseases has a long history and has been a major interest of mathematical biologists at least since the late 1980s, after the AIDS epidemic reminded the Western world that infectious diseases were still a force to be reckoned with[a]. Yet, the vast majority of epidemic models focus on disease transmission within a single population (although the population may be divided into subgroups), and the questions primarily center on which individuals within a community are at highest risk, when and how an epidemic spreads *within* the community, why it spreads the way it does, and what can be done to limit that spread. These are very important questions, but humans do not live in a single, isolated community — they live in innumerable distinct, but socially connected communities. Only a small proportion of epidemic models extend beyond a single community to look at how infectious diseases spread across space, even though this is exactly what is needed to understand the potential risks of global epidemics such as the 2003 SARS epidemic.

Mathematical models are ideally suited to exploring the potential impacts of new or re-emerging diseases, because in such cases data may not exist to allow for more traditional mathematical and statistical analyses. Models provide the framework with which to explore the consequences of certain hypothesized events and can be used to evaluate the effectiveness of potential control strategies, as long as they are based on an understanding of the fundamental biology of the host-pathogen interaction and as long as the new disease is relatively similar to known diseases.

Mathematical models for the geographic spread of infectious diseases are not only useful for understanding the potential impact of new diseases, however; they can also help us to understand more completely how different factors impacted the spread of past epidemics within and among communities. Such studies aid in understanding the basic nature of infectious disease transmission, and they provide insights into the human host-pathogen in-

[a]Outside of the Western world public health authorities have always had to deal with relatively high rates of mortality from infectious diseases — especially malaria and other mosquito-borne diseases and gastrointestinal illnesses spread through inadequate sanitary systems. Consequently, scientists and public health authorities in these countries never possessed the relaxed attitudes towards infectious diseases commonly seen in Western countries during the last half of the 20th century.

teraction that can be used to build more realistic models for the spread of future epidemics of both known diseases and new pathogens.

To illustrate this process, the remainder of this paper describes studies of the spread of the 1918-19 influenza pandemic through several small communities in central Canada. Two different approaches are used in these studies: a) a population-based model consisting of a system of ordinary differential equations and its analogous computer model, and b) an individual-based fully stochastic computer model. The basic structure of each of these models is presented, the major results derived from these models are discussed, and insights resulting from the two approaches are compared. Finally, some thoughts are presented on the advantages and limitations of each approach, the questions for which they are ideally suited, and the kinds of insights the study of past epidemics can provide for understanding future epidemics.

2. The epidemiological scenario being modeled

2.1. *The study communities*

The models to be described below are part of a long-standing collaboration between the author and Ann Herring, an anthropologist at McMaster University in Canada. The models were developed to help understand how and why the 1918-19 Spanish flu epidemic spread among three Aboriginal "communities" in central Manitoba — Norway House, Oxford House, and God's Lake (Figure 1). The "communities" each consisted of a fur trading post operated by the Hudson's Bay Company (HBC), a small number of permanent post residents of predominantly European or mixed Aboriginal-European ancestry, and a larger number of associated family groups living in camps of about 15 individuals — a father, his adult son(s), wives, and children or two brothers and their families [7]. During the winter, the family groups were dispersed over a fairly large geographic area. The adult males engaged in fur trapping, and when they accumulated enough furs, they would set out from the camp to the HBC post (which, at the minimum, consisted of a company store) where they would exchange furs for food and other supplies[b]. During the summer, families associated with a particular HBC post would aggregate in the vicinity of that post. At that time baptisms, weddings, visiting relatives, and other social activities would commonly occur.

The environment of the Norway House region was highly seasonal, and

[b]On rare occasions women would also be involved in these activities.

because of that, life during the winter differed from that of the summer in a number of additional ways. During the winter travel occurred predominantly by dogsled or snowshoes, while canoes or larger boats were the preferred mode of travel in the summer. Trips between posts took 1-2 days longer in the winter than in the summer, and the size of traveling groups was much smaller (1-2 individuals compared to 10 or more). Predominant food sources also varied seasonally.

Figure 1. Hudson's Bay Company fur trading districts in Canada. Inset in the upper left shows the locations of the three study communities within the Keewatin District.

2.2. *Characteristics of the flu in the Norway House region*

The 1918-19 flu (often called the Spanish flu) was a major worldwide pandemic that led to the deaths of upwards of 50,000,000 people [8]. Essentially every part of the world was affected by this pandemic (see [9] for examples from many different locations). Mortality rates from the disease averaged around 2.5-5 deaths per 1000 population [8], but some communities experienced much higher rates. For example, Hebron and Okak, two small communities on the northern coast of Labrador, suffered 68% and 78% mortality, respectively [10]. The mortality at Norway House, one of the three study communities, was close to 20% [11], one of the highest mortality levels worldwide. Other communities in the region, such as Fisher River and Berens River, also experienced higher than average mortality, but the levels in those communities were significantly lower than that at Norway House, and, in addition, the historic record indicates that the epidemic never got to Oxford House and God's Lake, the two nearest communities (and the other two study communities) (Figure 2).

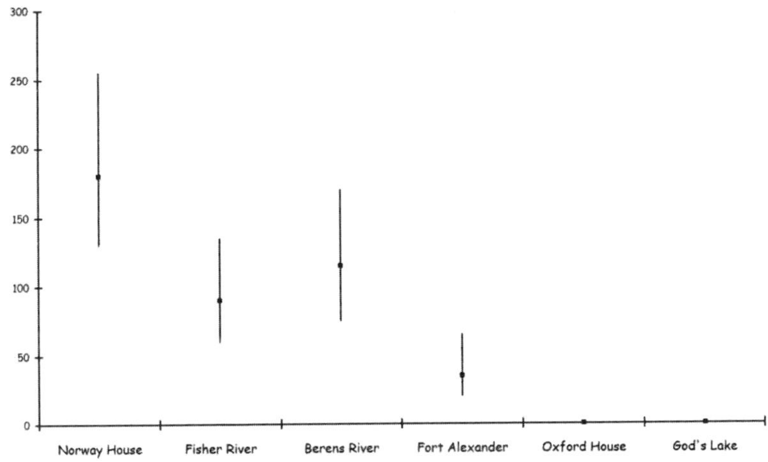

Figure 2. Mortality rates from the 1918-19 influenza epidemic in six Hudson's Bay Company post communities within the Keewatin District.

Besides this high level of inter-community heterogeneity, analysis of data from Norway House indicates extensive heterogeneity in mortality among the families within the community. For example, in seven of the 50 or so families everyone or nearly everyone died, while the majority of families had one or no deaths. Understanding the underlying reasons for such heterogeneity within and between communities has been the motivating factor underlying the mathematical modeling activities associated with this research. Studies have addressed the impact of changes in mobility rates and patterns, both seasonally and over time [11-12], how the social structure influences disease patterns [11-13], the possible impact of quarantine [14], and whether other diseases and other locations and time periods lead to similar or different insights [15-17]. Space constraints do not allow a review of most of this work. In addition, at this time only systematic studies of the impact of varying the rates and patterns of mobility have been completed using the individual-based models, although, as will be described in more detail below, the individual-based and population-based formulations also differ significantly in how they model the social and settlement structures of the study population. In order to facilitate discussion of the advantages and disadvantages of each approach, the focus in the remainder of this paper will be limited to discussion of how the different approaches have been used to address the impact of seasonal changes in rates and direction of travel and to elucidate ways that social structure impacts disease patterns.

3. The population-based model

3.1. *Structure of the model*

The population-based model used in this study combines a standard SIR epidemic model with a simple mobility model to describe the patterns of interaction among populations. A brief description of its structure is presented here; see [18] for details.

Consider a population consisting of n communities, each of constant size N_i, and with the residents of each community distributed into three disease classes — susceptible, infectious, and recovered. Since the model is used to describe a single, short-lived epidemic, it assumes that there are no births and deaths. Furthermore, it is assumed that the disease confers permanent immunity upon recovery. All residents, including infectious individuals, are allowed to travel among the communities, and in order to keep track of both a person's home and his or her location, two subscripts are used to identify individuals in the different disease classes, with the first subscript referring to the home community and the second referring to the visited community. Thus, I_{ij} represents the class of infectious residents of i who are visiting community j.

In the mobility submodel, residents are assumed to leave a community i at a constant rate, σ_i. The probability of traveling from the home community i to any other community j is given by ν_{ij}. A person from community i who travels to community j returns home at a rate ρ_{ij}.

The transmission term for the infection process in the i^{th}community of a population with n communities in this model is:

$$\sum_{k=1}^{n}\sum_{j=1}^{n} k_k \beta_{ijk} \frac{S_{ik} I_{jk}}{N_k^*}$$

where N_k^* is the number of individuals physically present in a community rather than the census population, and is found by summing the number of residents in each disease class who are at home and the total number of visitors to that community.

The full epidemic model adds terms representing the mobility process operating in the population to the epidemic terms corresponding to each disease status (S, I, and R). Because the return rates to region i from region k, ρ_{ik}, are not equal in general to the rates at which residents leave region i to travel to region k, $\sigma_i \nu_{ik}$, it is necessary to subdivide the population into residents who are present at their home region and residents who are visiting another region. This generates a model that consists of three sets of paired equations corresponding to the three disease states:

$$\begin{aligned}
\frac{dS_{ii}}{dt} &= \sum_k \rho_{ik} S_{ik} - \sigma_i S_{ii} - \sum_j \kappa_i \beta_{iji} \frac{S_{ii} I_{ji}}{N_i^*} \\
\frac{dS_{ik}}{dt} &= \sigma_i \nu_{ik} S_{ii} - \rho_{ik} S_{ik} - \sum_j \kappa_k \beta_{ijk} \frac{S_{ik} I_{jk}}{N_k^*} \\
\frac{dI_{ii}}{dt} &= \sum_k \rho_{ik} I_{ik} - \sigma_i I_{ii} + \sum_j \kappa_i \beta_{iji} \frac{S_{ii} I_{ji}}{N_i^*} - \gamma I_{ii} \\
\frac{dI_{ik}}{dt} &= \sigma_i \nu_{ik} I_{ii} - \rho_{ik} I_{ik} + \sum_j \kappa_k \beta_{ijk} \frac{S_{ik} I_{jk}}{N_k^*} - \gamma I_{ik} \\
\frac{dR_{ii}}{dt} &= \sum_k \rho_{ik} R_{ik} - \sigma_i R_{ii} + \gamma I_{ii} \\
\frac{dR_{ik}}{dt} &= \sigma_i \nu_{ik} R_{ii} - \rho_{ik} R_{ik} + \gamma I_{ik}
\end{aligned}$$

As in other epidemic models, γ is the rate of recovery from the disease. This model can be modified in standard ways to allow for vital dynamics, other disease states, or additional details specific to a particular disease.

3.2. *Results from simulations of the population-based model*

Initial simulations of the population-based model (hereafter designated as the NOG-ODE model) were designed to explore the effects of changing the rates of travel among the study communities (represented by σ in the model), changing the probabilities of traveling between particular communities (represented by ν), and changing the location of the initial case. Figure

3 illustrates the impact of summer vs. winter mobility patterns using mobility parameter estimates derived from Hudson's Bay Company post records. Notice that changing the rates and patterns of mobility from winter values to summer values changes the timing of the epidemic in both Oxford House and God's Lake, but does not influence the size of the epidemic peak, a measure of the severity of the epidemic. Simulations varying the location of the initial case had a similar effect — the timing of epidemics within communities was altered, but not the severity of those outbreaks (results not shown).

Because changing the rates and patterns of mobility had little or no effect on the size of epidemic peaks, this aspect of the model was unable to explain the observed heterogeneity in the incidence of flu among the study communities. The model contained one other social parameter that could be varied, however — κ, the rate of contact within a community. Consequently, a decision was made to see whether changes in this parameter could help explain the observed results.

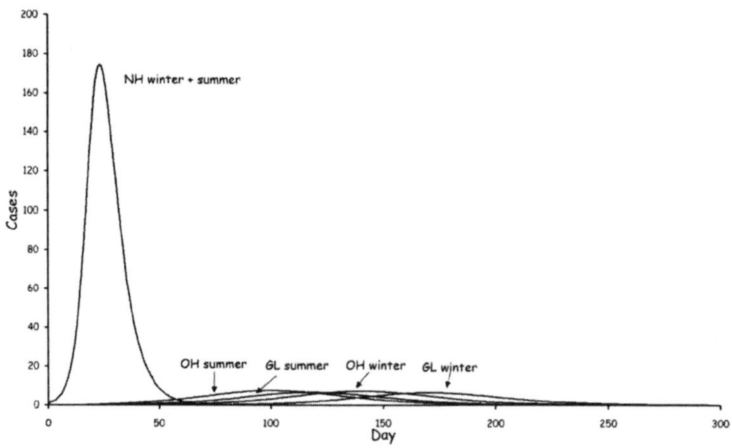

Figure 3. Epidemic curves resulting from summer and winter mortality rates and patterns at Norway House (NH), Oxford House (OH), and God's Lake (GL). Epidemics at Oxford House and God's Lake occur later in the winter than in the summer, but are of equal length and size.

Figure 4 shows that, in fact, changes in the rates of contact within communities can lead to significant changes not only in timing of epidemic peaks, but also in the size of those peaks. It appears from these results that within-community factors, such as the degree of contact among individuals at the local level, has a much stronger influence on the risk of disease spread than travel between communities. In other words, in these models, travel seeds communities with a disease, but once the disease enters a community, the within-community factors take over and determine ultimate risk.

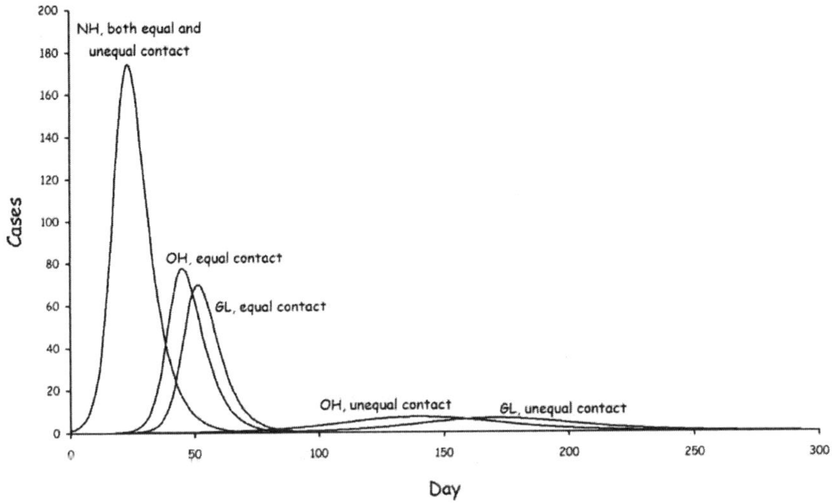

Figure 4. Effect of changes in rates of contact within communities. In simulations producing "unequal contact" curves, the number of contacts at Norway House was twice that at Oxford House and God's Lake; in simulations producing "equal contact" curves, contact rates were the same for all three communities and equal to the original Norway House contact rate.

The recognition of the relatively low importance of mobility and the high importance of within-community factors in determining the severity of an epidemic stimulated attempts to model in finer detail the social structures of the study communities. This goal faced one significant issue — whether deterministic differential equations were the appropriate modeling technique to use. The real study communities that are being modeled were small (the largest was about 750 individuals), which was of some concern in the initial modeling efforts. The ease of the ODE modeling technique, the relative lack of other studies focusing on the geographic spread of infectious diseases, and the quality of the results derived from the initial models justified their use in pursuing the initial questions. However, as the research began to focus on more detailed models of social structure it became clear that the population-based ODE models needed to be replaced by an individual-based model. In particular, newer models needed to be able to model a community dispersed over large areas into tiny groups of about 15 individuals — conditions that clearly invalidate the use of ODE models. Consequently, in the last few years research has shifted to developing an individual-based model that better reflects the real social structure and activities of the study communities and that can deal with the variability of individual behaviors and the stochasticity that results when populations are small.

4. The individual-based model

4.1. *Structure of the model*

An initial individual-based stochastic model (called hereafter the N-IB model), was developed by Carpenter[c] [19] using RePast, a Java-based computer simulation toolkit [20]. The structure of this model was designed to capture the essential social structure of the main HBC post community, Norway House. Because this model considers only a single post community, while the original ODE model (the NOG-ODE model) incorporates three linked communities, Carpenter's model was extended by Ahillen[d] [21] to consider all three HBC posts included in the ODE model. This 3-group individual-based model will be referred to as the NOG-IB model.

Both the N-IB and NOG-IB models separate individuals according to disease status, but they consider four disease classes: susceptible, infectious, recovered and exposed, with the latter consisting of individuals who have been infected but are not yet infectious. In addition the NOG-IB model incorporates death from influenza, a feature that has been incorporated into an extension of the N-IB model but that is not present in the original N-IB model. Epidemiological parameters include the probability of infection given contact, the length of the latency period, and the length of the infectious period of the disease. The length of the latency period is constrained to one day and the length of the infectious period is assumed to be constant, although the models incorporating mortality (NOG-IB and the extension of N-IB) have randomness built into the death process that alters

[c]Carpenter's model was developed under the guidance of Sattenspiel and with the help of Suman Kanuganti, Matthew Tanner, and Steven Tanner.
[d]Ahillen's model was developed under the guidance of Sattenspiel and with the help of Nate Green.

the length of the infectious period to some extent. Disease transmission following contact between a susceptible individual and an infectious individual occurs with constant probability. With the exception of the stochasticity introduced during disease transmission and the additions of the exposed class and mortality, the epidemiological structure of the individual-based models is similar to the NOG-ODE model.

The between-post mobility, which is only applicable to the NOG-IB model, is also modeled in a fashion similar to the NOG-ODE model. The rates of travel among communities and the probabilities of travel to particular communities (represented in the NOG-ODE model by σ and ν, respectively) were set at values similar to those in the NOG-ODE model, although they could vary stochastically. The length of stay in destinations, represented by ρ in the NOG-ODE model, was set at 1 day in both the NOG-ODE and NOG-IB models. One significant difference between the NOG-ODE model and the NOG-IB model is that the latter model incorporates the number of days needed to travel between posts (a measure of distance) and constrains movement along specific paths, while the NOG-ODE model does not incorporate explicit paths and assumes all trips take 1 day to complete.

Unlike the epidemiological and between-post mobility processes, the within-post mobility and overall settlement structure vary significantly between the individual-based models (N-IB and NOG-IB) and the NOG-ODE model. In the NOG-ODE model the population is divided into the three post communities only and all individuals within each community are assumed to be identical except for their classification by disease status. In addition, interactions among individuals do not have an explicit mobility component and are assumed to occur randomly within a community. The two individual-based models use a much more detailed and realistic population structure, which takes similar forms in both models — the difference between the two individual-based models is almost entirely related to the addition of two additional posts in the NOG-IB model; the within-post structure is not significantly different.

The basic structure of each community in the N-IB and NOG-IB models is a central post surrounded by four smaller camps. Each of the camps is composed of a number of family groups of about 15 individuals and consisting of adult males, adult females, and a composite group of young children and older individuals of both sexes. These groups were chosen to allow enough flexibility to use the model to address interesting and unusual characteristics of the populations and the Spanish flu epidemic, while keeping the number of age classes to a manageable level. In particular, it was essential to separate males and females because travel in the real study populations involved only adult males during the winter (when the epidemic hit), but all ages and both sexes during the summer. In addition, the 1918 flu epidemic was unusual worldwide because individuals in the prime of life were affected to a much greater extent than in normal flu epidemics, while young children and older individuals were impacted at more normal levels. Thus, it was essential to separate adults from individuals of other ages. In addition, the model assumes that individuals have more contact with their extended family members than with people from other family groups.

Population figures for each post were derived from the historic record, and were set at 750 for Norway House, 330 for Oxford House, and 300 for God's Lake. As suggested by ethnographic evidence, during the winter many residents were dispersed on the landscape into the associated camps, while during the summer they aggregated around the posts in extended family groups. Hence, in the winter model, the populations of Norway House, Oxford House, and God's Lake were distributed evenly among the post itself and the 4 associated camps (i.e., 150 individuals were assigned to each of these settlement groups (post + each of 4 camps) at Norway House, 66 individuals were assigned to each Oxford House settlement group, and 60 individuals were assigned to each God's Lake settlement group). Each of the individuals was also assigned to a family within his or her settlement group, with the average family size set at 15. To reflect population aggregation, in the summer model all individuals were assigned to their associated post, although they were assigned to family groups as well.

Consistent with evidence from the historic record, only permanent male residents of the posts (i.e., no camp residents) were allowed to engage in inter-post travel; this travel is described briefly above. Travel between the camps and their associated post was limited to adult males living in the camps and was assumed to follow specific paths, with camps varying in their distance from the post (Figure 5). The probability that a particular individual stayed in the camp on any particular day was set at 0.99 to reflect the rarity of any one individual traveling within the region during the winter. Because all family groups were aggregated at the post during the summer, there was no camp to post travel at that time.

Using this basic structure, the model was run for 200 days, which was sufficient for the epidemic to run its course. The first 20 days were run disease-free so that the travel patterns could settle into normal patterns before introducing the disease. One thousand simulations were run for each parameter set, and results were averaged to generate the overall patterns presented below. Only results from simulations of the NOG-IB model will be discussed; interested readers may consult [19] for results from simulations of the original N-IB model.

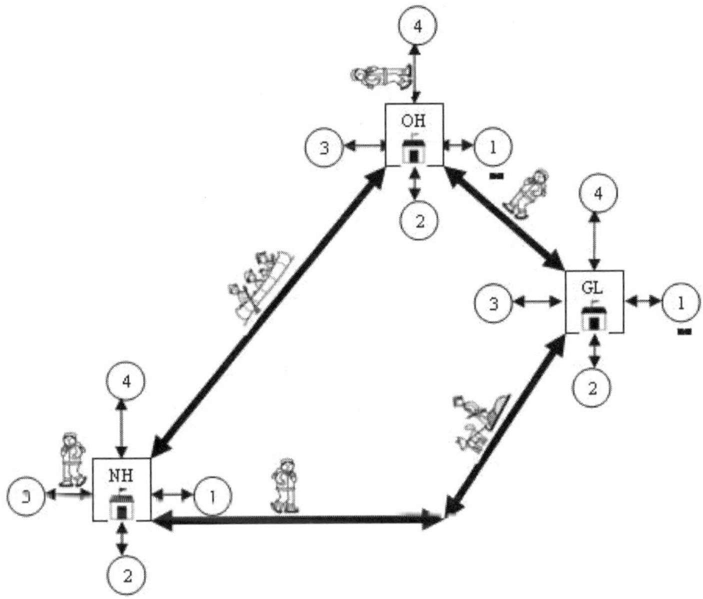

Figure 5. Diagram of the NOG-IB model landscape showing the three communities, Norway House (NH), Oxford House (OH), and God's Lake (GL), each with a central post and four camps (numbered). The inter-community pathways between posts and between each post and its associated camps are also indicated. (Original source: [21]).

4.2. *Results from simulations of the NOG-IB model and comparison with the NOG-ODE model*

Initial simulations of the NOG-IB model focused on understanding the impact of seasonal changes in both travel patterns and settlement and social structures. In some ways the simulated patterns were similar to those found with the NOG-ODE model, but in many ways they differed significantly. As Figure 6 shows, simulations using estimates of summer mobility rates and patterns lead to earlier, more quickly spreading, and shorter epidemics with higher peaks. They also result in higher overall death and infection rates (results not shown). As discussed above and illustrated in Figure 3, changing these parameters in the NOG-ODE model leads only to faster epidemics — both summer and winter epidemics are the same length and severity, but summer epidemics occur earlier than winter epidemics. In order to get the same kinds of results from the NOG-ODE model as those observed in the NOG-IB model, it was necessary to alter contact rates within communities.

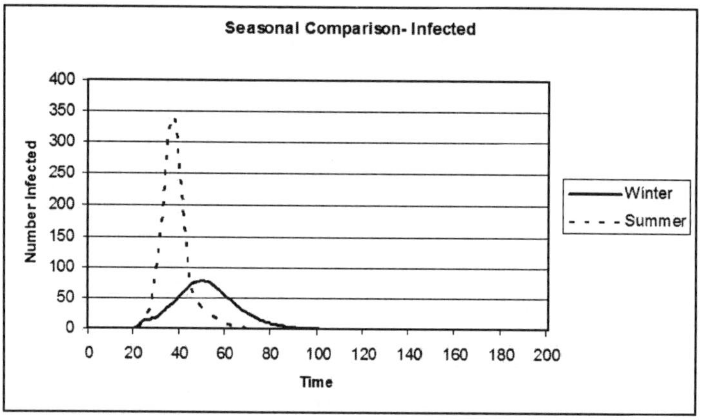

Figure 6. Average number of infected individuals for winter and summer simulations of the NOG-IB model. Because the simulated epidemics never reach God's Lake and only rarely reach Oxford House, graphs show case numbers for all three posts combined. (Original source: [21]).

This comparison between models is not as simple as it appears, however, because, unlike the NOG-ODE model, the summer formulation of the NOG-IB model differs significantly from the winter formulation not just in travel rates, but also in settlement structure and rates of contact among individuals. In order to determine the relative roles of mobility and settlement patterns in generating the observed differences in epidemic patterns, other sets of simulations were run using hybridized mobility patterns. In particular, simulations were run using both winter mobility rates with the summer settlement and contact structure and summer mobility rates with the winter settlement and contact structure. Analysis of these results indicates that most of the summer/winter differences in epidemic patterns can be explained by differences in settlement patterns (and hence contact within and among groups), and not by differences in mobility. Thus, the basic result of the NOG-ODE model — that epidemic severity is more strongly influenced by within-community factors such as settlement structure and contact than by between-community factors such as travel patterns — is consistent with simulation results from the individual-based, stochastic NOG-IB model.

Simulations of the NOG-IB model that center on exploring the influence of changes in the location of the starting point of the epidemic do lead to significant differences from results of similar simulations of the NOG-ODE model. In particular, changing the starting location in simulations of the NOG-ODE model affects the timing of epidemic curves, but the epidemic always reaches all three posts and always generates peaks that do not vary significantly in length or height. Results from simulations of the NOG-IB model are very different. In general, introducing the disease at Oxford House or God's Lake leads to shorter and milder epidemics that are limited to that community and do not spread further. In fact, changing the starting location affects almost every measure of the epidemic, not just the timing. Since these simulations did not involve a change in settlement structure and contact rates (i.e., comparisons involved simply changing the starting location while keeping all other parameters constant), they cannot be explained by differences in contact rates or other factors. Most likely they are related to the smaller size of the two outlying communities and the fact that smaller communities generate lower numbers of potentially infected travelers. This, combined with the stochastic nature of the NOG-IB model, is probably sufficient to explain observed differences.

One major feature of the actual 1918 flu epidemic that the NOG-ODE model was not able to explain is why the epidemic devastated Norway House, but apparently never reached Oxford House or God's Lake, even though they were two of the nearest communities to Norway House. Using realistic parameter estimates derived from the historic record, all simulations of the NOG-ODE model suggested that the epidemic should always reach all three communities, even though winter parameter estimates generated very small epidemics in the outlying communities. Results from simulations of the NOG-IB model do provide some explanation of the observed data. Simulations using winter parameter estimates always fail to reach God's Lake, and only reach Oxford House an average of 13 times in sets of 1000 runs. Furthermore, even simulations using the settlement structure and more frequent summer travel never reach God's Lake and still reach Oxford House only about 37% of the time. Thus, these simulations suggest that the observed between-community heterogeneity in impact in 1918 should actually be expected, rather than surprising. It is likely that Oxford House and God's Lake were spared the horrible effects of the worldwide pandemic because of a combination of small community size, low mobility, and geographic distance from Norway House, and the stochastic effects resulting from these conditions.

5. Implications of these results for modeling the geographic spread of infectious diseases

It is clear from the results of these studies that different modeling approaches can sometimes give different answers to the same questions. Thus, if a model is intended to be applied to real situations it is essential to choose the approach very carefully and to use multiple approaches whenever possible. In particular, the decision about whether to use a population-based model or an individual-based model depends on the questions the model is intended to address and on the nature of the population being modeled.

The methods used to develop and analyze population-based models are usually more familiar to mathematicians and the models are almost always much easier to implement. Furthermore, although the high quality data needed to estimate model parameters may be difficult to obtain, the simplifying assumptions inherent in considering populations of identical, ideal individuals reduce those data needs significantly. The significantly more intensive data requirements of most individual-based models may well prohibit their use in many situations — without adequate data with which to estimate model parameters, the results of even the most carefully formulated models may be called into question. Consequently, in situations with serious data limitations, population-based models may be the best approach. In addition, if the intent is to gain good, theoretical understanding of the impact of different factors on disease spread, then population-based models may well be adequate for the purpose.

The major limitation of most population-based models is that they usually do not incorporate enough of the stochasticity operating in real world populations. The underlying methods used in most population-based epidemic models are based on an assumption that the modeled populations are so large that random effects can be ignored (although there is a small literature using stochastic, population-based models). For models focusing on moderately sized cities or larger areas, population-based models may be adequate for their intended purposes, and this in combination with their ease of implementation and more modest data needs may justify their use.

On the other hand, if the intent of a model is to explore a situation where stochastic factors are likely to be significant (such as the 1918 flu epidemic in central Canada), then it is probably best to use a stochastic model, and if individual behavior is the focus of the research question, then an individual-based model is most appropriate.

Interesting insights may be possible using a population-based model as a first approximation, but the constraining large-population assumptions can lead to results that probably do not reflect reality well and that may need to be revised as stochasticity is brought into play. For theoretical questions this is not likely to be a major problem; the benefits of the initial insights from models that are simpler to use are usually greater than any costs related to ignoring the underlying random effects, as long as the population-based model is well formulated and incorporates a sufficient degree of biological reality. For practical questions, however, it is probably necessary, at the very least, to develop corresponding individual-based models to ensure that any policies and actions derived from analysis of the models will have the intended consequence.

The analysis of the NOG-ODE simulations led to important insights about the relative roles of movements across a landscape and the nature of social interactions in facilitating the spread of an infectious disease through a region. In particular, these insights helped to shift the focus of study away from a view that movement of individuals and the diseases they carry was the most important influence on geographic patterns to one that recognized the complexity of social structure and human social interactions within communities and the much more important role those factors play in determining the ultimate pathways of a disease through a set of linked communities. However, even the initial simulations of the individual-based model described above show clearly that the theoretical conclusions derived from the population-based models need to be modified in light of results from the more realistic individual-based model. This is not to say that the population-based model was erroneous or limited in its usefulness — it provided important insights that clearly advanced our understanding of how infectious diseases spread across time and space and that stimulated many years worth of productive research. But for the practical problem of understanding the spread of the 1918 flu in central Canadian fur-trading populations and its differential impact in different communities and family groups, it is necessary now to shift focus to the more realistic individual-based models.

Both population-based and individual-based models are being used to study the geographic spread of infectious diseases in modern populations, and for many diseases, such as SARS, influenza, and smallpox, both approaches have proven valuable. Continued effort to develop models using multiple approaches to address the same questions will help increase the modern world's much needed understanding of how infectious diseases spread across time and space.

Acknowledgment

This work was supported by NSF Grant No. SBR-0094449 and by the Social Sciences and Humanities Research Council of Canada. Carrie Ahillen and Connie Carpenter provided substantive and important comments on the description of the individual-based model structure and also were the primary forces leading to the development and application of those models. This paper would not have been possible without their help.

References

1. I. M. Longini, Jr., A. Nizam, S. Xu, K. Ungchusak, W. Hanshaoworakul, D. A. T. Cummings, and M. E. Halloran, *Science* **309**, 1083 (2005).
2. N. M. Ferguson, D. A. T. Cummings, S. Cauchemez, C. Fraser, S. Riley, A. Meeyai, S. Iamsirithaworn, and D. S. Burke, *Nature* **437**, 209 (2005).
3. R. F. Grais, J. H. Ellis, and G. E. Glass, *Eur. J. Epidemiol.* **18**, 1065 (2003).
4. S. Riley, C. Fraser, C. A. Donnelly, A. C. Ghani, L. J. Abu-Raddad, A. J. Hedley, G. M. Leung, L.-M. Ho, T.-H. Lam, T. Q. Thach, P. Chau, K.-P. Chan, S.-W. Lo, P.-Y. Leung, T. Tsang, W. Ho, K.-H. Lee, E. M. C. Lau, N. M. Ferguson, and R. M. Anderson, *Science* **300**, 1961 (2003).
5. G. Chowell, P. W. Fenimore, M. A. Castillo-Garsow, and C. Castillo-Chavez, *J. Theor. Biol.*, **224**, 1 (2003).
6. L. Hufnagel, D. Brockmann, and T. Geisel, *Proc. Natl. Acad. Sci., USA*, **101**, 15124 (2004).
7. A. I. Hallowell, *The Ojibwa of Berens River, Manitoba: Ethnography into History*, Fort Worth: Harcourt Brace (1992).
8. N. P. A. S. Johnson and J. Mueller, *Bull. Hist. Med.*, **76**, 105 (2002).
9. H. Phillips and D. Killingray (eds.), *The Spanish Influenza Pandemic of 1918-1919*, London: Routledge (2003).
10. N. Markham, *Them Days*, **11(3)**, 3 (1986).
11. D. A. Herring and L. Sattenspiel, In *The Spanish Influenza Pandemic of 1918-1919*, H. Phillips and D. Killingray (eds.), London: Routledge, p. 156 (2003).
12. L. Sattenspiel and D. A. Herring, *Hum. Biol.*, **70**, 91 (1998).
13. L. Sattenspiel, A. Mobarry, and D. A. Herring, *Am. J. Hum. Biol.*, **12**, 736 (2000)
14. L. Sattenspiel and D. A. Herring, *Bull. Math. Biol.*, **65**, 1 (2003)

15. L. Sattenspiel, In *The Changing Face of Human Disease*, N. Mascie-Taylor, J. Peters, and S. T. McGarvey (eds.), Boca Raton: CRC Press, p. 40 (2004).
16. E. Williams, *Whooping Cough Among Western Cree and Ojibwa Fur-trading Communities in Subarctic Canada: a Mathematical Modeling Approach,* unpublished MA thesis, University of Missouri-Columbia (2004)
17. M. Stoops, *Simulating the Spread of Smallpox in 19th Century Canadian Fur Trapping Communities*, unpublished Honors thesis, University of Missouri-Columbia (2004).
18. L. Sattenspiel and K. Dietz, *Math. Biosci.*, **128**, 71 (1995).
19. C. Carpenter, *Agent-based Modeling of Seasonal Population Movement and the Spread of the 1918-1919 Flu: the Effect on a Small Community*, unpublished MA thesis, University of Missouri-Columbia (2004).
20. M. J. North, N.T. Collier, and J.R. Vos, Experiences creating three implementations of the Repast agent modeling toolkit, *ACM Trans. Mod. Comp. Simul.*, **16**(1), 1 (2006).
21. C. Ahillen, *Agent-based Modeling of the Spread of the 1918-1919 Spanish Flu in Three Canadian Fur Trading Communities*, unpublished MA thesis, University of Missouri-Columbia (2006).

MATHEMATICAL MODELS OF TUBERCULOSIS: ACCOMPLISHMENTS AND FUTURE CHALLENGES

CAROLINE COLIJN, TED COHEN

Harvard School of Public Health, Kresge Bldg
677 Huntington Ave., Boston, MA 02138, U.S.A.
E-mail: ccolijn@hsph.harvard.edu

MEGAN MURRAY

Massachusetts General Hospital, Infectious Disease Unit
Gray 5, 55 Fruit St., Boston, MA 02114, U.S.A.

Tuberculosis is a leading cause of infectious mortality. Although anti-biotic treatment is available and there is vaccine, tuberculosis levels are rising in many areas of the world. The recent emergence of drug-resistant of TB is alarming, as are the potential effects of HIV on TB epidemics. Mathematical models have been used to study tuberculosis in the past and have influenced policy; there is renewed opportunity for mathematical models to contribute today. Here we review and compare the mathematical models of tuberculosis dynamics in the literature. We present two models of our own: a spatial stochastic individual-based model and a set of delay differential equations encapsulating the same biological assumptions. We compare two different assumptions about partial immunity and explore the effect of preventative treatments. We argue that seemingly subtle differences in model assumptions can have significant effects on biological conclusions.

1. Introduction

Despite many decades of study, the widespread availability of a vaccine, an arsenal of anti-microbial drugs and, more recently, a highly visable World Health Organization effort to promote a unified global control strategy, tuberculosis (TB) remains a leading cause of infectious mortality. It is responsible for approximately two million deaths each year. Although TB is currently well-controlled in most countries, recent data indicate that the overall global incidence of TB is rising as a result of resurgence of disease in Africa and parts Eastern Europe and Asia (Dye, 2006). In these regions, the emergence of drug-resistant TB and the convergence of the HIV (human immunodeficiency virus) and TB epidemics have created substantial new challenges for disease control.

Mathematical models have played a key role in the formulation of TB control strategies and the establishment of interim goals for intervention programs. Most of these models are of the SEIR class in which the host population is categorized by infection status as susceptible, exposed (infected but not yet infectious), infectious and recovered. One of the principle attributes of these models is that the force of infection (the rate at which susceptibles leave the susceptible class and move into an infected category, ie become infected) is a function of the number of infectious hosts in the population at any time t and is thus a nonlinear term. Other transitions, such as the recovery of infectious individuals and death, are modeled with linear terms with constant coefficients.

In many such models, there is a sharp threshold behavior, and the asymptotic dynamics are determined by a parameter R_0. When $R_0 < 1$, the disease-free equilibrium is (usually globally) asymptotically stable, and when $R_0 > 1$ there exists a unique endemic equilibrium, which is also (usually globally) stable. R_0 represents the of average number of new infectious cases caused by an infectious case in a fully susceptible population, over the course of the entire infectious period. However, with vaccination, relapse and reinfection there can be other threshold values and the interpretation of R_0 is not so clear. Sometimes the term "effective reproductive number", usually denoted R, is used in these situations. Not all SEIR-type models have an R_0 threshold; see Vynnycky and Fine (1998) for a discussion of R_0 in TB models, and Heffernan et al. (2005) for a review of R_0 estimation in different kinds of models.

While the mathematical models of TB in the literature share structural similarities, there are important, if sometimes subtle, differences in the way that the natural history of disease and the TB transmission process are represented. We suggest that some of these differences may have substantial effects and it behooves us to have a greater comparative understanding of the properties of these models. Here we present a brief historical review of the development of several dynamic tuberculosis models; we do not attempt to present an exhaustive survey of every model that has appeared in the literature, rather we highlight a subset which illustrate different important methodological approaches.

2. The natural history of tuberculosis

Tuberculosis is caused by infection with the bacterium *Mycobacterium tuberculosis*. It is primarily transmitted by the respiratory route; individuals with active disease may infect others if the airborne particles they produce when they cough, talk, or sing are inhaled by others. Once infected, an individual enters a period of latency during which he exhibits no symptoms and is not infectious to others. This latent period can be of extremely variable length and the great majority of those infected (~90%) will never have clinical tuberculosis. However, a small proportion of individuals progress to disease relatively rapidly, falling ill within months or several years after infection. Others may be asymptomatically infected for decades before they become sick. Once ill and infectious, individuals may recover without treatment, may be cured with antibiotics, or may die from TB. Recovered individuals may relapse to disease or be reinfected. The degree of protection afforded by a previous infection and mechanism by which individuals with partial immunity are protected are controversial.

3. Mathematical models of tuberculosis

For clarity, we group models by their structure: ordinary differential equations (SEIR-type models), age-structured and delayed models, comprising both partial differential equation and discrete-time compartment models, and spatially structured models. Within each category we describe contributions in roughly chronological order.

Because the natural history of tuberculosis is complex and difficult to accurately reflect in a simple model, modelers have made several different simplifying choices. The first relates to the progression of a latent infection to active disease. Observational data suggest that upon infection the risk of progression to active disease is much higher for the first few years than it is subsequently (Vynnycky and Fine, 2000). It is not clear whether this is due to innate protection in some individuals (who have reduced risk of progression), or if it is a time-varying risk of progression in all individuals. There are three ways that modelers have represented progression, which we have summarized schematically in Figure 1. The first, shown in Figure 1 A or B is to simply ignore the different rates of progression and to represent only one latent class which progresses at a single rate to the infectious class. Some such models include a direct transfer of some fraction of the susceptibles to the infectious class (A), representing essentially immediate progression. See Blower et al. (1995) and Gomes et al. (2004) for examples.

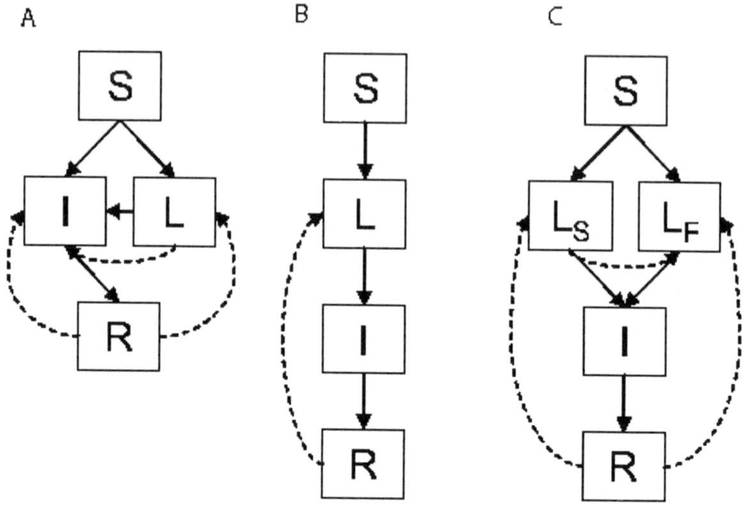

Figure 1. Three schematic models showing different representations of progression from latent infection to active disease. Dashed lines represent reinfection.

The second approach is to model progression probability as a function of the time since infection. This may be done either with explicit age and maturation (of the infection) structure, as is done by Waaler (1968c) and Vynnycky and Fine (1997b), or the explicit inclusion of variable latent periods in a delayed and/or integro-differential equation as was done by Feng et al. (2001). While this approach is more challenging mathematically, it removes the need to set as a parameter the fraction of susceptibles who, upon infection, are "predestined" to be fast progressors, as is done in the third approach. In the second approach, the latent class in Figure 1 B would progress at different rates to I depending on the age since infection; the rate could be any (usually decreasing) function of the time since infection.

The third approach is to explicitly represent two different latently infected classes: "fast progressors" and "slow progressors". This approach is summarized in Figure 1 C. In these models, newly infected individuals are assigned to one of these two compartments and experience the corresponding rate of progression. Typically individuals can move from the slow group to the fast group (if a re-infection event occurs), and not vice versa. Examples of models in this group are numerous, and include those of Dye et al. (1998), Murray and Salomon (1998), Ziv et al. (2001), Cohen and Murray (2004) and others.

Another more subtle division among TB models is whether, and how, they choose to represent reinfection. Some models do not permit reinfection, implying that previous infection confers complete immunity. However, more recent observation (Rie et al., 2005; Warren et al., 2004) shows that individuals may be infected with multiple circulating TB strains (cases of active disease with two strains have been observed, as well as individuals with sequential episodes of TB with different strains each time). Some models allow reinfection to move individuals from the recovered and latent class(es) to the infectious class, as indidicated by the dashed lines in Figure 1 A (Gomes et al., 2004), some allow reinfection to move individuals from the recovered and slow progressing classes to the latent class (dashed lines in Figure 1 B) or to the fast-progressing class (Figure 1 C; Cohen et al. (2006a)), and still others use explicit representation of multiple strains, where reinfection moves individuals from any compartment of one strain to another (Castillo-Chavez and Feng, 1997), or a combination (Cohen and Murray, 2004). As mentioned above, there is no consensus on how much protection, and what kind of protection, is gained through previous infection. Some authors allow partial immunity through reduction of the probability of reinfection (for example Cohen and Murray (2004); Dye et al. (1998) with a new strain while others specify that individuals with prior infection are as likely as fully susceptible people to be infected, but less likely to progress to disease upon reinfection (Vynnycky and Fine, 1997b).

3.1. *Ordinary differential equation models*

The first mathematical model of TB was presented by Waaler et al. (1962). Following this, there were several numerical studies, primarily focusing on cost-effectiveness of different interventions (Brogger, 1967; Revelle et al., 1969). Revelle et al. (1969) used a model with one progression rate and various latent classes representing different treatment and control strategies, and argued that vaccination was cost-effective in countries with high TB burdens. Waaler continued his work in Waaler (1968a), Waaler (1968b), Waaler and Piot (1969), Waaler (1970) and Waaler and Piot (1970). After the 1970's little work on models of tuberculosis appeared in the literature until the mid-1990's.

In 1995, Blower et al. presented two differential equation models of TB, a simpler model (see Eq. (1)) and a more detailed one. Both are SEIR-type models; the detailed model has both infectious and non-infectious active TB as well as recovery.

It is useful to give examples of several models that are proto-typical of models that come later. We use our own (and standard) notation, S for susceptible, L for latent, I for infectious, and R for recovered: The first model of Blower et al. (1995) assumes that susceptible people are born into the population at rate Λ. Susceptible individuals are infected at rate βIS and move either into the latent class L or directly into the infectious class I (fast progression). Latent individuals progress to active disease when they are infectious at a constant rate (slow progression). In the infectious state, individuals suffer an increased death rate due to disease.

$$\begin{aligned}\dot{S} &= \Lambda - \beta IS - \mu S \\ \dot{L} &= (1-p)\beta IS - (\nu + \mu)L \\ \dot{I} &= p\beta IS + \nu L - (\mu + \mu_T)I.\end{aligned} \tag{1}$$

Here β is the transmission parameter, μ is the natural death rate, μ_T is the rate of death due to TB, ν is the progression rate from latency to active disease and p is the portion "fast progressors".

In their second, more detailed, model, Blower et al. (1995) use a similar approach but specify two active TB classes (one infectious and one non-infectious), a recovered class (with entry cI for cure), and relapse into either active class. Note that, as mentioned above, a parameter (in this case p) sets the proportion of new infections that move directly into the infectious class.

In their analysis (Blower et al., 1996), R_0 is decomposed into a sum of R_0 values for each of three routes to disease (fast progression, slow progression and relapse), forming the basis for the authors' subsequent discussion of "linked TB epidemics". The model is matched to TB mortality data and R_0 is used to derive a population threshold below which the disease cannot take hold. An interesting aspect of this work is that the authors compute doubling times of TB epidemics (in their initial transient rise as the epidemic invades a population); these are very long ($\sim$ 100 years) for R_0 near 1, as it has been estimated to be by Salpeter and Salpeter (1998). The authors also compare aspects of young and old TB epidemics, which are different because of the long time scales inherent in the model.

In 1997, Castillo-Chavez and Feng (1997) present an SEIR model with one form of latency and one class of active TB. In this model, individuals can only move to the infectious class from the latent class, so there is only one progression rate, and there is recovery from latency and active disease back to the susceptible class. They find an R_0 for the model, and show global asymptotic stability of the disease-free equilibrium when $R_0 < 1$, and local asymptotic stability of the unique endemic equilibrium for $R_0 > 1$; global asymptotic stability of this was proved in Feng et al. (2001)). Their first model is analogous to Eq. (1) and is given by

$$\begin{aligned}\dot{S} &= \Lambda - \sigma IS/N + r_1 L + r_2 I - \mu S \\ \dot{L} &= \sigma IS/N - (\mu + \nu + r_1)L \\ \dot{I} &= \nu L - (\mu + \mu_T + r_2),\end{aligned} \tag{2}$$

where r_1 and r_2 are cure rates out of latent and active infection, respectively, and $N = S + L + I$ is the population size.

In the same paper, Feng and Castillo-Chavez present a two-strain model, in which the drug-resistant strain is not treated, and latent, infectious and treated individuals may be re-infected with the drug-resistant strain. Each strain has a different R_0, and there are 3 equilibrium points (no disease, coexistence of both strains, and only the drug-resistant strain). Without acquisition of drug resistance, there is an additional equilibrium with only the drug-sensitive strain. The authors discuss stability of the equilibria and find, interestingly, areas of parameter space of postive measure where coexistence of the strains is possible; they report that coexistence is rare when drug resistance is mainly primary (resulting from transmission) but almost certain if the resistant strain is the result of aquisition, for example under poor treatment. Neglecting disease-induced death and setting the transmission parameter equal for the two strains, they are able to prove that the disease-free equilibrium is globally asymptotically stable if both R_0's are less than 1. These results were extended in Mena-Lorca et al. (1999).

In Blower et al. (1996) and Blower and Gerberding (1998), Blower and colleagues discuss two additional models with a focus on chemophrophylactic treatment (which prevents progression from latency to active disease). These are both like Eq. (1) in that there is a direct transition from S to I. The second model contains two strains, drug-sensitive and drug-resistant. The drug-resistant phenotype may be acquired among those treated for drug-sensitive active disease, or directly transmitted to susceptible individuals. This was the first multi-strain model of TB. In Blower et al. (1996) the authors focus on the threshold R_0 and the development of drug resistance. They define a variable X to be the number of drug-resistant cases caused by the treatment of one drug-sensitive case. The authors then use the condition $R_0 < 1$ to compute $r_{\max}(X)$, the maximal acceptable probability of treatment failure. They conclude that control programs could become perverse (meaning $X > 1$), though this requires a rather high probability of aquisition of drug resistance due to treatment failure (up to 0.5. In countries with a high TB burden, they conclude that the efficacy of treatment combined with the effective overall treatment rates must be kept high in order to control TB. They estimate in model terms the World Health Organizations objectives for the year 2000, and report that these targets do not satisfy $R_0 < 1$.

Blower and Gerberding (1998) expand on this work; here the authors focus on trajectories in the two-strain model (rather than threshold conditions with R_0), and simulate specific control policies, numerically in the short term, and using R_0 analyses for long-term consequences. In their model, policies leading to the same long-term equilibrium can have very different transient approaches to that equilibrium, which is relevant in TB because transients can be very long (as the authors point out). Furthermore, some transients in their model show a decline in the portion of drug-resistant TB over a 10-year period, followed by a slow increase. The authors discuss this qualitatively in terms of TB's inherent fast and slow time scales; in conclusion they argue for vaccination, and warn about the consequences of focusing control measures only on drug-sensitive TB. In Blower et al. (1999) the authors comment that in their models, while increasing antibiotic use can contribute to the emergence of drug resistance, increasing treatments with higher efficacy can still have an overall beneficial effect.

Other papers by Blower and coauthors include, but are not limited to, Sanchez and Blower (1997) and Porco and Blower (1998), in which the authors perform a sensitivity analysis of the three R_0 values from Blower et al. (1995) (fast, slow and relapse), and of TB outcomes, respectively. They generally find $1 < R_0 < 9$, and that the most important parameters are the infection rate (β), the portion p "fast" progressors, the reactivation rate ν, and the death rates.

It is not always clear how to define R_0 for a given model. E. Vynnycky, more of whose work is discussed in Section 3.2 below, argues in Vynnycky and Fine (1998) that the interpretation and application of R_0 is difficult when the rate of transmission of disease changes in time (which they argue in Vynnycky and Fine (1999)), and when reinfection contributes to disease. Indeed, there are a variety of biological questions that have been asked of TB models that go beyond the threshold behavior in R_0.

Several TB models aim to investigate optimal treatment strategies. Lietman and Blower (2000) study pre- and post-exposure vaccines, using models with fast and slow progressors, and vaccines parametrized by their "take", "degree" and "duration", permitting various mechanisms by which these programs may be less than 100% effective. They find that even if a vaccine is only moderately effective, it may reduce TB epidemics if coverage is high. A strategy of continuous vaccination of newborns after a single mass vaccination of susceptibles appears to perform best. However the vaccines simulated are theoretical and estimates of the efficacy of the existing vaccine Bacille Calmette-Guérin (BCG) are highly variable (Colditz et al., 1994, 1995). In Ziv et al. (2001) the authors use an SEIR-model with fast and slow progression to numerically compare the effects of preventative treatment of those in the fast-progressing latent class with treatment of those with active, infectious disease; they conclude that contact tracing and preventative treatment compare quite favorably to treatment of those with disease.

Murray and Salomon (1998) present a model with 19 different states, which includes fast and slow progression, treatment with chemoprophylaxis (INH), superinfection (e.g. reinfection), three clinical categories of active disease, and "good" and "bad" treatment. The authors calibrate the model to five different regions of the world, and use it (numerically) to quantitatively predict that a major extension to DOTS could be quite effective, preventing millions of deaths.

Dye et al. (1998) present a model with explicit fast and slow progression from two latent classes. They study drug-resistant TB alone, representing treatment failures as potential transmitters of drug-resistant TB. Using Monte Carlo methods to estimate the model's R_0 for drug-resistant TB, they argue that short-course chemotherapy can bring resistant strains under control, preventing drug-resistant TB from emerging, and that this can probably be done by meeting the WHO targets for case detection and cure. This is in contrast to the conclusion of Blower et al. (1996), whose analysis was based on an R_0 threshold in a different model. However, Dye's result depends on the assumption that drug-resistant strains are less transmissable than drug-sensitive strains, and does not explicitly represent the dynamics of drug-sensitive disease. In a more recent paper, Dye and Williams (2000) used a model that allowed the relative fitness of drug-resistant strains to be as high as 1 (i.e. the same as drug-sensitive strains), and concluded that drug-resistant strains could threaten control of TB. Minimally, they estimate that 70% of drug-resistant cases must be detected and 80% of these must be cured in order to prevent drug-resistant TB outbreaks. In Espinal et al. (2001), however, the authors used data from one molecular epidemiology study and a ceiling of fitness of the drug-resistant strain of 0.7, and concluded that drug-resistant strains were likely to remain a localized problem rather than a problem for global control.

Differing conclusions about the potential threat that drug-resistant (DR) and multi-drug-resistant[a] (MDR) TB strains pose for the global control of TB emphasize the importance of estimates of the fitness costs associated with the acquisition of drug-resistant phenotypes. Citing an earlier review in which they note significant variation in the deleterious effects associated with resistance-conferring mutations (Cohen et al., 2003), Cohen and Murray (2004) expand earlier models to allow the fitness of drug-resistant strains to be heterogeneous. This model allows that while most mutations will come at a cost to the transmissability of the strain, some strains may acquire resistance with minimal negative effect. Thus even in the presence of a "good" control program, the average fitness of MDR strains will increase as more fit strains are preferentially transmitted. They find that short term trends in burden of MDR TB are relatively uninformative and should not be used to speculate that MDR TB can or will be contained.

Blower and Chou (2004) introduce a model which allows "amplification" of drug resistance. That is, they allow that strains may become resistant to increasing numbers of drugs if sub-optimal treatment regimens are used. They find, in agreement with results of Dye and Williams (2000) and Cohen and Murray (2004) that individuals with drug-resistant disease must be treated to avoid the further emergence of resistance.

In most of the models discussed thus far, the question of reinfection has not been a central one. However it has been included in various models including Vynnycky and Fine (1997b), Dye et al. (1998), Gomes et al. (2004), Cohen and Murray (2004) and Cohen and Murray (2005). In Feng et al. (2000) the authors add a transition from the latent class to the infectious class of the form $p\beta LI/N$, and show that this changes the qualitative dynamics of the model, allowing for the existence of multiple endemic equilibria even when $R_0 < 1$, if p is sufficiently large. Lipsitch and Murray (2003) argue that the required value of p is too large for the result to be meaningful for TB. However, the change in qualitative behavior of the model from the usual R_0-determined dynamics clearly indicates the necessity of establishing the existence of a threshold behavior with respect to R_0 before using basic definitions of R_0 (the average number of new infectious cases from one infectious case in a susceptible population) to derive treatment thresholds.

[a]A strain is considered multi-drug-resistant if it is resistant at least to isoniazid and rifampin

In Gomes et al. (2004), the authors detect a "reinfection threshold", which is a value of R_0, larger than one, after which reinfection is the predominant form of disease transmission. Breban and Blower (2005) correctly point out that this is not a sharp threshold, in that it does not correspond to a bifurcation point; $R_0 = 1$ is the only bifurcation point. However, Gomes et al. (2005) reply that there is a bifurcation in a submodel of the model of Gomes et al. (2004) which gives rise to what resembles threshold behavior.

Several of the authors referred to here, along with others have begun in more recent years to explore the impact of the high prevalence of HIV on TB epidemics. While a full review of this literature is beyond the scope of this paper, this is a topic of increasing importance in Africa where both diseases are at critically high levels and so we will include some examples of these models. In Porco et al. (2001) the authors use discrete event simulation model to predict HIV's impact on TB epidemiology. They find that HIV can significantly affect levels of TB, but that the system is sensitive to the TB treatment rate. In Currie et al. (2003) a difference equation model is developed representing both HIV and TB dynamics with simple submodels. The system is fitted to time series data, to compare TB treatment with combined HIV and TB prevention in high-burden areas. Prevention in their model is by reduction of HIV transmission, highly active antiretroviral therapy (HAART), and chemoprophylactic TB treatment (of latent infection). They conclude that prevention is insufficient and treatment of active TB disease should not be replaced with these measures. Cohen et al. (2006b) developed a dynamic model of linked TB and HIV epidemics which includes a state of mixed drug-resistant and drug-sensitive TB infection; the inclusion of states of multiple pathogen infection (i.e. TB/HIV co-infection as well as superinfection with multiple strains of TB) reflects recent recognition that individuals may simultaneously harbor several distinct *M. tuberculosis* strains. Using this model it was found that current recommendations for use of single drug preventive therapy to prevent progression from infection to TB disease among those with HIV co-infection may reduce TB prevalence in the short-term but may eventually be counterproductive without directed efforts to improve the diagnosis and treatment of those with drug-resistant TB.

3.2. *Age-structured and delayed models*

Waaler (1968c) presents an early numerical age structured model for TB.

At the time, a system whose explicit solution was available only numerically was reasonably novel, and the paper argues for its relevance more than it uses the model to derive results. The natural history of TB contains two latently infected classes (fast and slow progression), where all susceptibles are initially in the fast-progression class. It does not include reinfection.

A much cited age-structure model of TB is given by Vynnycky and Fine (1997b). Here, vaccination is assumed to completely protect 77% of those vaccinated. Upon infection individuals enter a "fast latency" which lasts 5 years, and move from there to a slow-progressing latency (if they do not develop active disease). It does include reinfection of those in the slow-progressing latent class. Reinfected individuals move into the faster-progressing latent class, but their previous infection affords them some protection from progression, *not* from transmission of the new infection. Vynnycky and Fine (1997b) model transmission in a way that differs from most SEIR models of infectious disease transmission, in that the force of infection is not modeled as a function of the prevalence of infectious cases (βIS). Rather, the "force of infection" term is a prescribed function of time, derived from data (Vynnycky and Fine, 1997a). In Vynnycky and Fine (2000), this model was used to argue that the lifetime risk of developing disease, the length of the latency period, and the time period between infection and transmission have changed significantly over the 20th century. In a later paper, Vynnycky and Fine (1999) argue that a major factor in the decline of TB has been the declining effective contact rates; in other words, that the number of people an infectious case contacts sufficiently to transmit disease has declined.

Castillo-Chavez and Feng (1998) also consider an age-structured model, in their case with age-dependent transmission rates. They have one form of latency, as in their earlier work, and this model does not contain reinfection. They study vaccination, and define a vaccine-dependent R_0 threshold $R(\psi)$. They prove stability of the disease-free equilibrium when $R(\psi) < 1$, and the existence of an endemic steady state when $R(\psi) > 1$, and discuss analytically-determined optimal vaccination strategies. These turn out to be either vaccination at a single age, or vaccination in precisely two age classes. It should be noted that they arrive at their conclusions by analytically minimizing an integral cost function, and not through numerical simulation.

Dye et al. (1998) also have developed an age-structured model, with the goal of exploring TB control under the DOTS strategy and under improved case detection and cure. Their model uses discrete time rather than a partial differential framework. It includes two latent classes (slow and fast) with a set portion of new infections moving into each (see Figure 1 C), as well as reinfection. They validate their model by comparing their results to annual risk of infection data from the Netherlands[b]. They then simulate different epidemiological situations. They conclude that if tuberculosis is stable, and if HIV is not included in the model, then WHO targets of 70% case detection and 85 % cure "would reduce the incidence rate by 11% (range 8-12) per year and the death rate by 12% (9-13) per year". The effect would be smaller if TB were already in decline. They conclude that DOTS has greater potential today (in developing countries) than it did 50 years ago in developed countries, but that case detection and cure rates must be improved. Salomon et al. (2006) extend this model, calibrating it to natural TB epidemics and then studying the effects of hypothetical new treatments with differing durations. In their model, drug resistance could result from transmission and acquisition. They first calibrate the model to represent TB in South-East Asia. They conclude that if more rapid treatments for TB were available, with reduced infectious periods, this could significantly reduce TB mortality and incidence, particularly if these become available soon.

There are, of course, other alternatively-structured models. For example, Debanne et al. (2000) presented a multivariate Markov model that projected different levels of TB disease among different ethnic groups in the US. Salpeter and Salpeter (1998) used TB incidence data to derive a function for the time delay from initial infection to active disease, and develop an integral equation for R_0. This is not a dynamic epidemiological model for TB, but it is relevant to TB models because the variable delays and widely ranging estimates of R_0 are important. They report R_0 near 1, which in Blower et al. (1995) was associated with the longest time scales. Feng et al. (2001) analyze the impact of variable latent periods, and show that their inclusion in the model of Castillo-Chavez and Feng (1997) does not change the qualitative dynamics of the model.

3.3. *Spatial and network models*

Schinazi (2003) presents a spatial model of tuberculosis which takes place on a lattice. Each site is in one of three states: susceptible, latent and infectious. Infection is only possible among nearest neighbors. Latent

[b]The annual risk of infection is defined as the per year probability that a susceptible individual becomes infected.

individuals may become infectious either through progression or through reinfection. An interesting aspect of this model, from our current point of view, is that the long-term accessible behaviors are quite different in the spatial version of the model than in the (homogeneously mixed) mean-field approximation. All of the models discussed above, except for Song et al. (2002), assume a fully homogeneously mixed population. Furthermore, in models for other diseases, spatial effects have been shown to be quite important Keeling et al. (1997); Keeling and Eames (2005); Eames and Keeling (2003); Pourbohloul and Brunham (2004).

In Cohen et al. (2006a), we present a spatial stochastic model for TB on spatially structured random graphs. In this model, susceptible individuals first enter a fast-progressing latent class which lasts 5 years, after which they move to the long-term latent class. As in Vynnycky and Fine (1997b), reinfection is possible, and previous infection partially protects reinfected individuals from progression to active disease. This model was used to examine the role of spatial structure on the relative importance of exogenous reinfection in TB epidemics, in the absense of treatment. Where in the models above (Blower and Gerberding, 1998; Vynnycky and Fine, 1997b) exogenous reinfection is thought to play a role only when disease burden is high, spatial effect such as the formation of clusters of locally high latency can result in exogenous reinfection being important even when disease incidence is low.

4. Our models

We have developed two models of TB transmission: a stochastic individual-based model and a delay differential equation model encapsulating the same biological assumptions. In this section we briefly introduce the two models, and compare their predictions of the relative contribution of exogenous reinfection to TB epidemics. This question is relevant for drug-sensitive TB, as we have modeled it here, but more importantly it will be critical in understanding the clustering and evolution of different strains of drug-resistant TB as they emerge. In each model, we then compare two different ways to represent the partial immunity that individuals gain after an initial TB infection: protection from the new infection itself, vs. protection from progression to active disease after the new infection. We compare the models' predictions about the effectiveness of chemoprophylactic treatment (prevention of primary progression). This is related to the relative contribution of reinfection, because where reinfection is common, one would expect that a treatment that prevents primary progression only would not be as effective as it would if reinfection were rare.

4.1. *Spatial model*

The first model we describe is an individual-based stochastic model. The population is represented by a spatially structured random graph and disease transmission may occur between two individuals i and j if there is an edge between vertices i and j. The graphs are created following Read and Keeling (2003), probabilistically and with a preference for shorter edges. Varying how much more likely short edges are than long ones allows graphs to be created with more local (short) or more global (long) edges. Because there are fewer vertices nearby than far away, graphs with primarily short links have higher clustering coefficients[c] than those with longer links.

Individuals may be in one (and only one) of the following states: susceptible, latently infected, infectious and recovered. The transitions in the model are described in detail elsewhere (Cohen et al., 2006a; Colijn et al., 2006); the essential features are that all newly latent individuals suffer from a "fast" progression rate, whereby they enter the infectious class. After 5 years, if an individual has not yet progressed to active disease, this rate drops. Reinfected latent individuals also suffer a faster rate of progression.

In Cohen et al. (2006a) we apply this model to the question of the relative importance of reinfection in TB epidemics. The central result is that on graphs with higher clustering coefficients, spatial clustering emerges during the natural fall-off of the TB epidemic between 1900 and 1990. As a result, there are pockets of relatively high latent infection, where reinfection can be an important contributor to disease. Furthermore, most TB incidence in the model takes place in such pockets. As a result, exogenous reinfection can contribute more to disease burden than was previously believed.

Here, we explore the two main ways to model the partial immunity that is conferred by a previous infection, and simulate chemoprophylactic treatment under the two representations. In our earlier work we have modeled partial immunity as protecting reinfected individuals from progressing to active disease; here we compare this to protection from gaining the second infection at all. More specifically, let τ be the per contact rate of transmission of disease, p_1 be the rate of primary (fast) progression, and f be the degree of partial immunity. Thus, where the reinfection probability

[c]The clustering coefficient of a graph is the likelihood that if vertices i and j are both connected to vertex k then they are connected to each other.

for a latent or recovered individual in contact with k neighbors in Cohen et al. (2006a) and Colijn et al. (2006) was $1-(1-\tau)^k$, and the subsequent progression rate was $p_I = (1-f)p_1$, we now set the infection probability to be $(1-f)(1-(1-\tau)^k)$ and the progression rate $p_I = p_1$. We make a similar extension in a delay differential equation model.

4.2. *Delay model*

Our delay differential equation model is presented in detail in Colijn et al. (2006). Its equations are given in Eq. (3). The notation is as follows: S is the fraction of the population in the susceptible state, L is the fraction in the slow-progressing latent state, I is the fraction that are infectious and the parameters are as defined in Appendix A.

$$\begin{aligned}
\frac{dS}{dt} &= \gamma - \beta I S - \mu S \\
\frac{dL}{dt} &= \beta_{\tau_1} e^{-(\mu+p_1)\tau_1} I_{\tau_1} S_{\tau_1} + \bar{\beta}_{\tau_1} e^{-(\mu+p_I)\tau_1} I_{\tau_1}(L_{\tau_1} + R_{\tau_1}) \\
&\qquad - \bar{\beta} I L - p_2 L - \mu L + r I \\
\frac{dI}{dt} &= \beta_{\tau_1} e^{-\mu\tau_1}(1 - e^{-p_1\tau_1}) I_{\tau_1} S_{\tau_1} + \bar{\beta}_{\tau_1} e^{-\mu\tau_1}(1 - e^{-p_I\tau_1}) I_{\tau_1}(L_{\tau_1} + R_{\tau_1}) \\
&\qquad + p_2 L + r_{\mathrm{rel}} R - r I - \mu_{\mathrm{TB}} I - r_{\mathrm{TR}} I_{\tau_2} \\
\frac{dR}{dt} &= r_{\mathrm{TR}} I_{\tau_2} - r_{\mathrm{rel}} R - \bar{\beta} I R - \mu R
\end{aligned} \tag{3}$$

Susceptible individuals first enter a fast-progressing latent class which lasts $\tau_1 = 5$ years, and from which they either die at rate μ or progress to active infection. If they do not die or progress (i.e. a portion $e^{-(\mu+p_1)\tau_1}$) they enter the slow-progressing latent class, L. A portion $(1 - e^{-p_1\tau_1})$ *did* progress to active infection and for simplicity these are simply placed in the infectious class, though this is an approximation of what would otherwise be an integral equation for I (such as the one given in Feng et al. (2001)). We ensure a constant population by setting

$$\gamma = \mu(1 - I) + \mu_{\mathrm{TB}} I. \tag{4}$$

As above, to model partial immunity through protection from progression we set $p_I = (1-f)p_1$, and to model protection from reinfection we set $\bar{\beta} = (1-f)\beta$, where f is the degree of partial immunity.

Figure 2 shows the disease trajectories generated by the two representations at the same parameter values, along with a trajectory from the spatial

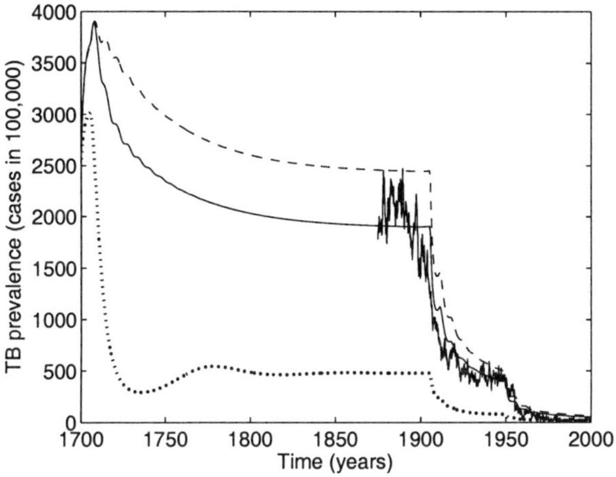

Figure 2. Tuberculosis disease trajectories showing the comparison between two different forms of partial immunity. The dashed line represents protection through β and the solid lines protection through p_I in the delay and spatial stochastic models. All other parameters are the same. The dotted line shows the model trajectory at the same parameter values with the assumption that there is no reinfection at all.

stochastic model for comparison. If partial immunity affects $\bar{\beta}$ (dashed line), disease levels are higher than if partial immunity reduces the progression p_I. The difference is most visible when disease prevalence is high, but it persists throughout the epidemic. The model trajectory under the assumption that there is no reinfection at all ($\bar{\beta} = 0$) is shown for comparison.

One implication of this observation is that if the model were fit to data, the two assumptions about how partial immunity acts would lead to different parameter estimates. Assuming that partial immunity affects $\bar{\beta}$, β, the overall transmission parameter, would need to be 20% lower than if one assumed that partial immunity affects progression. This in turn would affect intuitive understanding about the likelihood of new transmission (for example, of drug-resistant strains).

Even more notable are the parameter estimates that would be needed to match the model trajectory in the absense of reinfection (dotted line in Figure 2) to realistic levels. In our model, increasing β is not sufficient to accomplish this. In fact, β needs to be *decreased* by half, and p_1 must be increased by a factor of 4[d]. β is notoriously hard to measure, as it en-

[d]Further increases in β lead to very slow, non-monotonic approach to equilibrium between 1900 and 1950 with a sharp increase between 1935 and 1950 rather than a decrease

capsulates the per contact transmission rate as well as the overall "contact rate" in a population, and SIR-type models are usually quite sensitive to it. TB models are no exception (Sanchez and Blower, 1997; Porco and Blower, 1998). Our model and others are also very sensitive to p_1. Diverse estimates of β and p_1 would certainly have implications for determining what realistic interventions, if any, would lead to an $R_0 < 1$ and elimination of disease.

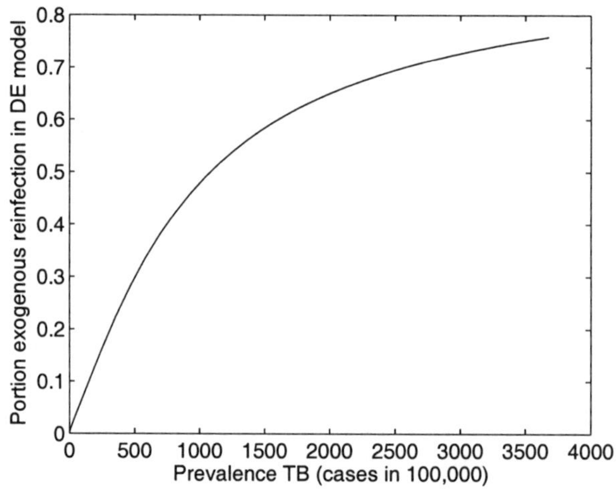

Figure 3. The contribution of exogenous reinfection to disease incidence at various prevalences

While the two forms of partial immunity differ in the resulting prevalence of disease under the same parameters, their prediction about the overall contribution of exogenous reinfection are essentially the same. That is, at a given prevalence of infectious TB, both models predict the same contribution to new disease from reinfection. Figure 3 shows the relationship between disease prevalence at equilibrium and the contribution to new disease due to exogenous reinfection, in the delay differential equation model (and consequently under the assumption of a homogeneously mixed population). In Cohen et al. (2006a) we show the contribution from reinfection in the spatial stochastic model for local and global graphs; reinfection is more important at low prevalences for more local graphs, as mentioned above.

Preventative treatment, such as isoniazid preventative therapy (IPT), can reduce the chance of progression from latency to active disease, if it is administered shortly after an initial infection. However, it is not possible to determine how recently a latent tuberculosis infection was acquired, except by serially testing individuals such as health care workers who are assumed to be at high risk. One would expect that such treatment would be quite effective when the risk of reinfection was small, and less so if individuals were likely to be reinfected subsequently. Figure 4 shows the effect of reducing the primary progression rate p_1 by a factor of 2 in the delayed model, without changing p_I.

Figure 4 also shows the model's threshold behavior. Near the critical value, the decrease in the primary progression rate can lead to elimination of disease under both forms of partial immunity. Just above the threshold, preventative therapy is still very effective, and it is less so as disease prevalence rises; this effect is due to the high levels of reinfection.

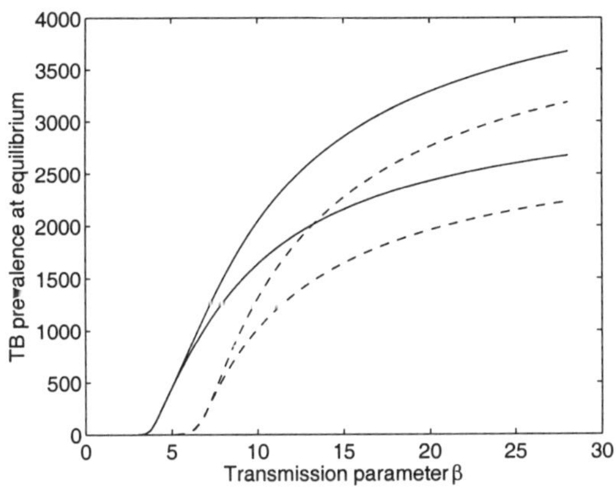

Figure 4. The effect of preventative therapy on equilibrium disease prevalence in the two forms of partial immunity. Upper solid line: partial immunity lowers $\bar{\beta}$. Upper dashed line: $\bar{\beta}$ with preventative therapy. Lower solid line: partial immunity lowers p_I. Lower dashed line: p_I protection with preventative therapy.

Figure 5 shows the results from the spatial model. The prevalence is shown as a function of the length scale of the graph and the per year, per contact transmission probability. In both plots, the higher surface represents the prevalence of TB without preventative treatment, and the lower surface shows the prevalence with preventative treatment. The figure shows that preventative treatment is more effective in the model for graphs with longer length scales; these graphs also have lower clustering coefficients. In local graphs, an infectious individual infects its neighbors and creates a localized pocket of latent infection. When one of the neighbors progresses, other members of the cluster are likely to be reinfected. It is therefore more likely a given individual will be repeatedly reinfected on a graph with a higher clustering coefficient. This in turn reduces the overall effectiveness of preventative therapy.

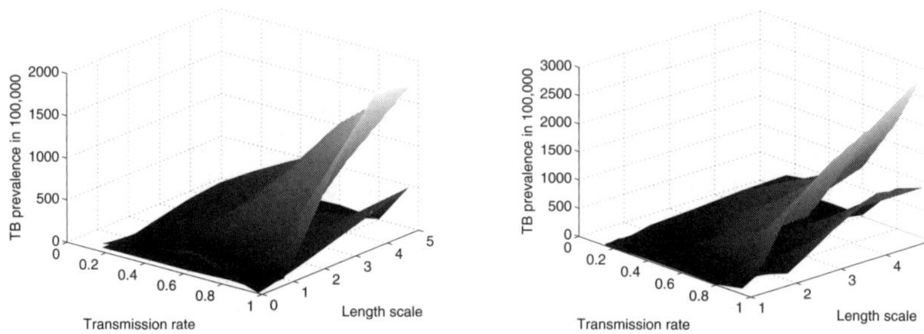

(a) p_I protection

(b) $\bar{\beta}$ protection

Figure 5. The effect of preventative therapy in the spatial model, showing the dependence on the length scale of the underlying graph.

5. Conclusions

We have reviewed the literature on the mathematical modeling of tuberculosis dynamics. Multiple models exist, encapsulating different assumptions about the dynamics of progression from latent infection to active disease, the nature of reinfection and the subsequent partial immunity, and the complexities of different TB strains as well as HIV. Because of the rising importance of drug-resistant TB and HIV, the modeling literature is moving steadily in these directions. With complexities inherent in such modeling efforts, it is particularly important to clearly specify and compare model assumptions about the biological processes driving the TB epidemic, and their ultimate consequences for the more complex dynamics of interest.

Here, we describe results from two new models of TB, a spatial stochastic model and a delay differential equation model. The stochastic model indicates that if disease transmission is indeed local, this may reduce the effectiveness of widely applied preventative treatment. Both models allow us to examine the portion of new disease that is due to exogenous reinfection without having to specify a portion of new infections destined to be "fast progressors". The specific implementation of partial immunity does not affect the estimated contribution of reinfection to disease levels, but spatial effects do. However, the choice of mechanism for partial immunity would significantly affect parameter estimation, differences that will persist as various authors extend their models to include different strains and HIV, and may significantly affect the conclusions.

References

S. Blower, K. Koelle, and T. Lietman. Antibiotic resistance–to treat... *Nat Med*, 5(4):358, Apr 1999. doi: 10.1038/7328.

S. M. Blower and J. L. Gerberding. Understanding, predicting and controlling the emergence of drug-resistant tuberculosis: a theoretical framework. *J Mol Med*, 76(9):624–636, Aug 1998.

S. M. Blower, A. R. McLean, T. C. Porco, P. M. Small, P. C. Hopewell, M. A. Sanchez, and A. R. Moss. The intrinsic transmission dynamics of tuberculosis epidemics. *Nat Med*, 1(8):815–821, Aug 1995.

S. M. Blower, P. M. Small, and P. C. Hopewell. Control strategies for tuberculosis epidemics: new models for old problems. *Science*, 273(5274): 497–500, Jul 1996.

Sally M Blower and Tom Chou. Modeling the emergence of the 'hot zones': tuberculosis and the amplification dynamics of drug resistance. *Nat Med*, 10(10):1111–1116, Oct 2004. doi: 10.1038/nm1102.

Romulus Breban and Sally Blower. The reinfection threshold does not exist. *J Theor Biol*, 235(2):151–152, Jul 2005. doi: 10.1016/j.jtbi.2004.12.026.

S. Brogger. Systems analysis in tuberculosis control: a model. *Am Rev Respir Dis*, 95(3):419–434, Mar 1967.

C. Castillo-Chavez and Z. Feng. To treat or not to treat: the case of tuberculosis. *J Math Biol*, 35(6):629–656, Jun 1997.

C. Castillo-Chavez and Z. Feng. Global stability of an age-structure model for tb and its applications to optimal vaccination strategies. *Math Biosci*, 151(2):135–154, Aug 1998.

T. Cohen, C. Colijn, B. Finklea, and M. Murray. Exogenous re-infection in tuberculosis: local effects in a network model of transmission. 2006a.

Ted Cohen and Megan Murray. Modeling epidemics of multidrug-resistant m. tuberculosis of heterogeneous fitness. *Nat Med*, 10(10):1117–1121, Oct 2004. doi: 10.1038/nm1110.

Ted Cohen and Megan Murray. Incident tuberculosis among recent us immigrants and exogenous reinfection. *Emerg Infect Dis*, 11(5):725–728,

May 2005.

Ted Cohen, Ben Sommers, and Megan Murray. The effect of drug resistance on the fitness of mycobacterium tuberculosis. *Lancet Infect Dis*, 3(1):13–21, Jan 2003.

Ted Cohen, Marc Lipsitch, Rochelle P Walensky, and Megan Murray. Beneficial and perverse effects of isoniazid preventive therapy for latent tuberculosis infection in hiv-tuberculosis coinfected populations. *Proc Natl Acad Sci U S A*, 103(18):7042–7047, May 2006b. doi: 10.1073/pnas.0600349103.

G. A. Colditz, T. F. Brewer, C. S. Berkey, M. E. Wilson, E. Burdick, H. V. Fineberg, and F. Mosteller. Efficacy of bcg vaccine in the prevention of tuberculosis. meta-analysis of the published literature. *JAMA*, 271(9): 698–702, Mar 1994.

G. A. Colditz, C. S. Berkey, F. Mosteller, T. F. Brewer, M. E. Wilson, E. Burdick, and H. V. Fineberg. The efficacy of bacillus calmette-gurin vaccination of newborns and infants in the prevention of tuberculosis: meta-analyses of the published literature. *Pediatrics*, 96(1 Pt 1):29–35, Jul 1995.

C. Colijn, T. Cohen, and M. Murray. Spatial structure and contact networks in modeling tuberculosis. 2006.

Christine S M Currie, Brian G Williams, Russell C H Cheng, and Christopher Dye. Tuberculosis epidemics driven by hiv: is prevention better than cure? *AIDS*, 17(17):2501–2508, Nov 2003. doi: 10.1097/01.aids.0000096903.73209.ac.

S. M. Debanne, R. A. Bielefeld, G. M. Cauthen, T. M. Daniel, and D. Y. Rowland. Multivariate markovian modeling of tuberculosis: forecast for the united states. *Emerg Infect Dis*, 6(2):148–157, 2000.

C. Dye and B. G. Williams. Criteria for the control of drug-resistant tuberculosis. *Proc Natl Acad Sci U S A*, 97(14):8180–8185, Jul 2000. doi: 10.1073/pnas.140102797.

C. Dye, G. P. Garnett, K. Sleeman, and B. G. Williams. Prospects for worldwide tuberculosis control under the who dots strategy. directly observed short-course therapy. *Lancet*, 352(9144):1886–1891, Dec 1998.

Christopher Dye. Global epidemiology of tuberculosis. *The Lancet*, 367 (9514):938–940, 2006.

Ken T D Eames and Matt J Keeling. Contact tracing and disease control. *Proc Biol Sci*, 270(1533):2565–2571, Dec 2003. doi: 10.1098/rspb.2003.2554.

Marcos A. Espinal, Adalbert Laszlo, Lone Simonsen, Fadila Boulahbal,

Sang Jae Kim, Ana Reniero, Sven Hoffner, Hans L. Rieder, Nancy Binkin, Christopher Dye, Rosamund Williams, Mario C. Raviglione, the World Health Organization-International Union against Tuberculosis, and Lung Disease Working Group on Anti-Tuberculosis Drug Resistance Surveillance. Global trends in resistance to antituberculosis drugs. *N Engl J Med*, 344(17):1294–1303, April 2001.

Z. Feng, C. Castillo-Chavez, and A. F. Capurro. A model for tuberculosis with exogenous reinfection. *Theor Popul Biol*, 57(3):235–247, May 2000. doi: 10.1006/tpbi.2000.1451.

Zhilan Feng, Wenzhang Huang, and Carlos Castillo-Chavez. On the role of variable latent periods in mathematical models for tuberculosis. *Journal of Dynamics and Differential Equations*, 13(2):425–452, 2001.

M. Gomes, A. Franco, M. Gomes, and G. Medley. The reinfection threshold promotes variability in tuberculosis epidemiology and vaccine efficacy. *Proc. R. Soc. B*, 271(1539):617–623, March 2004.

M. Gabriela M Gomes, Lisa J White, and Graham F Medley. The reinfection threshold. *J Theor Biol*, 236(1):111–113, Sep 2005. doi: 10.1016/j.jtbi.2005.03.001.

J. M. Heffernan, R. J. Smith, and L. M. Wahl. Perspectives on the basic reproductive ratio. *J R Soc Interface*, 2(4):281–293, Sep 2005. doi: 10.1098/rsif.2005.0042.

M. J. Keeling, D. A. Rand, and A. J. Morris. Correlation models for childhood epidemics. *Proc Biol Sci*, 264(1385):1149–1156, Aug 1997.

Matt J Keeling and Ken T D Eames. Networks and epidemic models. *J R Soc Interface*, 2(4):295–307, Sep 2005. doi: 10.1098/rsif.2005.0051.

T. Lietman and S. M. Blower. Potential impact of tuberculosis vaccines as epidemic control agents. *Clin Infect Dis*, 30 Suppl 3:S316–S322, Jun 2000.

Marc Lipsitch and Megan B Murray. Multiple equilibria: tuberculosis transmission require unrealistic assumptions. *Theor Popul Biol*, 63(2): 169–170, Mar 2003.

J. Mena-Lorca, J. X. Velasco-Hernandez, and C. Castillo-Chavez. Density-dependent dynamics and superinfection in an epidemic model. *IMA J Math Appl Med Biol*, 16(4):307–317, Dec 1999.

Christopher J.L. Murray and Joshua A. Salomon. Modeling the impact of global tuberculosis control strategies. *PNAS*, 95(23):13881–13886, November 1998.

T. C. Porco and S. M. Blower. Quantifying the intrinsic transmission dynamics of tuberculosis. *Theor Popul Biol*, 54(2):117–132, Oct 1998. doi:

10.1006/tpbi.1998.1366.

T. C. Porco, P. M. Small, and S. M. Blower. Amplification dynamics: predicting the effect of hiv on tuberculosis outbreaks. *J Acquir Immune Defic Syndr*, 28(5):437–444, Dec 2001.

Babak Pourbohloul and Robert C Brunham. Network models and transmission of sexually transmitted diseases. *Sex Transm Dis*, 31(6):388–390, Jun 2004.

Jonathan M Read and Matt J Keeling. Disease evolution on networks: the role of contact structure. *Proc Biol Sci*, 270(1516):699–708, Apr 2003. doi: 10.1098/rspb.2002.2305.

Charles Revelle, Floyd Feldmann, and Walter Lynn. An optimization model of tuberculosis epidemiology. *Management Science*, 16(4):B190–B211, December 1969. ISSN 0025-1909.

A. Van Rie, V. Zhemkov, J. Granskaya, L. Steklova, L. Shpakovskaya, A. Wendelboe, A. Kozlov, R. Ryder, and M. Salfinger. Tb and hiv in st petersburg, russia: a looming catastrophe? *Int J Tuberc Lung Dis*, 9(7): 740–745, Jul 2005.

Joshua A Salomon, James O Lloyd-Smith, Wayne M Getz, Stephen Resch, Mara S Snchez, Travis C Porco, and Martien W Borgdorff. Prospects for advancing tuberculosis control efforts through novel therapies. *PLoS Med*, 3(8):e273, Aug 2006. doi: 10.1371/journal.pmed.0030273.

E. E. Salpeter and S. R. Salpeter. Mathematical model for the epidemiology of tuberculosis, with estimates of the reproductive number and infection-delay function. *Am J Epidemiol*, 147(4):398–406, Feb 1998.

M. A. Sanchez and S. M. Blower. Uncertainty and sensitivity analysis of the basic reproductive rate. tuberculosis as an example. *Am J Epidemiol*, 145(12):1127–1137, Jun 1997.

Rinaldo B. Schinazi. On the role of reinfection in the transmission of infectious diseases. *Journal of Theoretical Biology*, 225(1):59–63, November 2003.

Baojun Song, Carlos Castillo-Chavez, and Juan Pablo Aparicio. Tuberculosis models with fast and slow dynamics: the role of close and casual contacts. *Math Biosci*, 180:187–205, 2002.

E. Vynnycky and P. E. Fine. The annual risk of infection with mycobacterium tuberculosis in england and wales since 1901. *Int J Tuberc Lung Dis*, 1(5):389–396, Oct 1997a.

E. Vynnycky and P. E. Fine. The natural history of tuberculosis: the implications of age-dependent risks of disease and the role of reinfection. *Epidemiol Infect*, 119(2):183–201, Oct 1997b.

E. Vynnycky and P. E. Fine. The long-term dynamics of tuberculosis and other diseases with long serial intervals: implications of and for changing reproduction numbers. *Epidemiol Infect*, 121(2):309–324, Oct 1998.

E. Vynnycky and P. E. Fine. Interpreting the decline in tuberculosis: the role of secular trends in effective contact. *Int J Epidemiol*, 28(2):327–334, Apr 1999.

E. Vynnycky and P. E. Fine. Lifetime risks, incubation period, and serial interval of tuberculosis. *Am J Epidemiol*, 152(3):247–263, Aug 2000.

H. Waaler, A. Geser, and S. Andersen. The use of mathematical models in the study of the epidemiology of tuberculosis. *Am J Public Health*, 52: 1002–1013, Jun 1962.

H. T. Waaler. Cost-benefit analysis of bcg-vaccination under various epidemiological situations. *Bull Int Union Tuberc*, 41:42–52, Dec 1968a.

H. T. Waaler. The economics of tuberculosis control. *Tubercle*, 49:Suppl:2–Suppl:4, Mar 1968b.

H. T. Waaler. Model simulation and decision-making in tuberculosis programmes. *Bull Int Union Tuberc*, 43:337–344, Jun 1970.

H. T. Waaler and M. A. Piot. The use of an epidemiological model for estimating the effectiveness of tuberculosis control measures. sensitivity of the effectiveness of tuberculosis control measures to the coverage of the population. *Bull World Health Organ*, 41(1):75–93, 1969.

H. T. Waaler and M. A. Piot. Use of an epidemiological model for estimating the effectiveness of tuberculosis control measures. sensitivity of the effectiveness of tuberculosis control measures to the social time preference. *Bull World Health Organ*, 43(1):1–16, 1970.

Hans Waaler. A dynamic model for the epidemiology of tuberculosis. *American Review of Respiratory Disease*, 98:591–600, 1968c.

Robin M Warren, Thomas C Victor, Elizabeth M Streicher, Madalene Richardson, Nulda Beyers, Nicolaas C Gey van Pittius, and Paul D van Helden. Patients with active tuberculosis often have different strains in the same sputum specimen. *Am J Respir Crit Care Med*, 169(5):610–614, Mar 2004. doi: 10.1164/rccm.200305-714OC.

E. Ziv, C. L. Daley, and S. M. Blower. Early therapy for latent tuberculosis infection. *Am J Epidemiol*, 153(4):381–385, Feb 2001.

Appendix A

The parameters for Eq. (3) are listed in the table below. Where two values are listed, these are pre-1900 and post-1900 values.

Table 1. Model parameters.

Parameter	Description	Value	Unit
β	transmission parameter	10.4, 5.3	year^{-1}
$\bar{\beta}$	reinfection transmission	$(1-f)\beta$	year^{-1}
μ	natural death rate	0.02	year^{-1}
μ_{TB}	TB mortality rate	0.35	year^{-1}
f	partial immunity	0.4	none
r_{rel}	relapse rate	0.05	year^{-1}
r	self-recovery rate	0.22	year^{-1}
r_{TR}	treatment rate	0, 0.9	year^{-1}
p_1	fast progression rate	0.033	none
p_I	reinfected progression rate	$(1-f)p_1$	year^{-1}
p_2	slow progression rate	0.0003	none
τ_1	duration of fast latency	5	year
τ_2	treatment delay	0.1	year

A SPACE-TIME SCAN STATISTIC FOR DETECTION OF TUBERCULOSIS OUTBREAKS IN THE SAN FRANCISCO HOMELESS POPULATION

BRANDON W. HIGGS, MOJDEH MOHTASHEMI

MITRE Corporation, 7515 Colshire Dr. McLean, VA, 22102 USA
and
MIT Department of Computer Science
77 Mass. Ave. Cambridge, MA 02139-4307 USA
E-mail: bhiggs@mitre.org

JENNIFER GRINSDALE, L. MASAE KAWAMURA

TB Control Section, San Francisco Dept. of Public Health
Ward 94, 1001 Potrero Ave., San Francisco, CA 94110 USA

San Francisco (SF) has the highest rate of TB in the US. Although in recent years the incidence of TB has been declining in the general population, it appears relatively constant in the homeless population. In this paper, we present a spatio-temporal outbreak detection technique applied to the time series and geospatial data obtained from extensive contact and laboratory investigation on TB cases in the SF homeless population. We examine the sensitivity of this algorithm to spatial resolution using zip codes and census tracts, and demonstrate the effectiveness of it by identifying outbreaks that are localized in time and space but otherwise cannot be detected using temporal detection alone.

1. Introduction

Tuberculosis (TB) is one of the top four diseases for infection-induce mortality in the world today. There are currently about 54 million people infected with the bacterium *Mycobacterium tuberculosis* with approximately 8 million new infections occurring each year. TB kills nearly 2.4 million people annually. In the U.S. alone, there are currently about 12.5 million people who have been infected by TB (Ginsberg, 2000).

Though advances in health and medicine had considerably reduced the incidence of TB after the "mid" 20th century, there was an increase in cases in many parts of the United States in the mid-1980s, and early-1990s, in part due to increased homeless individuals and prevalence of AIDS (which compromises the immune response). In San Francisco, annual TB cases peaked in 1992 with "51.2 cases per 100,000 persons and decreased significantly thereafter to 29.8 cases per 100,000 persons in 1997" (Jasmer et al., 1999). Currently TB case rates are approximately 20 per 100,000 persons. An important characteristic of TB is that once droplet nuclei containing bacteria are expelled into the air, such droplets are able to circulate in confined spaces so that direct and prolonged contact with an infectious person is no longer a prerequisite to acquire infection (Wells 1934; Wells et al., 1934). This factor is of particular importance in the homeless population, where individuals typically seek care at locations of limited space, such as shelters or single room occupancies (SRO)s, which can amplify a single exposure quickly.

With such an efficient mode of transmission in the homeless population, strategies to mitigate the spread of TB are important. Surveillance systems should be equipped to target the spatial spread within and between shelters and SROs over time to reduce the likelihood of potential outbreaks. Combining this information across different dimensions, such as time and space (Klovdahl et al., 2001), significantly enhances our understanding of the underlying transmission dynamics, and hence will improve upon existing public health intervention policies for control and prevention of TB.

In this paper, we propose the scan statistic first examined in 1965 (Naus, 1965) and later implemented in other work (Kleinman et al., 2005; Kulldorff, 1997; Kulldorff et al., 2005; Wallenstein, 1980; Weinstock, 1981) for detection of potential TB outbreaks in the homeless population of San Francisco from the years of 1991-2002. We demonstrate that the scan statistic is a sensitive measure for identifying aberrant frequencies (from normal trends) of TB cases within certain time and spatial distributions that would otherwise go undetected using deviations from a global average or methods that depend only on temporal patterns.

We find that the distribution of TB cases within zip code regions follows a power law distribution, where many individuals with TB reside in a select few zip codes and few individuals with TB reside in many different zip codes. We show that a more resolved clustering of region into census tracts can reduce this skewness in the spatial distribution of TB cases and improve detection sensitivity.

2. Methods

2.1. *San Francisco Department of Public Health TB Data*

TB case data is kept electronically in a patient management database maintained by the San Francisco Department of Public Health, TB Control Section. All case information, including address of residence and homeless status at the time of diagnosis, was downloaded directly from the database. Census tract information was obtained from the 2000 census.

2.2. *Space-Time Permutation Scan Statistic Applied to TB Data*

A variation to the scan statistic introduced by Kulldorff (Kulldorff et al., 2005) was implemented here as a suitable method for early detection of TB outbreaks in the San Francisco homeless population, particularly for those time/region-specific increases in case frequency that are too subtle to detect with temporal data alone. Similar to the scan statistic proposed by Kulldorff et al, the scanning window utilizes multiple overlapping cylinders, each composed of both a space and time block, where time blocks are continuous windows (i.e. not intermittent) and space blocks are geographic-encompassing circles of varying radii. Briefly explained here (see Kulldorff et al., 2005 for a more in depth account of the algorithm), for each cylinder, the expected number of cases, conditioned on the observed marginals is denoted by μ where μ is defined as the summation of expected number of cases in a cylinder, given by

$$\mu = \sum_{(s,t)\in A}^{\mu_{st}} \tag{1}$$

where s is the spatial cluster and t is the time span used (e.g., days, weeks, months, etc.) and

$$\mu_{st} = \frac{1}{N}(\sum_{s} n_{st})(\sum_{t} n_{st}) \tag{2}$$

where N is the total number of cases and n_{st} is the number of cases in either the space or time window (according to the summation term). The observed number of cases for the same cylinder is denoted by n. Then the Poisson generalized likelihood ratio (GLR), which is used as a measure for a potential outbreak in the current cylinder, is given by

$$\left(\frac{n}{\mu}\right)^{n}\left(\frac{N-n}{N-\mu}\right)^{(N-n)} \quad \text{(Kleinman et al., 2005).} \tag{3}$$

Since the observed counts are in the numerator of the ratio, large values of the GLR signify a potential outbreak. To assign a degree of significance to the GLR value for each cylinder, Monte Carlo hypothesis testing (Dwass, 1957) is conducted, where the observed cases are randomly shuffled proportional to the population over time and space and the GLR value is calculated for each cylinder. This process of randomly shuffling is conducted over 999 trials and the random GLR values are ranked. A p-value for the original GLR is then assigned by where in the ranking of random GLR values it occurs.

For our space window, we restricted the geographic circle to three radius sizes of small, medium, and large: 0.23 miles (0.37 km), 0.44 miles (0.70 km), and 0.56 miles, (0.90 km), respectively. For our time window, the TB case count is much lower than the daily data feeds typical of surveillance systems used to monitor emergency room visits or pharmacy sales, for example. To compensate for the smaller proportion of total cases, monthly case counts were used (with a time window of 2 months) spanning the years of 1991-2002. We observed that the homeless population that we surveyed is a closer approximation to a uniform population at risk as compared to the general population and has less dependence on certain social patterns. For example, unlike associations between emergency department visits and specific days of the week, or medicine sales promotions and increases in sales of medication targeting a specific ailment (Kulldorff et al., 2005), the TB-infected homeless population in not affected by these variables and exhibits greater uniformity in case counts with time. Many of the confounding factors that can influence these case count fluctuations in the general population such as socio-economic status, days of the week, holidays, and season, do not affect the true case counts of TB. We conducted the scan statistic across all 144 months as well as stratifying across seasons, to adjust for the largest confounding variable, but did not observe a large difference in the top scoring GLR space/time combinations between the two methods.

2.3. *Data*

The dataset consists of 392 individuals that have been diagnosed by the San Francisco Department of Public Health with active TB and identified as homeless over the time period of 1991-2002. The primary residences of these individuals have also been identified, where a residence is defined as either a shelter, a single room occupancy (SRO), or a county prison. The total number of zip codes that account for these residences include 22, with

80 census tracts.

3. Results

3.1. *Scale-free property of TB cases in the homeless population (1991-2002)*

When examining the frequencies of TB cases across the 12 year time period within each geographically partitioned region (zip codes) in the San Francisco homeless population, there are a small number of hub zip codes. These hubs are defined as regions with a large density of TB cases as compared to the other regions. Some of these hubs tend to fluctuate in density (i.e., become smaller/larger and less/more dense) in certain years, while others persist as regions of highly dense case counts throughout the 12 year time period, where highly dense is defined as an apparent increase. A viable assumption that explains the existence of these highly dense hubs is the relationship between region population and number of TB cases. One would assume that the more populated regions would have more TB cases, simply as a function of a larger population size. However, when utilizing the 2000 population census data for each zip code, the correlation between these two variables is very low (r¡0.15), so adjusting for population (i.e. per capita statistic) does not alter the observed hub regions.

More interesting than the observation of these hub regions is the degree of self-organization that exists, resembling large-scale properties of complex networks (Barabasi et al., 1999). When examining the total number of TB cases in the homeless population for each region across the 12 year time period, the number of cases and regions follow a power law for zip codes (Figure 1 blue points). Similar to large networks that exhibit a power-law distribution for self-organizing into a scale-free state, the average number of TB cases across regions demonstrate the same characteristic of this "universal architecture" (Keller 2005). That is to say, as new connecting edges (k), are added to a node, the edges attach with a probability $P(k)$ proportional to the number of edges already connected to the current node. This decay rate is represented by $P(k) \sim k^{-\gamma}$ where γ is the exponent constant. The rate at which a node acquires edges is given by $\frac{\partial k_i}{\partial t} = \frac{k_i}{2t}$ which gives $k_i(t) = m\left(\frac{t}{t_i}\right)^{0.5}$, where t_i is the time at which node i was added to the system and m is the number of nodes. So the probability that a node i has a connectivity smaller than k, $P[k_i(t) < k]$ can be written in terms of m as $P(t_i > \frac{m^2 t}{k^2})$ or $1 - P(t_i \leq \frac{m^2 t}{k^2})$, which is equal to $1 - \frac{m^2 t}{k^2}(t + m_0)$ (Barabasi et al., 1999). The number of cases (c) is analogous to the edge degree or k, and the cumulative number of regions with c cases is analogous to $P(k)$. The degree of decay for the plot of the cumulative number of regions with c cases versus c represents the distribution property, or how evenly distributed the number of cases are over the regions. A function that decays quickly has a more even distribution of cases over the regions than a function that decays slower. From Figure 1, it is apparent that the distribution of cases over the zip codes is distributed much less evenly than the distribution of cases over census tracts.

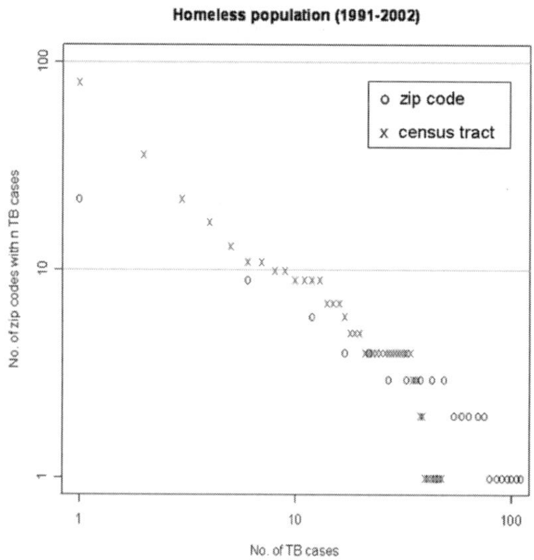

Figure 1. Cumulative distribution of TB cases in the homeless population within both zip codes (circles) and census tracts (crosses) in the San Francisco area.

Based on this skewed distribution of the hub zip codes, a simple surveillance system might target such TB outbreak "hotspots" for mitigation with the assumption that the primary mode of transmission can be greatly reduced through moderation of the most TB case-rich zip codes. However, as illustrated in Figure 2, the zip codes where individuals with TB reside are more spatially indistinct than clear. There is an averaging effect for each zip code when summating the total number of cases for each, where individuals with TB that reside on the border of two or more zip codes can create a hub that is not identified by a single zip code. For example, within the space between zip codes 1, 2, and 3, there is a large density of residences for TB cases, however, attempts at targeting (for purposes of mitigation) each zip code separately would be inappropriate since the primary density of cases occur within the space between the three zip codes. A scan statistic can address this issue with the addition of more cylinders with finer overlapping space windows. However, the space window will always have to include at least two zip codes to include the density between the regions. To account for this problem and to work with spatial clusters that have TB cases more evenly distributed throughout, we found it more useful to use the smaller units to partition the space, such as census tract.

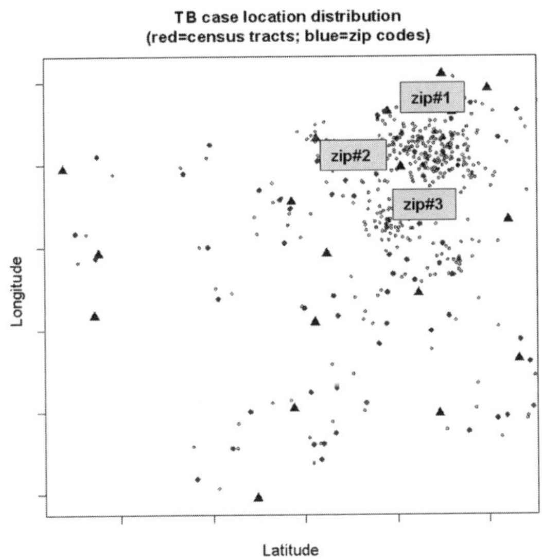

Figure 2. Distribution of TB cases in the homeless population (1991-2002). Small open points represent TB cases, closed points represent census tracts, and triangles represent zip codes. Dense regions of TB cases are not evenly accounted for with zip code partitioning as illustrated in the region between zip codes 1-3.

3.2. *Analysis of homeless population outbreaks (1991-2002)*

Table 1 lists the homeless TB case occurrences for the most significantly scoring space-time windows. Out of the 12 year time period analyzed, specific spatial regions in the years of 1991, 1992, 1993, and 1997 produce significant signals for a high count of TB cases, relative to normal variability. This result of potential TB outbreaks within these years is consistent with previous reports that have documented the early-1990s as the time period where there was a resurgence of TB cases in the San Francisco area (Jasmer et al., 1999).

Table 1. Scan statistic results for most significant GLR values.

GLR	p-value	Expected	Observed	Radius*	Circle†	Date1	Date2
629.716	$p < 0.0001$	0.32	4	s,m	1	Mar-91	Apr-91
85831.34	$p < 0.0001$	0.36	6	s,m,l	1	Apr-91	May-91
2868.833	$p < 0.0001$	0.21	4	s,m	1	May-91	
155956.7	$p < 0.0001$	0.32	6	s,m,l	1	May-91	Jun-91
3010.488	$p < 0.0001$	1.28	8	l	2	Aug-91	Sep-91
143.755	$p < 0.0001$	1.18	6	s,l	3	Aug-91	Sep-91
77.675	$p < 0.0001$	0.09	2	s	4	Jul-92	Aug-92
297.477	$p < 0.0001$	0.04	2	s	4	Aug-92	
297.477	$p < 0.0001$	0.04	2	s	4	Aug-92	Sep-92
297.477	$p < 0.0001$	0.04	2	m	5	Feb-93	Mar-93
749.225	$p < 0.0001$	0.03	2	m	5	Mar-93	Apr-93
376.38	$p < 0.0001$	0.64	5	m	6	Sep-97	Oct-97

* Size of space component of cylinder used to group regions: small (s), medium (m) and large (l).

† Unique circle surrounding the regions of significant signal.

The top GLR values (most significant signals) correspond to different successive combinations of the months in the range of March to June in 1991 for a small region of census tracts that is contained by all three different sized circles (Figure 3). All three of these circles intersect at some point to share similar information, which explains why the same region exhibits a high signal in multiple time windows. For the regions defined by these 3 circles of increasing radii that do not intersect, one can observe how far the signal extends. When examining only the temporal plot of TB case frequencies in the homeless population, stratified by month for the years 1991-2002 (Figure 4), there are apparent peaks observed in 1991 that correspond to the scan statistic significantly scoring time windows. For example, within the range of March 1991 to October 1991, the number of TB cases in the homeless population never drops below 3 cases. In fact, in 4/12 months for this year there were at least 5 cases of TB. For this example, the frequency peaks are large (compared to normal trends across the 12 year period), such that the temporal information alone could potentially be used to identify the peak signal months in the year 1991, once the cases were mapped to the general residential area.

However, for the other significantly scoring space-time windows in the table within the years of 1992, 1993, 1997, the temporal information alone does not intuitively infer a potential signal. In fact, for some of the significantly scoring space-time windows, the temporal information can be

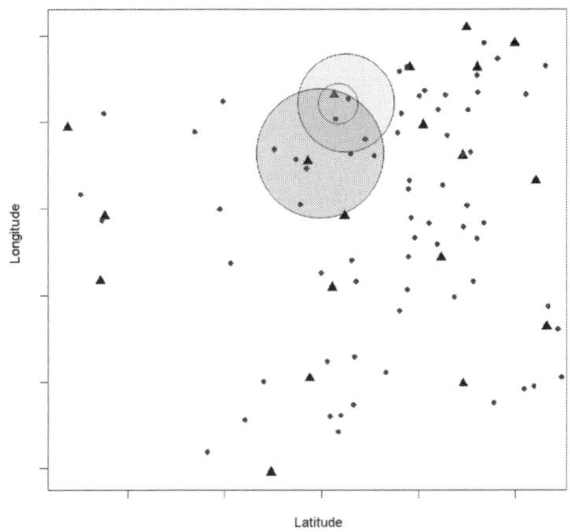

Figure 3. Identified region of significant alarm in 1991. Points represent census tracts and triangles represent zip codes. Three overlapping circle sizes (small, medium, and large) represent the spatial window for different time windows including the months of March to June in 1991.

counterintuitive. For example, if one examines the temporal information alone for the year 1992, there are 4 months where 2 cases of TB and 1 month where 1 case of TB is documented (Figure 4). This year accounts for a small number of cases for any one month, such that if one relies only on the temporal information for detection of a significant signal, none of the months in this year would raise a flag. So what information causes this time window for the 3 month combinations of July 1992 to September 1992 to appear as a significant signal? The answer is in the number of TB cases for a particular census tract relative to the total number of cases over the entire 12 year period (Figure 5). That is to say, for the 2 TB cases that are spatially mapped within the same cylinder (1992/smallest circle), there are only 8 total TB cases within this same region over the entire 12 year period. So, 25% of the total TB cases in this region over the 12 year time span occurred in 1992, and as a result, produce a potential signal.

All of the significant signals are plotted in Figure 6 with the dates of case detection labeled. The numbers in the plot correspond to the unique signals denoted in the 'Circle' column of Table 1, to better illustrate the spatial trend of significant frequencies of TB cases over the time period. It is interesting to note the spatial distribution of the significant TB case counts for the 4 years determined. Through the months of March to June in 1991, the significant signals occur in the north central region (represented

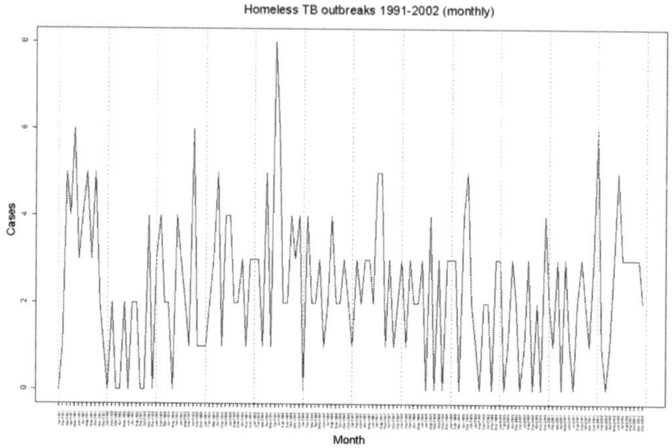

Figure 4. Temporal plot of TB cases in the homeless population for the years of 1991-2002. Data is stratified by months, where the dashed vertical lines divide years.

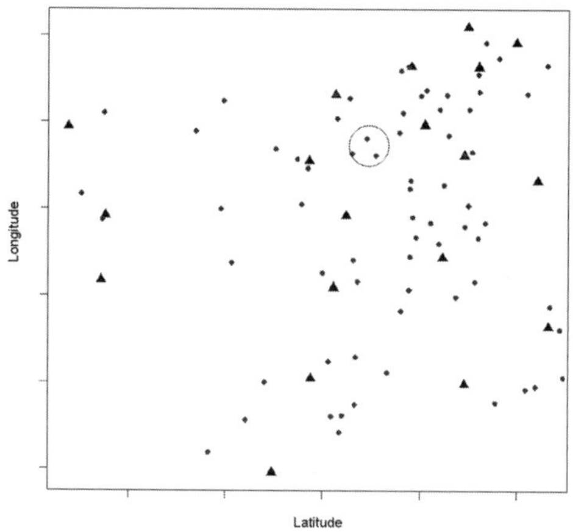

Figure 5. Identified region of significant alarm in 1992. Points represent census tracts and triangles represent zip codes. Small circle size represents the spatial window for different time windows spanning the months of July to September in 1992.

by circles labeled 1), then from August to September of the same year, the signal predominates slightly northwest (represented by circles labeled 2 and 3), but in the same general northern region. Then in the months of July to September of 1992, the signal occurs back in the northern central region (represented by circle labeled 4), as it had originated in the months of May to June in 1991. In the months of February to April in 1993, the significant signal occurs in a completely separate region in the south central area (represented by circle labeled 5). This is interesting since the signal seems to deviate from the other years in a completely separate region for this year. Finally, in the months of September and October in 1997, the significant signal occurs back near the north central region (represented by circle labeled 6) as had originally occurred for significant signals in the years of 1991 and 1992.

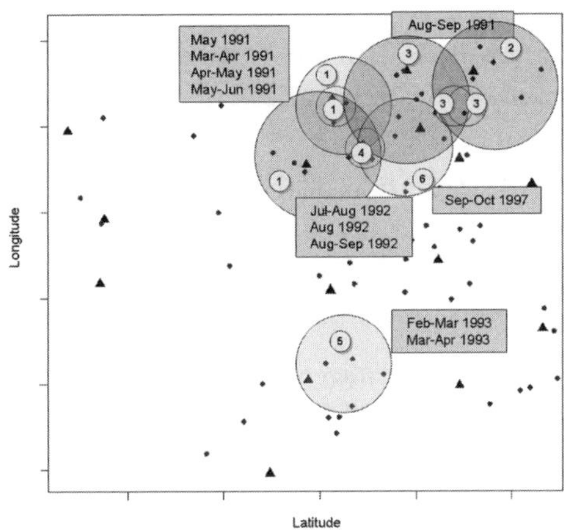

Figure 6. Spatial plot of top scoring signals from Table 1 with different sized circles representing spatial windows (small, medium, and large). Dates for time windows are provided next to each space window.

References

Barabasi AL, Albert R. (1999) Emergence of Scaling in Random Networks. *Science*, **286**:509-512.

Dwass M. (1957) Modified randomization tests for non-parametric hypotheses. *Ann Math Statist*, **29**:181187.

Ginsberg, A. (2000) A Proposed National Strategy for Tuberculosis Vaccine Development. *Clinical Infectious Diseases,* **30**:S233-242.

Jasmer RM, Hahn JA, Small PM, Daley CL, Behr MA, Moss AR, Creasman JM, Schecter GF, Paz EA, Hopewell PC. (1999) A Molecular Epidemiologic Analysis of Tuberculosis Trends in San Francisco, 1991–1997. *Ann Intern Med,* **130**:971-978.

Keller EF. (2005) Revisiting "scale-free" networks. *BioEssays*, **27**:1060-1068.

Kleinman KP, Abrams AM, Kulldorff M, Platt R (2005) A model-adjusted space-time scan statistic with application to syndromic surveillance. *Epidemiol. Infect.*, 000:1-11.

Klovdahl AS, Graviss EA, Yaganehdoost A, Ross MW, Wanger A, Adams GJ, Musser JM. (2001) Networks and tuberculosis: an undetected community outbreak involving public places. *Soc Sci Med.*, **52**(5):681-694.

Kulldorff M. (1997) A spatial scan statistic. *Commun Stat A Theory Methods*, **26**:1481-1496.

Kulldorff M, Heffernan R, Hartmann J, Assuncao R, Mostashari F (2005) A space-time permutation scan statistic for disease outbreak detection. *PLOS*, 2(3).

Naus J. (1965) The disctribution of the size of maximum cluster of points on the line. *J Am Stat Assoc*, **60**:532-538.

Wallenstein S. (1980) A test for detection of clustering over time. *Am J Epidemiol*, **111**:367-372.

Weinstock MA. (1982) A generalized scan statistic test for the detection of clusters. *Int J Epidemiol*, **10**:289-293.

Wells WF. (1934) On airborne infection. Study II. Droplets and droplet nuclei. *Am J Hyg*, **20**:611–618.

Wells WF, Stone WR. (1934) On air-borne infection. Study III. Viability of droplet nuclei infection. *Am J Hyg*, **20**:619–627.

DYNAMICS OF TUBERCULOSIS UNDER DOTS STRATEGY

PATRÍCIA D. GOMES, REGINA C. P. LEAL-TOLEDO, C. E. C. CUNHA
Universidade Federal Fluminense, Instituto de Computação
Rua Passo da Pátria 156, Bloco E, sala 301 - São Domingos
Niterói-R.J.- Brazil - CEP-24210-240
E-mail: leal@ic.uff.br

Tuberculosis still remains as a serious public health problem. Thinking about more efficient interventions for its combat and control WHO advocates that DOTS (Directly Observed Short-time Treatment Strategy) can improve cure rates and case detection. In this context, mathematical modeling can be used to evaluate the behavior of tuberculosis under DOTS, supplying informations to optimal action strategies. The model presented in this work, describes the dynamic of 4 individuals groups: susceptibles, never before infected; latently-infected or cured of TB under chemotherapy; infectious individuals with pulmonary TB and sputum smear positive; and noninfectious with pulmonary TB but sputum-smear negative or extra-pulmonary TB. Individuals, who complete treatment successfull, are cured of tuberculosis but remain infected and individuals, who do not complete treatment, continue with the illness (infectious or noninfectious). The model, as considered here allows us to evaluate the effect of improve cure rates and case detection. The Effective Reproductive Rate (R_0^e) was found for this model and used as epidemiological measure of severity of an epidemics. If $R_0^e > 1$, an epidemics occurs and if $R_0^e < 1$, it is eradicated. The time-dependent uncertainty analysis is presented using the Monte Carlo Method.

1. Introduction

The Tuberculosis is among the diseases called "neglected diseases" Cohen[22] because this disease have been neglected by the government and other society segments per decades. New cases of tuberculosis (TB) in developing countries and its resurgence in developed countries, shows that this disease still constitute a big challenge in defining public politics to health. Thus, the control of this disease demands not only from government as society, health professionals and another sectors, the developing and implantation of action strategies, with the main objective of information and disease erradication.

One estimates that around 1/3 of the worldwide population (about 1.7 billion individuals) is infected by the Mycobacterium tuberculosis. Latin America and Caribbean countries are responsible for 400.000 cases and 50% of those occurred in Brazil, Peru and Mexico with a number of deaths, caused by this illness in Brazil, between to 4 to 5 thousands a year.

Nowadays, this disease is directly related with poor people who have not access to information or adequate treatment. Moreover, the tuberculosis still represents a serious social problem, because it is the disease that represents the greater "causa-mortis" than another diseases, among people with HIV SPS[23].

The main factors that harm the possibility of control and erradication of the Tuberculosis is the abandonment of the treatment as observed in a large percentage of patients. For this reason, the WHO (World Health Organization) declare the DOTS strategy the best intervention to combat this disease because it is based on the notification of cases and the accompaniment of the patients when these pleople took the medicine which guarantee the conclusion of the treatment.

According to the WHO, the implementation of DOTS is of great importance mainly in countries or regions where the tuberculosis has alarming values and the incidence of the disease characterize the tuberculosis as a health public problem Neto[12].

Among the twenty two countries which concentrate 80% of the cases of TB in all the world, Brazil occupies the fifteenth place, having 110,000 new cases registered each year. The treatment of this disease in this country presents a percentage of cure is 72% and the percentage of abandonment is 12% and in some regions this value can reach between 30% and 40% Neto[12].

In this context, a lot of models has been proposed mainly since 1993 when WHO declares TB a world-wide emergency Blower[3],Blower[4],Blower[5], Blower[6], Chavez[8], Chavez[9], Chavez[10], Chavez[11], Dye[14], Dye[15], Murphy[19], Murray[20], Sanchez[21]. Christopher Dye, WHO researcher and consultant about Tuberculosis, presented in Dye[14], a detailed mathematical model composed by nine categories of individuals which has as the main objective to evaluate as much the effect of DOTS treatment policy in the evolution of tuberculosis as the improvement in the detection of the cases and the cure rates in the epidemics. However, the model includes disease complexities which make difficult the analysis of the system of differential equations associated, as well as, the attainment of the Effective Reproductive Rate which is of great importance when the objective is making the qualitative analysis of the model presented.

In this work, we present a model with four categories of individuals gotten of simplification of the model of Dye et al. Dye[14]. This model, which was proposed in Gomes[17], allows to calculate the Effective Reproductive Rate (R_0^e) and with this, to evaluate the importance of successfull treatment and the consequences of the abandonment of the treatment, in the evolution of the disease, as established in DOTS strategy. This paper also presents the stability analysis of the critical points of the system of equations associated to the model using the Effective Reproductive Rate.

Examples are presented to evaluate the importance of the sucessfull treatment based in DOTS strategy.

2. The Mathematical Model

The proposed mathematical model consists of four ordinary differential equations that describe the temporal dynamics of four groups of individuals. The population $N(t)$ is divided in following groups: susceptible $S(t)$; infected individuals $L(t)$; diseased and infectious $T_i(t)$; diseased and non-infectious $T_n(t)$.

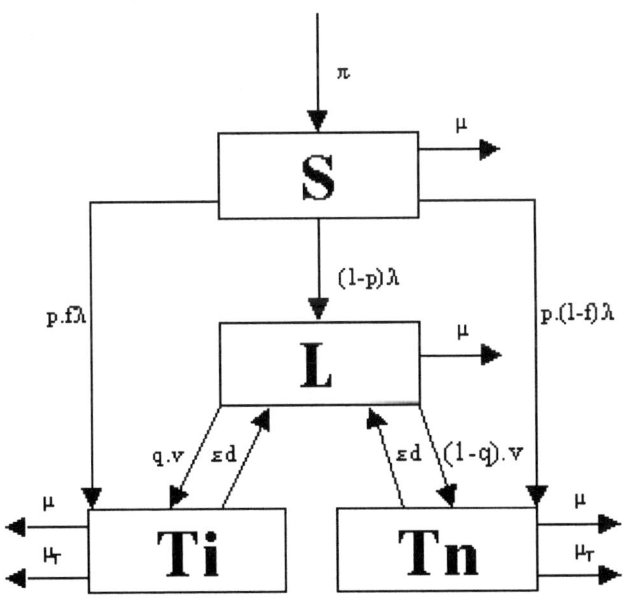

Figure 1. Flow diagram of model for tuberculosis

As showed in the flow diagram in Figure 1, after infection, a fraction (p) of susceptible individuals progresses to disease (until one year); of these, a fraction (f) progresses to infectious tuberculosis, while $(1-f)$ progresses to non-infectious pulmonary and extra-pulmonary tuberculosis. Of the individuals that do not progress to disease until one year after infection, $(1-p)$ stay infected; and only a small fraction (v) progresses to disease. In this case, a fraction (q) is associated with infectious disease and $(1-q)$ with another types of tuberculosis.

As suggested in Dye[15], we include in this model the possibility to evaluate how the effective treatment can affect the transmission dynamic of tuberculosis epidemics considering that the fraction of individuals who was diagnosed and treated (d) only a fraction (ϵ) is considered cured while the other remain sick. The infected individuals include those that are cured by treatment, because they stay with the bacillus in their organisms Dye[15].

As defined, the model presented is described for this nonlinear system of differential equations:

$$\begin{cases} \frac{dS(t)}{dt} = \pi - (\mu + \lambda)S(t) \\ \frac{dL(t)}{dt} = (1-p)\lambda S(t) + \epsilon d(T_i(t) + T_n(t)) - (v + \mu)L(t) \\ \frac{dT_i(t)}{dt} = pf\lambda S(t) + qvL(t) - (\mu + \mu_t + \epsilon d)T_i(t) \\ \frac{dT_n(t)}{dt} = p(1-f)\lambda S(t) + (1-q)vL(t) - (\mu + \mu_t + \epsilon d)T_n(t) \end{cases} \tag{1}$$

where $\lambda = \beta T_i$ is the risk of infection and β is a transmission coefficient given by: $\beta = (ECR)\frac{\mu}{\pi}$; ECR is the average number of new infectious cases caused by one case of infectious tuberculosis per unit time; π is the number of individuals migrate or born over time and μ_t, μ are tuberculosis death rate and death rate for other causes respectively.

The model presented has two equilibrium points: the no-disease equilibrium point: $S^t = \frac{\pi}{\mu}, L^t = {T_i}^t = {T_n}^t = 0$ and the endemic equilibrium point given by:

$$S^{nt} = \frac{\pi}{\mu R_0^e} \tag{2}$$

$$T_i^{nt} = \frac{\mu}{\beta}(R_0^e - 1) \tag{3}$$

$$L^{nt} = \left(\frac{a}{(v+\mu)a - \epsilon d(1-q)v}\right)\left[\epsilon d\left(\frac{p(1-f)\pi}{aR_0^e} + \frac{\mu}{\beta}\right) + \frac{(1-p)\pi}{R_0^e}\right](R_0^e - 1) \tag{4}$$

$$T_n^{nt} = \left\{\frac{p(1-f)\pi}{aR_0^e} + \frac{(1-q)v}{(v+\mu)a - \epsilon d(1-q)v} \times \left[\frac{(1-p)\pi}{R_0^e} + \epsilon d\left(\frac{p(1-f)\pi}{aR_0^e} + \frac{\mu}{\beta}\right)\right]\right\}(R_0^e - 1) \tag{5}$$

An important parameter when we evaluate qualitatively a mathematical model that describe the dynamic of a disease is the Effective Reproductive Rate (R_0^e). This rate is the average number of secondary infectious cases that are generated when one infectious case is introduced in a population completely susceptible in which exists a treatment.

In Gomes[17] is demonstrated more deeply as is possible to get R_0^e for this model, which is given by:

$$R_0^e = \left(\frac{\beta\pi}{\mu}\right)\left[\frac{pf}{a} + \frac{qv}{a(v+\mu) - \epsilon dv} \times \left((1-p) + \frac{\epsilon dp}{a}\right)\right], \tag{6}$$

where: $a(v+\mu) - \epsilon dv > 0$, $a = \mu + \mu_t + \epsilon d.$

3. Stability

3.1. *Stability of the no-disease equilibrium point*

The Jacobian matrix associated to system of differential equations (1) and evaluated at the trivial equilibrium point is given by:

$$\mathbf{J_t} = \left(\begin{array}{c|ccc} -\mu & 0 & -\frac{\beta\pi}{\mu} & 0 \\ \hline 0 & -(v+\mu) & \epsilon d + (1-p)\frac{\beta\pi}{\mu} & \epsilon d \\ 0 & qv & -a + pf\frac{\beta\pi}{\mu} & 0 \\ 0 & (1-q)v & p(1-f)\frac{\beta\pi}{\mu} & -a \end{array}\right) = \left(\begin{array}{c|c} J_{11}^* & J_{12}^* \\ \hline J_{21}^* & J_{22}^* \end{array}\right) \tag{7}$$

There is stability in trivial equilibrium point when all eigenvalues associated with this Jacobian matrix in this point have negative real parts. The matrix J_t has a upper block triangular form. Thus, the eigenvalues of matrix J_t can be given by the union of the eigenvalues of the square matrices: J_{11}^* and J_{22}^*. Therefore, the one-by-one matrix has eigenvalue $\lambda_1 = -\mu$, lacking to verify the sign of the eigenvalues of three-by-three matrix J_{22}^*. For this, as suggested in Blower[4] we use the following theorems Berman[1]:

Theorem 3.1. *Let* $\mathbf{C}$ *be a non-singular matrix, then all eigenvalues of matrix* $\mathbf{C}$ *have positive real parts.*

Theorem 3.2. *Let* $\mathbf{C}$ *be a non-singular matrix with nonpositive off-diagonal elements. The matrix* $\mathbf{C}$ *is a M-matrix if and only if:*

(a) The matrix $\mathbf{C}$ *is a positive inverse matrix.*
(b) All elements of main diagonal of matrix $\mathbf{C}$ *are positive and there is a positive diagonal matrix* $\mathbf{D}$*, where the product matrix* $\mathbf{CD}$ *is strictly dominant diagonal.*

To use the Theorem 3.1 and the Theorem 3.2 to evaluate the sign of eigenvalues of matrix $\mathbf{J}^*_{22}$ we define the matrix $\mathbf{A} = -\mathbf{J}^*_{22}$ with all the off-diagonal elements nonpositives and suppose the restriction:

$$a > \frac{pf\beta\pi}{\mu} \quad , \qquad a = \mu + \mu_t + \epsilon d \; . \tag{8}$$

For all eigenvalues of the matrix $\mathbf{A}$ have positive real parts, the matrix $\mathbf{A}$ must be a M-matrix. Thus, if exist the inverse matrix of the matrix $\mathbf{A}$, $\mathbf{A}^{-1} = \frac{\mathbf{P}}{det\mathbf{A}}$, where all elements are positive, the matrix $\mathbf{A}$ will satisfy the first item of Theorem 3.2. For this, $det\mathbf{A}$ (determinant of the matrix $\mathbf{A}$) and $\mathbf{P}$ (adjunct matrix of matrix $\mathbf{A}$) must have the same sign.

Therefore, we calculate $\mathbf{A}$ through expansion in cofators for last column and we get:

$$det\mathbf{A} = a\; det\mathbf{A_1} - \epsilon d\; det\mathbf{A_2} \; , \tag{9}$$

where A_1 and A_2 are given by:

$$\mathbf{A_1} = \begin{pmatrix} v + \mu & -\left(\epsilon d + (1-p)\frac{\beta\pi}{\mu}\right) \\ -qv & a - pf\frac{\beta\pi}{\mu} \end{pmatrix} , \tag{10}$$

$$\mathbf{A_2} = \begin{pmatrix} -qv & a - pf\frac{\beta\pi}{\mu} \\ -(1-q)v & -p(1-f)\frac{\beta\pi}{\mu} \end{pmatrix} . \tag{11}$$

Expanding the expression (9) and putting it in function of R^e_0, we get:

$$det\mathbf{A} = a[(\mu + \mu_t)(v + \mu) + \mu\epsilon d](1 - R^e_0) \; . \tag{12}$$

Concluding, we have two cases:
Case 1:

$$det\mathbf{A} > 0 \; if \; R^e_0 < 1 \; . \tag{13}$$

Case 2 :

$$detA < 0 \, if \; R_0^e > 1 \; . \tag{14}$$

Knowing the relation between the sign of $det\mathbf{A}$ and R_0^e, we need to evaluate the sign of the terms of the matrix **P**. Calculating the matrix **P** and supposing all the parameters of the system are positive, we can easily verify all the elements are positive with exception of the term P_{33} that is given by:

$$P_{33} = (v + \mu)\left(a - \frac{pf\beta\pi}{\mu}\right) - \left(\epsilon d + (1-p)\frac{\beta\pi}{\mu}\right)(qv) \; . \tag{15}$$

Comparing (10) and (15) we observe $P_{33} = det\mathbf{A_1}$. Thus, to $P_{33} > 0$, that is, all elements of the adjunct matrix of matrix **A** are positive, the strict inequality has to be verified:

$$(v + \mu)\left(a - \frac{pf\beta\pi}{\mu}\right) > \left(\epsilon d + (1-p)\frac{\beta\pi}{\mu}\right)(qv) \; . \tag{16}$$

Supposing the Case 1 in (13), where $R_0^e < 1$ and $det\mathbf{A} > 0$ and verifying in (11) that $detA_2 > 0$, we get of the expression (9):

$$P_{33} - det\mathbf{A_1} > \frac{\epsilon \, d \; det\mathbf{A_2}}{a} > 0 \; . \tag{17}$$

Therefore, if $det\mathbf{A} > 0$, then the matrix $\mathbf{A^{-1}}$ is positive and the first item of the Theorem 3.2 is proved.

To verify the second item of the Theorem 3.2, we suppose the existence of a 3-by-3 diagonal matrix **D**. For the product matrix **AD** can be dominant diagonal, the following relations must be verified:

$$(v + \mu)\, d_{11} > \left(\epsilon d + (1-p)\frac{\beta\pi}{\mu}\right) d_{22} + \epsilon d \; d_{33} \; , \tag{18}$$

$$\left(a - pf\frac{\beta\pi}{\mu}\right) d_{22} > qv \; d_{11} \; , \tag{19}$$

$$a \; d_{33} > (1-q)v \; d_{11} + p(1-f)\frac{\beta\pi}{\mu} \; d_{22} \; . \tag{20}$$

By successive substitutions, we get the following strict inequality:

$$\left(\frac{det\mathbf{A}}{det\mathbf{A1}}\right) d_{33} > 0 \; . \tag{21}$$

As $det\mathbf{A}$ and $det\mathbf{A_1}$ are positive when $R_0^e < 1$, we can conclude there is a positive diagonal matrix $\mathbf{D}$ where $\mathbf{AD}$ is stricly dominant diaginal since $d_{33} > 0$.

Thus, the matrix $\mathbf{A}$ is a M-matrix, that is, the matrix $\mathbf{J_{22}^*}$ has all of the eigenvalues with negative real part when $R_0^e < 1$. In this case, the disease will not be able to be established if it was introduced. Conversely, in the case where $R_0^e > 1$, there is, at least, one eigenvalue which do not have negative real part. Therefore, the endemic equilibrium point have to be evaluated because there is a possibility of disease be established if it was introduced.

4. Stability of Endemic Equilibrium Point

Evaluating the Jacobian matrix in the non-trivial equilibrium point, we get:

$$\mathbf{J} = \begin{pmatrix} -\mu R_0^e & 0 & -\frac{\beta\pi}{\mu R_0^e} & 0 \\ (1-p)\mu(R_0^e-1) & -(v+\mu) & \epsilon d + (1-p)\frac{\beta\pi}{\mu R_0^e} & \epsilon d \\ pf\mu(R_0^e-1) & qv & -a + pf\frac{\beta\pi}{\mu R_0^e} & 0 \\ p(1-f)\mu(R_0^e-1) & (1-q)v & p(1-f)\frac{\beta\pi}{\mu R_0^e} & -a \end{pmatrix} . \tag{22}$$

Calculating the determinant of the matrix $(\mathbf{J} - \lambda\mathbf{I})$ through the expansion in cofators for the last column, after massive calculations, we can reduce it to the following expression:

$$det(\mathbf{J} - \lambda\mathbf{I}) = \lambda^4 + \lambda^3\, a_3 + \lambda^2\, a_2 + \lambda\, a_1 + a_0 \,. \tag{23}$$

Using the definitions of $R_0^e(6)$, a (8) and $\beta' = \frac{\beta\pi}{\mu R_0^e}$, we have:

$$i) \quad a_3 = 2\,a + \mu R_0^e + v + \mu - pf\beta' \,, \tag{24}$$

$$ii) \quad a_2 = (\mu R_0^e)a_2^1 + \mu(R_0^e-1)a_2^2 + a_2^3 \,, \tag{25}$$

with: $a_2^1 = 2a + (v+\mu) - pf\beta'$, $\quad a_2^2 = pf\beta'$,

$$\begin{aligned} a_2^3 = a&\left[(v+\mu) + \left(a - pf\beta'\right)\right] - \left(\epsilon d + (1-p)\beta'\right) qv \\ &- \epsilon d(1-q)v + \left[(v+\mu)\left(a - pf\beta'\right)\right] , \end{aligned} \tag{26}$$

$$iii) \quad a_1 = (R_0^e - 1)\, a_1^1 + (R_0^e)\, a_1^2 + k_0 \,, \tag{27}$$

with:

$$\begin{aligned} k_0 = -\epsilon d&\left[\left(a - pf\beta'\right)(1-q)v + qvp(1-f) \times \beta'\right] \\ &+ a\left[(v+\mu)\left(a - pf\beta'\right) - \left(\epsilon d + (1-p)\beta'\right) qv\right] , \end{aligned} \tag{28}$$

$$a_1^1 = \beta' \mu \{pf[a + (v + \mu)] + (1 - p)qv\} \, , \tag{29}$$

$$a_1^2 = \mu \left\{ a \left[(v + \mu) + \left(a - pf\beta' \right) \right] - \left(\epsilon d + (1 - p)\beta' \right) qv \right.$$
$$\left. -\epsilon d(1 - q)v + \left[(v + \mu) \left(a - pf\beta' \right) \right] \right\} \, , \tag{30}$$

$$iv) \quad a_0 = Z_0 + K_0 \, , \tag{31}$$

where:

$$Z_0 = \beta' z_0 \, , \tag{32}$$

$$z_0 = (\mu(R_0^e - 1) \{\epsilon d [p(1 - f)qv - pf(1 - q)v]$$
$$+ a \times [(1 - p)qv + pf(v + \mu)]\}) \, , \tag{33}$$

$$K_0 = \mu R_0^e k_0 \tag{34}$$

To evaluate the signs of the eigenvalues associated to the matrix **J** in (22), we use the Routh-Hurwitz Theorem Frazer[16] which relate them to the sequence of test determinants. In this way, we can concude about the stability of the system.

Theorem 4.1. *A necessary and sufficient condition for the roots of the characteristic polynomial with real coefficients:*

$$a_4\lambda^4 + a_3\lambda^3 + ... + a_2\lambda^2 + a_1\lambda + a_0 = 0 \tag{35}$$

have negative real parts is that all determinants given by:

$$M_1 = \left[a_1 \right] \tag{36}$$

$$M_2 = \begin{bmatrix} a_1 & a_0 \\ a_3 & a_2 \end{bmatrix} \tag{37}$$

$$M_3 = \begin{bmatrix} a_1 & a_0 & 0 \\ a_3 & a_2 & a_1 \\ 0 & 1 & a_3 \end{bmatrix} \tag{38}$$

$$M_4 = \begin{bmatrix} a_1 & a_0 & 0 & 0 \\ a_3 & a_2 & a_1 & a_0 \\ 0 & 1 & a_3 & a_2 \\ 0 & 0 & 0 & 1 \end{bmatrix} \tag{39}$$

are strictly positive.

Thus, to verify the sign of eigenvalues of the matrix **J** in (22) for Theorem 4.1, we use the following result as suggested in Yang[24,25]:

Lemma 4.1. *If the independent term (a_0) of the characteristic polynomial given by the expression (35) is positive then all coefficients a_i $i = 1, 3$, are also positive.*

Proof. At first, we verify the conditions to the independent term (a_0) of the characteristic polinomial defined in (23) is strictly positive. Substituting the expressions of Z_0 and K_0 given in (32) and (34) in the expression (31), we get:

$$a_0 = \mu R_0^e k_0 + \beta' z_0 \,. \tag{40}$$

Adding and subtracting the part μk_0 in the expression (40), we observe that a_0 can be written as a function of R_0^e:

$$a_0 = \beta' z_0 + \mu(R_0^e - 1)k_0 + \mu k_0 \tag{41}$$

or:

$$a_0 = \mu(R_0^e - 1)\, a_0^1 + \mu\, a_0^2 \tag{42}$$

with:

$$a_0^1 = \{a\,[\mu\epsilon d + (\mu + \mu_t)(v + \mu)]\}\ , \tag{43}$$

$$\begin{aligned} a_0^2 = -\epsilon d &\left[\left(a - pf\beta'\right)(1-q)v + p(1-f) \times \beta' qv\right] + \\ a &\left[(v+\mu)\left(a - pf\beta'\right) - \left(\epsilon d + (1-p)\beta'\right) qv\right] . \end{aligned} \tag{44}$$

Expanding and simplifying the expressions of a_0^1 and a_0^2 we can show that $a_0^1 > a_0^2 > 0$. In this case, a_0 will be strictly positive if $R_0^e - 1 > 0$ in (42). After that, we need to verify if the coefficients a_1, a_2, a_3 are positive. By the restriction $a > \frac{pf\beta\pi}{\mu}$ (8), starting from $R_0^e > 1$, we have the restriction $a > pf\beta'$ and we get directly that $a_3 > 0$. As we can observe in the equation (25), the coefficient a_2 is composed by three parts, where:

i) a_2^1 is positive (due to restriction (8)) ;

ii) a_2^2 is positive by definition;

iii) a_2^3 is positive because $a > pf\beta'$, $R_0^e > 1$ and the strict inequality is valid (16).

Starting from this, we can conclude:

$$(v+\mu)\left(a-\frac{pf\beta\pi}{\mu R_0^e}\right)>\left(\epsilon d+(1-p)\frac{\beta\pi}{\mu R_0^e}\right)(qv) \tag{45}$$

and $a_2 > 0$.

It remains to prove that the coefficient a_1 *composed by 3 parts, is positive. For this, we analyze each part:*

i) the term a_1^1 *is positive for definition.*

ii) Assuming $R_0^e > 1$*, the restriction (8) and strict inequality (45) valid as described in the previous case, we can show that* k_0 *(28) and* a_1^2 *(30) are also positive.*

In this way, we can conclude all the coefficients of the characteristic polynomial (23) are strictly positive if a_0 *is. Thus, the lemma is proved.*

Verified the sign of all the coefficients of the characteristic polynomial, the following conditions must be satisfied to guarantee that the determinants described in the expressions (36), (37), (38), (39), are strictly positive:

$$a_1 > 0\,, \tag{46}$$

$$a_1\,a_2 - a_0\,a_3 > 0\,, \tag{47}$$

$$a_3\,[a_1\,a_2 - a_0\,a_3] - a_1^2 > 0\,, \tag{48}$$

and the previous conditions can be reduced to the following unique condition:

$$a_3\,[a_1\,a_2 - a_0\,a_3] - a_1^2 > 0\,. \tag{49}$$

To show the validity of the strict inequality (49) when $R_0^e > 1$, we apply the Rejection Technique Kalos[18] with 5000 simulations. Results are presented in Figure 2, using as possible range for each parameter, the values given by literature Blower[5] to tuberculosis, Table 1, where: ECR is average number of new infections caused by one infectious case; $\frac{1}{\mu}$ is life expectancy; p is proportion of new infections that become TB disease until 1 year; v is progression rate to TB; f is probability of developing infectious TB (fast

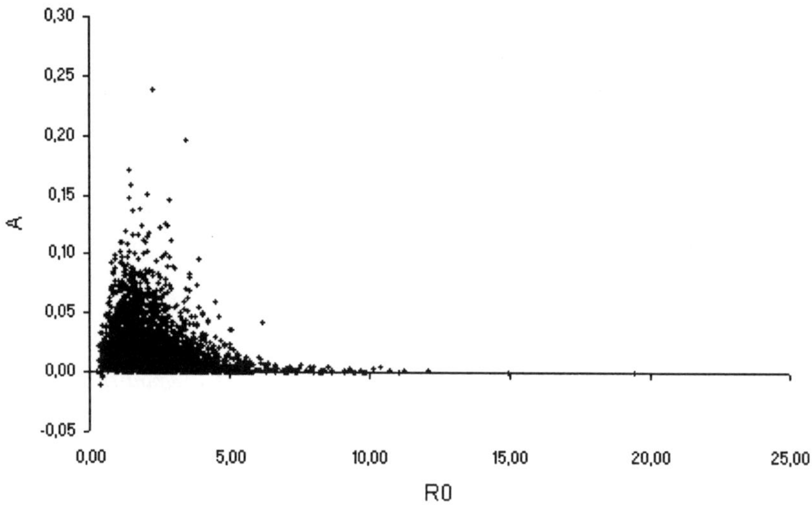

Figure 2. R_0^e x A

TB); q is probability of developing infectious TB (slow TB); μ_t is mortality rate due to TB (per capita).

We observe, in Figure 2, that $a_3\,[a_1\,a_2 - a_0\,a_3] - a_1^2 > 0$ when $R_0^e > 1$. In this way, we can conclude that when $R_0^e > 1$, the matrix **J** has all of its eigenvalues with negative real parts, and exists stability in the non-trivial equilibrium point.

5. Example

5.1. *Importance of the successfull treatment*

One of the biggest problems for the erradication of the epidemics is the abandonment of the treatment, where the individual continues to being a source of the transmission of the disease. In the presented model, to evaluate the influence of the successful treatment, the cure rate c is substituted for the product ϵd, where d is the detection rate and treatment of the cases of tuberculosis and ϵ is the efetivity of this treatment as suggested in Dye[14]. Thus, varying ϵ we can evaluate the efect of successfull treatment. In sequence, we present a simulation, where $\pi = 1500$, $\beta = 0.00018$, $p = 0.1$, $f = 0.66$, $\mu = 0.04$, $\mu_t = 0.461$, $q = 0.87$, $v = 0.005$ and we

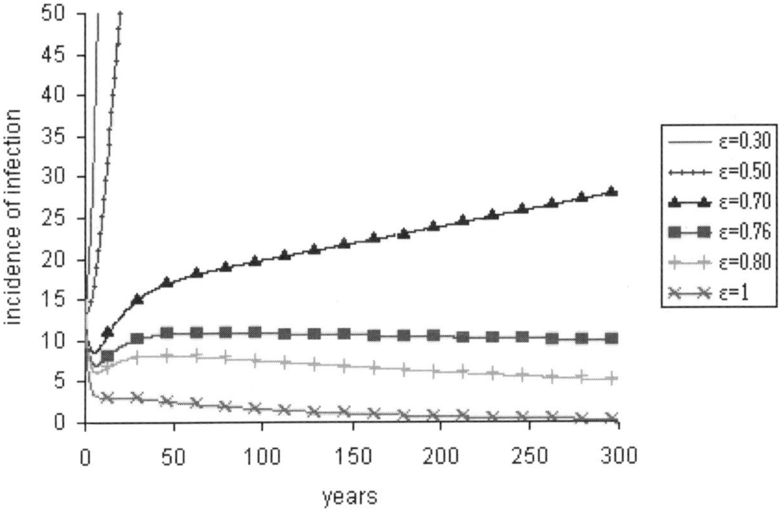

Figure 3. Incidence of Infection

can get by Effective Reproductive Rate that if more than 76% of the cases are detected and treated, the disease will not become established, that is, if $d > 0.76$ and 100% of the cases get successfull treatment ($\epsilon = 1$). To evaluate the influence of the successfull treatment, we present in the Figures from 3 to 6 grafics of Incidence of Infection ($II = \beta ST_i$); Incidence of Disease ($ID = p\beta ST_i + vL$); Prevalence of Infection ($PI = \frac{L+T_i+T_n}{N}$); Prevalence of Disease ($PD = \frac{T_i+T_n}{N}$), when $R_0^e > 1$, ($\epsilon = 0.30; 0.50; 0.70$); when $R_0^e = 1$, ($\epsilon = 0.76$) and when $R_0^e < 1$, ($\epsilon = 0.80; 1$).

Observing Figures 3 to 6, it is easy to understand why the WHO considers DOTS (the notification of cases and the accompaniment of sick people in the medicine ingestion), the more efficient strategy of treatment.

6. Uncertainty Analysis - an aplication

If we wish to do a quantitative analysis to the behaviour of the epidemics, where the uncertainty of the input data is considered, there arises the necessity to do an uncertainty analysis. This analysis allows us to evaluate the effect of uncertainty of these input parameters in the epidemiological outcome variables: Incidence of Infection, Incidence of Disease, Prevalence of Infection, Prevalence of Disease.

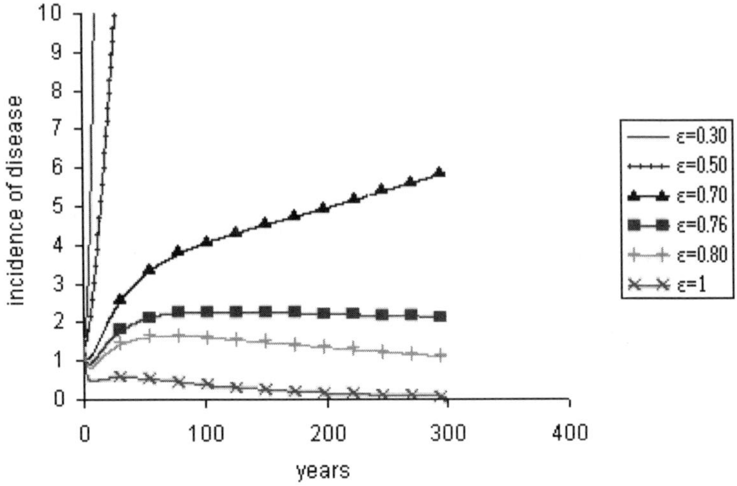

Figure 4. Incidence of Disease

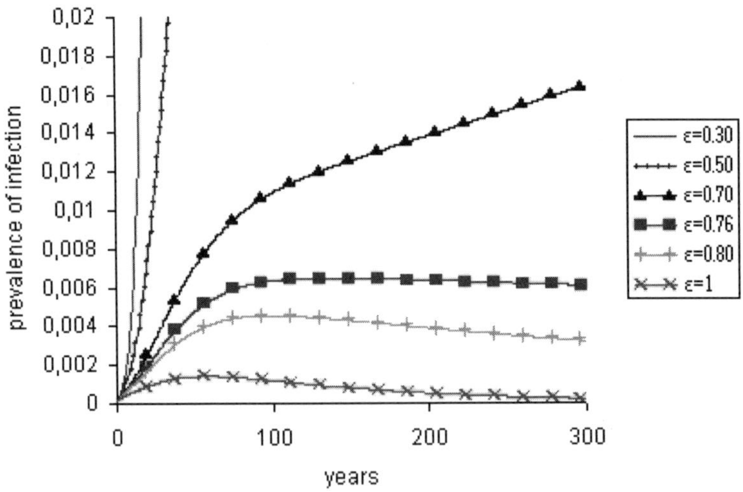

Figure 5. Prevalence of Infection

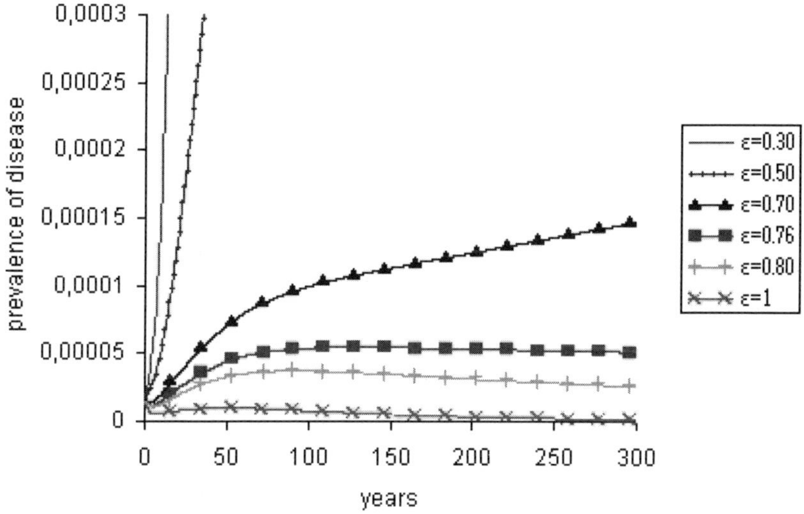

Figure 6. Prevalence of Disease

Table 1.

Symbol	Units	Lower	Mode	Upper
ECR	$year^{-1}$	3.0	7.0	13.0
$1 / \mu$	year	25.0	-	75.0
p	-	0.0	0.05	0.30
v	$year^{-1}$	0.00256	-	0.00527
f	-	0.50	0.70	0.85
q	-	0.50	0.85	1.0
μ_t	$year^{-1}$	0.058	0.139	0.461

In this work, we use a Technique of Monte Carlo Method with called Rejection Technique Kalos[18] to choose parameter values used to generate each simulation. The choise is based on theirs respectives PDFs and range of values, Table 1, for each one of the n simulations. As outcome values we can get measures of position Murray[20] as median, 1st quartile and 3rd quartile Rodrigues[2] for example.

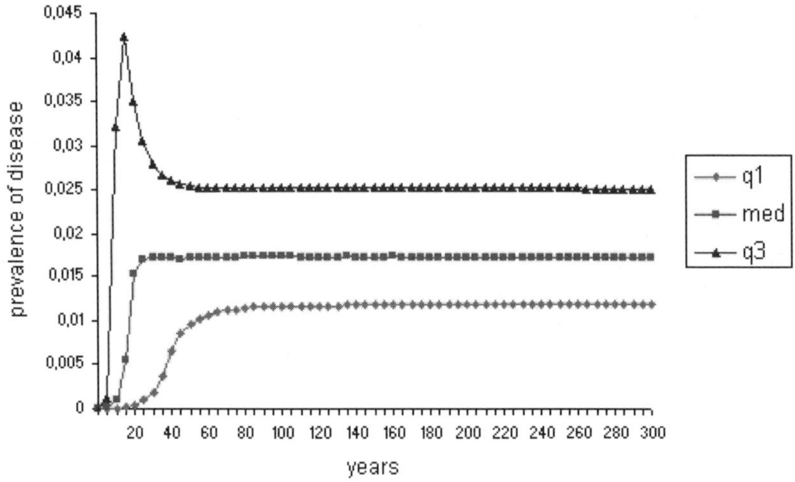

Figure 7. Prevalence of Disease - [ϵ: 0.021 to 0.086]

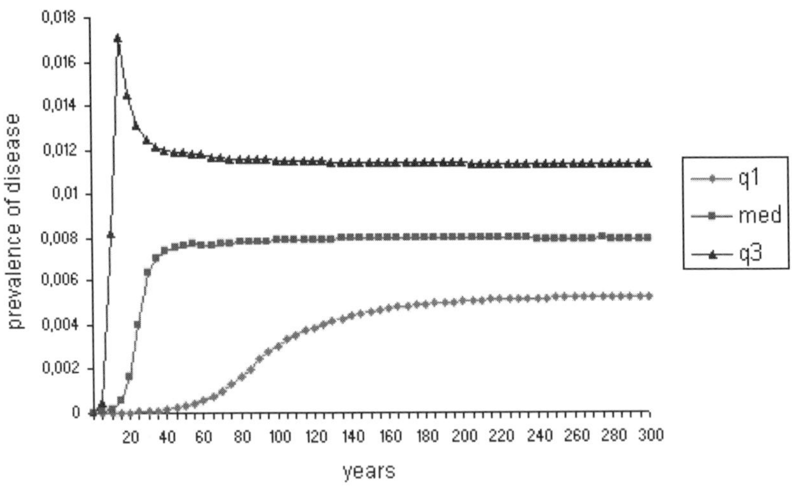

Figure 8. Prevalence of Disease - [ϵ: 0.086 to 0.40]

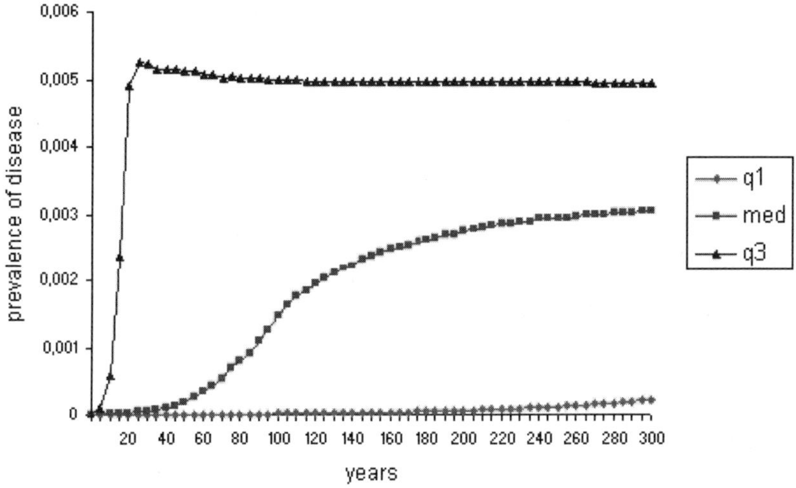

Figure 9. Prevalence of Disease - [ϵ: 0.40 to 0.70]

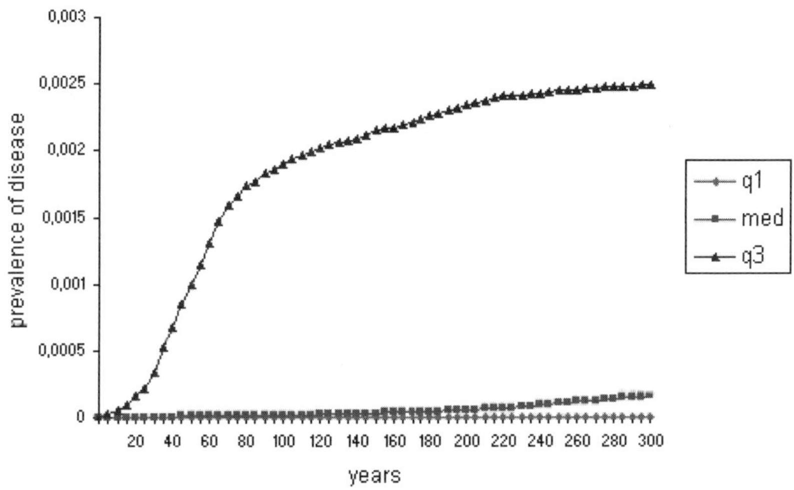

Figure 10. Prevalence of Disease - [ϵ: 0.70 to 1]

The following simulation presents the uncertainty analysis to Prevalence of Disease using the range of parameters supplied in Table 1. Supposing that 100% of the cases of TB are detected, we analize four situations (see figures 7, 8, 9 and 10):

(i) There is not treatment instituted. In this case, we use for ϵ the values suggested in Blower[5] for natural cure rate [0.021 - 0.086].
(ii) The effectiveness of treatment vary from 8.6% to 40%.
(iii) The effectiveness of treatment vary from 40% to 70%.
(iv) The effectiveness of treatment vary from 70% to 100%.

7. Conclusions

Mathematical model are powerful tools for studying the complex non linear dynamics of TB epidemics and transmission model can be extremely useful in designing epidemic control strategies.

In this work we presented a model which captures the essence of the complexity of the transmission dynamics of TB under DOTS strategy, evaluating the influence of abandonment of the treatment in the evolution of the disease.

As proposed, the model allows to calculate R_0^e (Effective Reproductive Rate) which is the parameter of great importance because it indicates whether or not the disease will became established within the population. Furthermore, the R_0^e turns the uncertainty analysis gotten, an instrument of decision making.

References

1. BERMAN A., PLEMMONS, R.J., Nonnegative Matrices in the Mathematical Sciences, SIAM, 1994.
2. RODRIGUES, P. C., Bioestatística, Ed. UFF, 2002.
3. BLOWER S.M., DOWLATABADI H., Sensivity and Uncertainty Analysis of Complex disease models: an HIV, as an example. International Statistical Review 62 (1994), 229-243.
4. BLOWER S.M., MC LEAN, A., PORCO T ET AL., Control Strategies for Tuberculosis Epidemics 8, Nature Medicine (1995), 815-821.
5. BLOWER S.M., MC LEAN, A., PORCO ET AL., Quantifying the Intrinsic Transmission Dynamics of Tuberculosis, Theorical Population Biology 54 (1997), 117-132.

6. BLOWER S.M., MC LEAN, A., PORCO T et al., The Intrinsic Transmission Dynamics of Tuberculosis Epidemics, Academic Press (1995).
7. BLOWER S.M., ZIV E., DALEY C.L., Early Therapy for Latent Tunberculosis Infection, American Journal of Epidemiology 153 (2001), 381-385.
8. CHAVEZ C., FENG Z., To treat or not to treat: the case of tuberculosis, Journal of Mathematical Biology 35 (1997), 629-659.
9. CHAVEZ C. ,FENG Z., Global stability of an age-structure model for TB and its applications to optimal vaccination strategies, Mathematical Biosciences 151 (1998), 135-154.
10. CHAVEZ C., CAPURRO A. F., APARICIO J.P., Transmission and Dynamics of Tuberculosis on Generalized Households, J. Theor. Biol. 206 (2000), 327-341.
11. CHAVEZ C., FENG Z., CAPURRO A. F., A model for tuberculosis with Exogenous Reinfection, J. Theor. Biol. 57 (2000), 235-247.
12. NETO, A. R., VILLA, T. C. S., Tuberculose - Implantação do DOTS em algumas regiões do Brasil - Histórico e peculiaridades regionais, Instituto Milênio Rede-TB Rede de Pesquisas em TB, (2006).
13. CUNHA C.E, Simulação Numérica de Epidemias de Tuberculose, Projeto Final do Curso de Ciência da Computação - UFF, 2002.
14. DYE C., GARNETT G.P., SLEEMAN K., WILLIANS B.G., Prospects for Worldwide Tuberculosis Control Under the WHO DOTS strategy, Lancet 352 (1998), 1886-1891.
15. DYE C., WILLIANS B.G., Criteria for the control of drug-resistant tuberculosis, PNASOnline 97 (2000); 8180-8185.
16. FRAZER, R. A.; DUNCAN, W. J.; COLLAR A. R., Elementary matrices, Cambridge, 1957.
17. GOMES, P. D., Um modelo matemático para a epidemia de TB, UFF, 2004.
18. KALOS M.H., WHITLOCK P.A., Monte Carlo Methods. John Wiley and Sons, 1986.
19. MURPHY B.M., SINGER B.H., ANDERSON S., KIRSCHNER D., Comparing epidemics tuberculosis in demographically distinct heterogeneous populations, Mathematical Biosciences 180 (2002), 1-24.
20. MURRAY, C.J.L., Salomon, J.A., Using Mathematical Models to Evaluate Global Tuberculosis Control Strategies. Harvard Center for Population and Development Studies, Cambridge, MA, (1998).
21. SANCHEZ M.A., BLOWER S.M., Uncertainty and Sensivity Analysis of the Basic Reprodutive Rate: Tuberculosis as an Example, American Journal of Epidemiology 12 (1997), 1127-1137.
22. COHEN, J., GLOBAL HEALTH: The New World of Global Health, Science, 311, 2006, 162–167
23. SECRETARIA DE POLÍTICAS DE SAÚDE, Situação da Tuberculose no Brasil, Série G: estatística e informação em saúde. Ministério da Saúde, 2002.
24. YANG H. M. Epidemiologia Matemática. UNICAMP, 2001.

25. YANG H. M., BASSANEZI, R. C.; LEITE, M. B. F., The basic reproduction ratio for a model of directly transmitted infections considering the virus charge and the immunological response, IMA Journal of Mathematics Apllied in Medicine and Biology, 17, 2000, 15–31.

MATHEMATICAL AND COMPUTATIONAL MODELING OF PHYSIOLOGICAL DISORDERS: A CASE STUDY OF THE IUPS HUMAN PHYSIOME PROJECT AND ANEURYSMAL MODELS

TOR A. KWEMBE

Department of Mathematics, Jackson State University,
P. O. Box 17610, Jackson, MS 39217.
E-mail: tor.a.kwembe@jsums.edu

The International Union of Physiological Sciences (IUPS) has undertaken a project called the Physiome Project with the goal of developing a comprehensive framework for modeling the human body using computational techniques which incorporate the biochemistry, biophysics and anatomy of cells, tissues and organs. The project aims to establish a web-accessible physiological databases dealing with model-related data which includes bibliographic information, at the cell, tissue, organ and organ system levels. The databases are intended to provide a quantitative description of physiological dynamics and functional behavior of the intact organism. The long-range objective is to understand and describe the human organism, its physiology and to use this understanding to improve human health. In this survey, we will give an overview of the Physiome Project and an analysis of the collection of mathematical and computational models aimed at detection, prevention and treatment of physiological disorders such as aneurysms. In conclusion, we will show the connection between the aneurysm database and the Physiome Project.

1. Introduction

The Physiome Project of the International Union of Physiological Sciences (IUPS) is attempting to provide a comprehensive framework for modeling the human body using computational methods which can incorporate the biochemistry, biophysics and anatomy of cells, tissues and organs. A major goal of the project is to use computational modeling to analyze integrative biological function in terms of underlying structure and molecular mechanisms. The project has established centers around the world to develop web-accessible physiological databases consisting of model-related data which includes bibliographic information, cell, tissue, organ and organ system levels.

The idea of the project was initiated by the IUPS following the successful completion of the first draft of the 13-year international Human Genome Project (Genome) which begun in October 1990 and completed in 2003. By the first draft a good number of the estimated 20,000-25,000 human genes were discovered and made accessible for further biological study. The other Genome project goal was to determine the complete sequence of the 3 billion DNA bases in the human genome. The sequencing of the 3 billion base pairs in the human genome, and the discovery of 17,000 of the likely 35,000 genes is only the start. A more tasking challenge of integrating this incredible wealth of information to allow the determination of structure and function at all levels of biological organisms is now the undertaking of the Physiome project.

The completion of the human genome sequence demonstrates the power of international and interdisciplinary sciences coordination and cooperation that the IUPS intends to duplicate in executing the Human Physiome project. Physiology has always been concerned with the integrative function of cells, organs and whole organisms. However, the explosion of scientific data at the molecular level produced by biomedical scientists have made it difficult for physiologists to relate integrated whole organ function to underlying biophysically detailed mechanisms[2]. The only means of coping with this explosion in complexity is mathematical and computational modeling. Biological systems are very complex and to understand them requires specially designed instrumentation, databases and software. It also requires a high degree of both international and interdisciplinary collaboration that has never been experienced. Another aim of the project is to develop a framework for handling the hierarchy of computational models, and associated experimental data, which will help integrate knowledge at the genomic and proteomic levels into an understanding of physiological function for intact organisms.

Organ and whole organism behavior needs to be understood in terms of systems, subcellular function and tissue properties. For example, understanding a reentrant arrhythmia in the heart depends on knowledge of numerous cellular ionic current mechanisms and signal transduction pathways as well as larger scale myocardial tissue structure and the spatial distribution of ion channel and gap junction densities1[1,2,5,6]. Such a model demands the knowledge of anatomy, cell and tissue properties of the heart for it to be able to reveal the integrated physiological function of the electrical activation, mechanics and metabolism of the heart under a variety of normal and pathological conditions[9]. The computational techniques and software tools developed for the hart project are applicable to other organs and systems of the body[2]. A number of universities in the UK, the US and New Zealand[3,10,11] have in collaboration developed a model of the lungs that encompass gas transport and exchange, pulmonary blood flow and soft tissue mechanics[2].

The Physiome Project as currently presented does not seems to pose many of the ethical, legal, and social issues that surrounded the human genome project. Issues such as the availability of private personal information that are inherent in a person's gene to the general public may not be a factor here. Nevertheless as rightly pointed out by Hunter[2] the project posed a lot of interesting scientific questions that may lead to the evolution of new science disciplines in the same fashion as the Genome project. Scientific questions such as (1) can the effect of a gene mutation be modeled all the way from its effect on protein structure and function to the way the altered properties of the protein affects a cellular process such as signal transduction, (2) how can the changed properties of the process alter the function of tissues and organs.

If the project is successfully completed as intended, there will be many benefits from this integrative framework to the academics and medical communities. Benefits such as (1) model simulation to determine the effects of parameter variation on the modeled physiological phenomenon or disorder such as embryological growth, ageing and disease, (2) the design of medical devices , (3) the diagnosis and treatment of diseases such as aneurysms and (4) the development of new drugs.

2. The IUPS Physiome Project

The fundamental concept of the physiome project is to understand and describe the human organism, its physiology and pathophysiology, and to use this understanding to improve human health[2]. For these and other benefits of the project to the environmental health issues and medical education, the project's other major objective is to develop computer models to integrate the observations from many laboratories into quantitative, self-consistent and comprehensive descriptions in meta data format accessible via standard data query platforms. As with the Human Genome Project, the internet is the medium through which the organizers are bring together a growing number of Physiome Centers that are developing and providing databases on the functional aspects of biological systems, covering the genome, molecular form and kinetics, cell biology, up to intact functioning organisms[2]. These databases will provide the raw information for users to use in developing physiological systems models that simulate the whole body organs. Aided by technical advances such as improved biological imaging techniques, a wealth of data on cell and tissue structures and physiological functions exists and are growing rapidly. Similarly, modelling resources and software are developing at a sufficient pace to allow the development of realistic computer simulations of whole organs to commence. In addition to working toward the long-term strategic goals, the project intends to facilitate and enhance biomedical research by continuously making well-documented and refereed models available to a wider range of non-mathematical users[2].

Thus, IUPS Physiome project is an effort to collect mathematical models of structure and function at spatial scales ranging from nano-scale molecular events to meter-scale intact organ systems up to the order of 10^9, and at temporal scales ranging from Brownian motion (microseconds) to a human lifetime (10^9s) up to the order of 10^{15}. In whatever way or method, for the project to be a success, IUPS must provide these services to the public in a form that is both human-readable and optimized for computer manipulation. Hence, the project is organizing the data in a hierarchy of models and modeling approaches requiring the model parameters at one scale to be linked to detailed models of structure and function at a smaller spatial scale[2].

The computational approach for developing the project data bank is a multi-scale modeling framework for understanding physiological function that allows models to be combined and linked in a hierarchical fashion. Models can be defined at various levels of abstraction: (i) The conceptual level where words are used to describe the model; (ii) The mathematical level where equations and boundary conditions are defined in standard mathematical notation; (iii) The formulation level, the modeling stage where equations are formulated in terms of the solution methodology; (iv) The solution level which involves numerical or closed form solutions or the development of algorithms for solving the parameterized equations on parameterized domains. To facilitate communication between research groups across these levels of abstraction, markup languages are needed to encapsulate the mathematical statements of the governing equations (MathML) and the way in which spatially and temporally dependent continuum fields are parametrized (FieldML). Scripting languages (such as Perl, Python or Matlab) are then needed to create modules which implement the mathematical operations on these fields and call libraries of numerical solution algorithms. The organizers proposed and developed the framework which includes MathML, FieldML, the scripting modules and a grouping construct called 'ModelML'. When domain specific ontologies are added to link this framework to biological entities such as cells, tissues or organs, the markup language framework is extended to become 'CellML', 'TissueML' or 'AnatML', etc. Computed variables are used to construct the functions that need to be minimized at the formulation level. It is intended that these computed variables will be constructed from MathML.

The primary sponsor of the Physiome Project is the International Union of Physiological Sciences (IUPS) under the auspices of the IUPS Physiome and the Bioengineering Committee: cochaired by Peter Hunter (University of Auckland, NZ) and Aleksander Popel (Johns Hopkins University, USA). Details about the Physiome project can be obtained from their web site cited in the references[2] or can accessed directly at www.physiomeproject.org.

This survey is an effort by the author to inform researchers with interest in biomathematics or mathematics and computational physiology of the existence of the IUPS Human Physiome Project. The intentions of the project are noble and these kinds of projects are not only good for the scientific world, but they also aid to enhance our understanding of the evolution of the human whole organ. If and when the project is completed, I think many researchers with interest in the biomedical areas will be influenced by many of the discoveries from the information gathered by the project. In the remainder of the paper, we shall present a survey of mathematical and computational methods of the abdominal aorta aneurysms in the hope to initiate work on the disease level of the human physiome project, the DiseaseML.

3. Aneurysmal Models

The last decade has generated extensive information on the genetic and molecular basis of disease. A major challenge remains the integration of this information into the physiological environment of the functioning cell and tissue. This paper illustrates how a completed Human Physiome project can enhance the use of computational biology in meeting this challenge in the context of abdominal aorta aneurysm (AAA). An AAA is characterized by a bulge in the abdominal aorta. The development of the AAA is associated with a weakening and dilation of the arterial wall and the possibility of rapture. The evolution of an aneurysm is assumed to be a consequence of the remodeling or a rebuilt of the artery's material constituents. The principal objective of mathematical modeling of aneurysms is to gain a greater insight in understanding the pathogenesis of the disease and to improve the criteria for predicting rapture and the decision as to when to operate on a AAA patient

In[12], Watton et al presented a first mathematical model that accounts for the evolution of the AAA. They modeled the artery as a two-layered cylindrical membrane using nonlinear elasticity and a physiologically realistic constitutive model. The model is subjected to two fundamental parameters, systolic pressure and physiological axial pre-stretch ratio. The model also takes into account of the way an aorta aneurysm is known to develop. It incorperates the microstructural recruitment and fibre density variables for the collagen into the Fung type strain energy density function. In this way they are able to address the rebuilt (or re-development) of the collagen as the aneurysm enlarges. An axisymmetric aneurysm with axisymmetric degradation of elastin and linear differential equations for the remodeling of the fibre variables is simulated numerically. They used physiologically determined parameters to model the abdominal aorta and realistic remodeling rates for its constituents to predict the dilations of the aneurysm. They obtained results that are consistent with in vivo observations. We will recast the model and the simulation here. However, to fully understand the model we will give a brief description of the structure of an abdominal aorta.

The abdominal aorta is a large elastic artery. It consists of three layers, the intima, media and adventitia. The intima is the inner most layer, the adventitia the outermost. The media is sandwiched between these two layers and is separated from them by the internal elastic lamellae which are thin elastic sheets composed of elastin. A substantial part of the arterial wall volume consists of an intricate network of macromolecules making up the extra-cellular matrix (ecm). This matrix is composed of a variety of versatile proteins that are secreted locally by fibroblast cells and assembled into an organized meshwork in close association with the surface of the cell that created them. The ecm determines the tissue's structural properties. For an artery, it consists primarily of elastin, collagen, smooth muscle and ground substance. The ground substance is a hydrophilic gel within which the collagen and elastin tissues are embedded. Collagen and elastin bear the main load. Collagen is the stiffer and most nonlinear of the two materials. However, the elastin bear the most load at low strains. This is due to the fact that at physiological strains the collagen is tortuous in nature and is crimped in the unstrained artery[8,12]. The elastin gives rise to the artery's isotropic and rubber-like behavior for small deformations. On the other hand, the collagen gives rise to the anisotropy and high nonlinearity at large deformations. In this model the mechanical contribution from the smooth muscle is ignored and the dilation of an aneurysm is by a loss of elastin and a weakening of the arterial wall[12].

Using the assumption that the strain field in the aneurysm tissue is uniform through the thickness of the arterial wall implies that the strain energy density functions (SEDFs), ω, are independent of the third coordinate x^3, after much work Watton et al arrived at the governing equation[12]

$$\int_0^{2\pi R} \left[\delta\left(h_M\omega_M + h_A\omega_A\right) - p\left(\mathbf{A}_1 \wedge \mathbf{A}_2\right) . \delta\nu^i\mathbf{a}_i\right] dx^1 dx^2 = 0\ , \qquad (1)$$

where the abdominal aorta is modeled as a thin cylinder of undeformed radius R, length L, and thickness h. The thickness of the media and adventitia are denoted by h_M and h_A respectively. $\mathbf{a}_\alpha = \frac{\partial \mathbf{q}^0}{\partial x^\alpha} = \mathbf{q}^0_{,\alpha}$ where $\mathbf{q}^0(\mathrm{x}^1,\ \mathrm{x}^2)$ is the position of the material point on the mid-plane of the membrane. $\mathbf{A}_\alpha = \mathbf{Q}^0{}_{,\ \alpha} = \mathbf{a}_\alpha + \mathbf{v}_{,\ \alpha}$ where $\mathbf{v}(\mathrm{x}^\alpha)$ is the mid-plane displacement field and $\mathbf{A}_\alpha$ is the tangent vector to the deformed mid-plane. They then used the appropriate functional forms for the SEDFs given by Holzapfel et al[4] for the media ω_M, and adventitia ω_A, as

$$\omega_M = c_M(\varepsilon_{11} + \varepsilon_{22} + \varepsilon_{33}) + \sum_{\varepsilon^2_{M_p} \geq 0} K_M\left(\exp\left[a_x\varepsilon^2_{M_p}\right] - 1\right)\ , \qquad (2)$$

$$\omega_A = c_A(\varepsilon_{11} + \varepsilon_{22} + \varepsilon_{33}) + \sum_{\varepsilon^2_{A_p} \geq 0} K_A\left(\exp\left[a_x\varepsilon^2_{A_p}\right] - 1\right)\ , \qquad (3)$$

where

$$\varepsilon_{J_p} = \varepsilon_{11}\sin^2(\gamma_{J_p}) + \varepsilon_{22}\cos^2(\gamma_{J_p}) + 2\varepsilon_{12}\sin(\gamma_{J_p})\cos(\gamma_{J_p}) \qquad (4)$$

is the Green's strain resolved in the direction of a collagen fibre which has an orientation of γ_{J_p} to the azimuthal axis. The constants C_M and C_A are associated with the non-collagenous matrix of the material, and a_x, K_M and K_A with the collagenous part. The concentration of elastin in the arterial wall $c_E(x^1, x^2, t)$ is taken to be a double Gaussian exponential

$$c_E(x^1, x^2, t) =$$
$$1 - \left(1 - (c_{\min})^{t/T}\right)\exp\left[-m_1\left(\frac{L_1 - 2x_1}{L_1}\right)^2 - m_2\left(\frac{L_2 - 2x_2}{L_2}\right)^2\right] \qquad (5)$$

where L_1 and L_2 denote the Lagrangian lengths of the membrane in the axial and circumference directions, m_1, $m_2 > 0$ are parameters that control the degree of localization of the degradation, and c_{min} is the minimum concentration of elastin at time $t = T$. For axisymmetric degradation, equation (5) reduces to

$$c_E(x^1, t) = 1 - (1 - (c_{\text{min}})^{t/T}) \exp\left[-m_1 \left(\frac{L_1 - 2x_1}{L_1}\right)^2\right] . \tag{6}$$

The SEDF is then the product of the concentration of elastin within the tissue and the neo-Hookean SEDF for the elastin. That is,

$$\omega_{\text{elastin}} = K_E c_E(x^1, x^2, t)(\varepsilon_{11} + \varepsilon_{22} + \varepsilon_{33}) , \tag{7}$$

where K_E is a material parameter to be determined that represents the mechanical behavior of elastin for the healthy abdominal aorta.

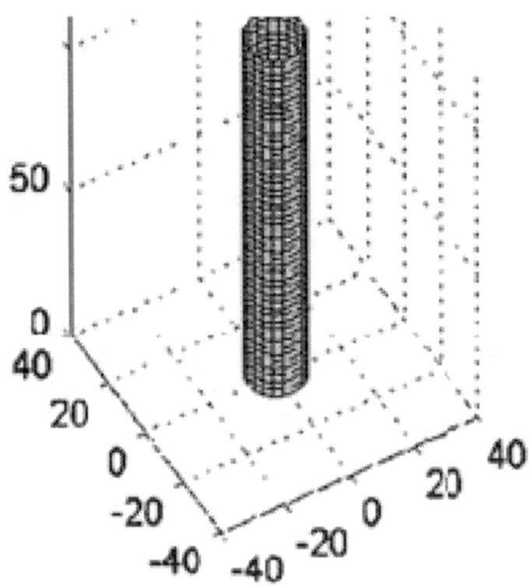

Figure 1. Undeformed Geometry

We will now consider the development of an aneurysm that arises due to an axisymmetric degradation of elastin. The degradation of elastin is given by equation (5) with $m_1 = 20$. The 3D figures given below gives the initial undeformed geometry at $t = 0$ in Figure (1) and the axisymmetric solution at 10 years using $m_1 = 20$ and $m_2 = 0$ in Figure (2). The Lagrangian mesh superimposed on the mid-plane of the aneurysm is attached to the material points and illustrates the deformation from the initial state.

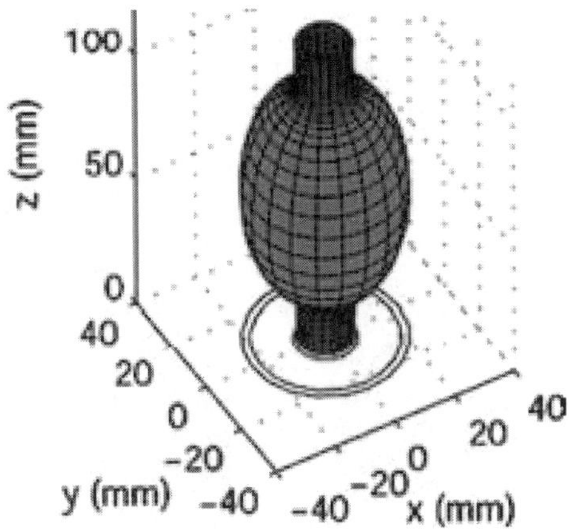

Figure 2. Deformed Geometry

4. Stented Abdominal Aortic Aneurysms

The Aortic Stent Graft interventional radiology technique, involves making a small nick in the groin and, under X-ray guidance, inserting a catheter into a blood vessel that leads to the aorta. A collapsed stent-graft, also known as an endograft is inserted through the catheter and moved to the site of the aneurysm, where it is deployed, reinforce the aorta and create a stronger pathway for the blood. When in place, blood flowing through the stent-graft no longer puts pressure on the ballooning walls of the aneurysm that are outside of the graft. Typically the patient is lightly sedated and has been given epidural anesthesia.

The stent graft is a piece of graft material, within which there are metal stents to support and secure the device to the wall of the aorta.

Utilizing a surgical access in the groin, the interventional radiologist and vascular surgeon work together to place the stent graft within the aorta at the location of the aneurysm to create a new channel for blood flow which effectively excludes the aneurysm from the circulation. The aneurysm clots off, leaving blood flowing through the stent graft in the same fashion as if a "vascular graft" had been placed during the routine type of surgical procedure. After placing the stent graft, the surgeon closes the access site in the groin and the patient is taken to the recovery room.

Realistic simulations of fluid-structure interactions in stented AAAs are important for AAA rupture prediction, optimal stent-graft placement, new stent-graft designs, and quantitative Endo-vascular Aortic Repair recommendations.

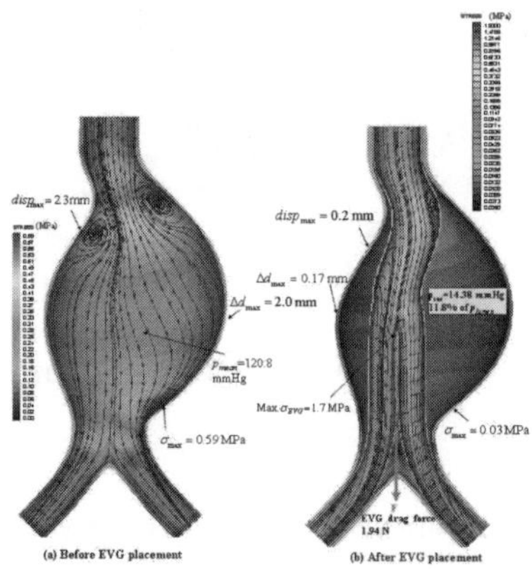

Figure 3. Unstented and Stented Aneurysm.

Insertion of a stent-graft into an aneurysm protects the aneurysm wall from high-pressure, pulsatile blood flow. Specifically, the pressure level in the aneurysm cavity drops by a factor of almost 10 and the maximum wall stress can be reduced by a factor of 20. Stented aneurysm simulation and analysis require the solution of coupled transient nonlinear fluid flow and solid structure equations to visualize the blood flow field interacting with the elastic stent-graft which, in turn, influences the stagnant blood in the aneurysm cavity, which transmits pressure waves to the distensible aneurysm wall. Professors C Kleinstreuer & Zhonghua Li, of the MAE Department at NC State University, Raleigh , NC are working on fluid structure interaction project to simulate fluid flows through AAAs to visualize the blood flow field interacting with aneurysmal wall and to ascertain the optimal time for inserting the stent graft. There experimentally validated results of the transient 3-D computer simulation model provide physical insight and quantitative recommendations for both endovascular surgeons for optimal stent-graft placement and implant manufacturers for improved stent-graft designs. With their permission, I am including here a graphic depiction of their results in Figure 3.

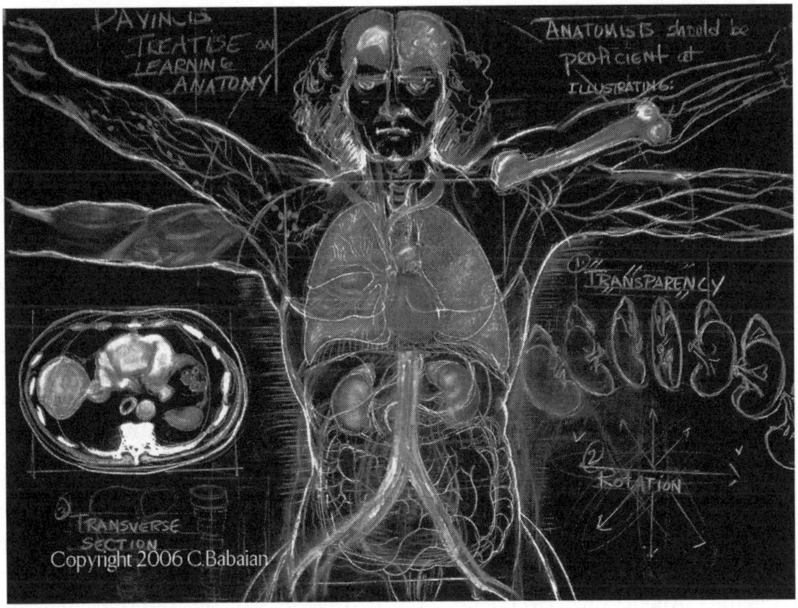

Figure 4. Da Vinci Vitruvian

5. Discussions

A major goal of the IUPS human physiome project is to use computational modeling to analyze integrative biological function in terms of underlying structure and molecular mechanisms. If the project were to be complete, then given a constituent parametric data of a AAA patient, such as DNA, blood type, blood pleasure, tissue samples an interventional radiologist or cardiovascular doctor would be able to use the project data to create a virtual carbon copy of this patient. Using the e-patient, Doctors would be able to accurately predict the rapture potential of the AAA and make an informed decision on type of treatment for the living patient. This is a wonderful project and I think we should all participate in it to make it a success. Leonardo da Vinci was ahead of us in his depiction of Vitruvius' perfect man. His version is considered the most accurate depictions of the human body. In the era of high tech we should better Da Vinci with a perfect depiction of a 3D functional e-person through the IUPS human physiome project. The da Vinci Vitruvian in figure (4) is courtesy of NSF science information site.

References

1. Glass L, Hunter PJ, McCulloch AD (1991) (eds) Theory of Heart. Springer-Verlag, New York
2. Hunter P, Robbins P, Noble D (2002) The IUPS human physiome project. Pflugers Arch - Eur J Physiol 445: 1 - 9
3. Howatson M, Pullan AJ, Hunter PJ (2000) Generation of an anatomically based three-dimensional model of the conducting airways. Ann Biomed Eng 28:793–802
4. Holzapel G A, Gasser T C, Ogden R W (2000) A new constitutive framework for arterial wall mechanics and a comparative study of material models. J Elasticity 61: 1 - 48
5. Kohl P, Noble D, Winslow RL, Hunter PJ (2000) Computational modelling of biological systems: tools and visions. Philos Trans R Soc Lond A 358: 579–610
6. Kohl P, Noble D, Hunter PJ (2001) (eds) The Integrated Heart: Modelling Cardiac Structure and Function. Philos Trans R Soc Lond A 359:1047–1337
7. Kwembe T A, Jones S A (2006) A Mathematical Analysis of Cylindrical Shaped Aneurysms. BIOMAT 2005, Proceedings of the International Symposium on Mathematical and Computational Biology, Rio de Janeiro, Brazil, World Scientific Publishers, 35 – 48
8. Raghavan M L, Webster M, Vorp D A (1999) Ex-vivo biomechanical behavior of AAA: assessment using a new mathematical model. Ann Biomed Eng 24: 573 - 582
9. Smith N P, Mulquiney PJ, Nash MP, Bradley CP, Nickerson DP, Hunter PJ (2001) Mathematical modelling of the heart: cell to organ. Chaos, Solitons and Fractals 13:1613–1621
10. Tawhai M H, Hunter PJ (2001) Characterising respiratory airway gas mixing using a lumped parameter model of the pulmonary acinus. Respir Physiol 127:241–248
11. Tawhai M H, Hunter PJ (2001) Multibreath washout analysis: modeling the influence of conducting airway asymmetry Respir Physiol 127:249–258
12. Watton P N, Hill N A, Heil M (2004) A mathematical model for the growth of the abdominal aortic aneurysm. Biomechan Model Mechanoniol 3: 98 - 113

LINEAR FEEDBACK CONTROL FOR A MATHEMATICAL MODEL OF TUMOR GROWTH

JEAN CARLOS SILVEIRA, ELENICE WEBER STIEGELMEIER, GERSON FELDMANN, MARAT RAFIKOV

DeFEM, Unijuí, 501 San Francisco Street, Ijuí,RS, Brazil
E-mail: jean.silveira@unijui.tche.br, elenice.stiegelmeier@unijui.tche.br, feldmann@unijui.tche.br, rafikov@unijui.tche.br

In this work we propose an application of the optimal control theory to the planning of tumor treatments. The tumor growth is represented by a system of three differential equations that considers the dynamics and interactions of three types of cells: normal, immune and tumoral. The problem of the tumor treatment was formulated in terms of the optimal control theory as the state regulator problem, aiming the reduction of the tumor cells population. The linear feedback regulator, that stabilized the nonlinear system with tumor around the globally stable tumor-free equilibrium point, was found. The numerical simulations show that the proposed optimal strategies can be accomplished by existing methods of cancer treatment, including radiotherapy.

1. Introduction

The principal therapy modalities for cancer treatment are surgery, chemotherapy, immunotherapy and radiotherapy. The use of one or another depends on a variety of factors, such as the state and severity of the tumor, the state of the patient immune system and the tumor site. A thorough revision of the contribution of mathematical modeling in the study of tumor growth was presented by Araujo and McElwain [1]. In the last few years, several papers have been published about the optimal tumor treatment planning strategies, based on mathematical models [2-4]. Mathematical modeling of this process is viewed as a potentially powerful tool in the development of improved treatment scheduling for chemotherapy, where the strength and toxicity of the drugs should be considered. In some works (for example, [2]) the control function appears in the model in explicit form before the formulation of the optimization problem. In the case of radiotherapy, the situation is more complex because one needs to consider the effects of radiation on normal cells as well as the so-called "latte effects" (those that appear only long times after irradiation) on the cells. In addition, the effects of radiation are different for normal and tumor cells. Therefore, currently used protocols for radiotherapy treatments are based on three main factors: total dose delivered to the tumor, number of fractions of application and fraction dose. To the best of our knowledge, little work has been done for models that also takes into account the effects of the patient immune system.

In the present work, we propose an application of the control theory to tumor dynamics. Tumor growth is described by a mathematical model that considers tumor cell growth, immune response and normal cell growth by a system of three differential equations [2]. We formulate the problem of tumor control in terms of the optimal control theory as a state regulator problem, aiming at the reduction of the tumor cells population. We found the linear feedback regulator that stabilizes the nonlinear system around the globally-stable tumor-free equilibrium point.

2. 2. The Mathematical Model

We consider a tumor growth model in which tumor cell growth, immune response and normal cell growth are represented by a system of three differential equations [2]. This model is an improved version of the prey-predator-protector model for cancer, proposed in [5]. It includes immune cells whose growth may be stimulated by the presence of the tumor and that can destroy tumor cells through a kinetic process. Normal cells and tumor cells compete for available resources, while immune cells and tumor cells interact in predator-prey fashion. The reaction of the immune cells with tumor cells is modeled in the same manner as that described in [6]. The growth of the tumor and normal cells is described by a logistic growth term.

We let $I(t)$, $T(t)$ and $N(t)$ denote the number of immune, tumor and normal cells at time t, respectively. The mathematical model is given by the following system of ordinary differential equations:

$$\begin{aligned} \dot{N} &= r_2 N\,(1 - b_2 N) - c_4 T N \\ \dot{T} &= r_1 T\,(1 - b_1 T) - c_2 I T - c_3 T N \\ \dot{I} &= s + \frac{\rho I T}{\alpha + T} - c_1 I T - d_1 I \end{aligned} \tag{1}$$

where the parameters r_1, r_2, b_1 and b_2 represent the *per capita* growth rates and reciprocal carrying capacities of normal and tumor cells, respectively. The positive nonlinear growth term for immune cells $\dfrac{\rho I T}{\alpha + T}$ represents the immune response, stimulated by the presence of tumor cells, where ρ and α are positive constants. The source of the immune cells is considered to be outside of the system so it is reasonable to assume a constant influx rate s. In the absence of any tumor, immune cells will die off at a rate d_1, resulting in a long-term population size of s/d_1 cells. The reaction of immune cells and tumor cells can result in either the death of tumor cells or the inactivation of the immune cells. The coefficients c_1 and c_2 represent this process whilec_3 and c_4 are coefficients that characterize the competition between normal and tumor cells.

There are three categories of equilibrium points that occur in this model:tumor-free, dead and coexistence. Bellow we briefly describe each one.

- *Tumor-free.* In this category, the tumor cell population is zero but the normal cells survive. The equilibrium point has the form
$$\left(\frac{1}{b_2}, 0, \frac{s}{d_1}\right). \tag{2}$$
If the following condition is satisfied
$$r_1 < \frac{c_2 s}{d_1} + c_3 \tag{3}$$
then the tumor-free equilibrium (2) is stable. This relates the *per capita* growth rate of the tumor $r_1$to the "resistance coefficient", $c_2 s/d_1$, which measures how efficiently the immune system competes with the tumor cells.
- *Dead.* An equilibrium point is classified as "dead" if the normal cell population is zero. There are two possible types of "dead" equilibria.
i) Type 1. $(0, 0, s/d1)$ in which both the normal and tumor cell populations are zero. This dead equilibrium is always unstable.
ii) Type 2. $(0, a, f(a))$ where the normal cell population is zero and tumor cells have survived. Here, a is a nonnegative solution to
$$a + \left(\frac{c_2}{r_1 b_1}\right) f(a) - \frac{1}{b_1} = 0,$$
where
$$f(a) \equiv \frac{s(\alpha + a)}{c_1 a(\alpha + a) + d_1(\alpha + a) - \rho a}.$$
This dead equilibrium can be either stable or unstable, depending on the parameters of the system.

- *Coexistence.* Here, normal and tumor cells coexist with nonzero populations. The equilibrium point is given by $(g(b), b, f(b))$, where b is a nonnegative solution of

$$b + \left(\frac{c_2}{r_1 b_1}\right) f(b) + \left(\frac{c_3}{r_1 b_1}\right) g(b) - \frac{1}{b_1} = 0,$$

where

$$g(b) \equiv \frac{1}{b_2} - \left(\frac{c_4}{r_2}\right) b$$

Depending on the values of these parameters, there could be zero, one, two, or three of these equilibriums which can be either stable or unstable.

Figure 2 shows an example that characterizes the system (1) with a stable coexisting equilibrium point.

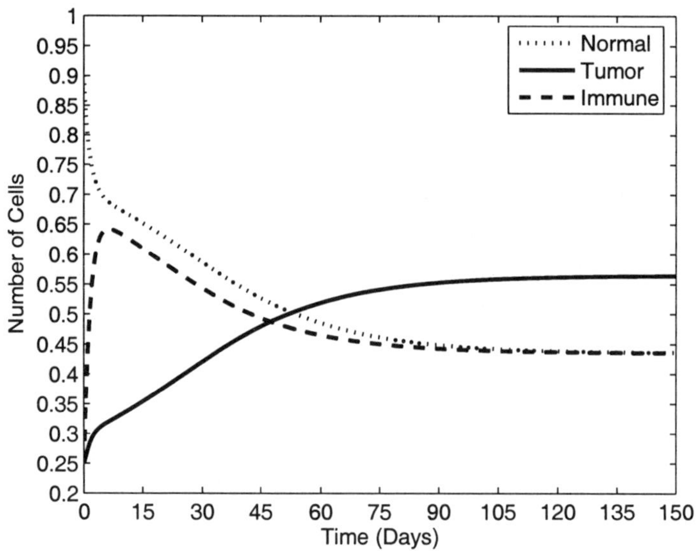

Figure 1. Evolution of cell populations, showing the stable coexisting equilibrium.

From Figure 2, we see that when the system progresses naturally without any intervention, the tumor burden overwhelms the system, while the normal cell population quickly decreases to below survival levels.

The equilibrium state that the system should ideally approach, in context of developing treatment therapy, is the tumor-free equilibrium.

3. Optimal Treatment Strategy

In this section, we consider the effect of therapy introduced in the system, and we use optimal control theory to look for an improved administration protocol. Our control problem consists of determining the function $u(t)$ (representing the chemotherapy or radiotherapy administration schedule) that will kill off the tumor cell population as effectively as possible, with the constraint that we do not also kill too many normal cells.

The optimal treatment problem is resolved by using Dynamic Programming reducing this problem to a solution of the Hamilton-Jacobi-Bellman equation. The fact that the steady-state solution of the Hamilton-Jacobi-Bellman equation is a Lyapunov function for the nonlinear system guarantees both stability and optimality. In the next subsection we present the supporting results of the optimal control design.

3.1. *Linear Design for the Nonlinear System*

In [7] a unified framework for continuous-time nonlinear-nonquadratic problems was presented in a constructive manner. The basic underlying ideas of the results in [7] are based on the fact that the steady-state solution of the Hamilton-Jacobi-Bellman equation is a Lyapunov function for the nonlinear system, thus guaranteeing both stability and optimality. In [8] these ideas were used to design linear feedback control for nonlinear systems. We consider the nonlinear control system

$$\dot{y} = Ay + h(y) + Bu \tag{4}$$

where $y \in R^n$ is a state vector, $A \in R^{n \times n}$ and $B \in R^{n \times m}$ are constant matrices, $u \in R^m$ is a control vector, and $h(y)$ is a vector, continues nonlinear functions. Assuming that

$$h(y) = G(y)\, y, \tag{5}$$

the dynamic system (4) has the following form

$$\dot{y} = Ay + G(y)y + Bu \tag{6}$$

Next, we present an important result, concerning a control law.

Theorem [6]. If there exist positive definite matrices Q and R, being Q symmetric, such that the matrix

$$\tilde{Q} = Q - G^T P - PG \tag{7}$$

is positive definite for the limited matrix G, then the linear feedback control

$$u = -R^{-1} B^T P y \tag{8}$$

is optimal, minimizing the functional

$$J = \int_0^\infty (y^T \tilde{Q} y + u^T R u) dt \tag{9}$$

and transferring the nonlinear system (6) from an initial to a final state

$$y(\infty) = 0. \tag{10}$$

In the functional (9), the symmetric matrix P is evaluated through the solution of the Ricatti matrix algebraic equation

$$PA + A^T P - PBR^{-1}B^T P + Q = 0. \tag{11}$$

3.2. *Formulation of the Optimal Treatment Problem*

We assume that the control $u(t)$ kills all types of cells, but that the kill rate differs for each type of cell. We denote by a_1, a_2 and a_3 the three different response coefficients. Then, system (1) including control is given by

$$\begin{aligned} \dot{N} &= r_2 N (1 - b_2 N) - c_4 T N - a_3 u \\ \dot{T} &= r_1 T (1 - b_1 T) - c_2 I T - c_3 T N - a_2 u \\ \dot{I} &= s + \frac{\rho I T}{\alpha + T} - c_1 I T - d_1 I - a_1 u \end{aligned} \tag{12}$$

Our goal is to direct the system (12) from an initial state to a desired regimen: the tumor-free equilibrium

$$\left(\frac{1}{b_2}, 0, \frac{s}{d_1}\right) \Rightarrow \tilde{N} = \frac{1}{b_2}; \tilde{T} = 0; \tilde{I} = \frac{s}{d_1}. \tag{13}$$

Defining

$$\begin{aligned} y_1 &= N - \tilde{N} \\ y_2 &= T - \tilde{T} \\ y_3 &= I - \tilde{I} \end{aligned} \tag{14}$$

as the deviation of the trajectory of system (12) from the desired one, one will obtain the following equation in form (6)

$$\dot{y} = Ay + G(y)y + Bu$$

where

$$A = \begin{bmatrix} r_2(1 - 2b_2\tilde{N}) & -c_4\tilde{N} & 0 \\ 0 & r_1 - (c_2\tilde{I} + c_3\tilde{N}) & 0 \\ 0 & -c_1\tilde{I} & -d_1 \end{bmatrix}; \quad B = \begin{bmatrix} -a_3 \\ -a_2 \\ -a_1 \end{bmatrix}; \tag{15}$$

$$G = \begin{bmatrix} -r_2b_2y_1 - c_4y_2 & 0 & 0 \\ 0 & -r_1b_1y_2 - c_3y_1 - c_2y_3 & 0 \\ 0 & \frac{\rho(y_3+\tilde{I})}{\alpha+y_2} - c_1y_3 & 0 \end{bmatrix} \tag{16}$$

Due to the above formulated theorem, the optimal control strategy is given by

$$u = -R^{-1}B^TP \begin{bmatrix} N-\tilde{N} \\ T-\tilde{T} \\ I-\tilde{I} \end{bmatrix} \tag{17}$$

The symmetric matrix P is evaluated through the solution of the matrix Ricatti algebraic equation (11). This control guarantees the stability of the tumor-free equilibrium if the matrix $\tilde{Q} = Q - G^TP - PG$ is positive definite.

3.3. *Numerical Simulations*

The parameters of the controlled system (12) are considered as in [2]: $a_1 = 0.2$, $a_2 = 0.3$, $a_3 = 0.1$, $b_1 = 1.0$, $b_2 = 1.0$, $\alpha = 0.3$, $c_1 = 1.0$, $c_2 = 0.5$, $c_3 = 1.0$, $c_4 = 1.0$, $d_1 = 0.2$, $r_1 = 1.5$, $r_2 = 1.0$, $s = 0.33$ and $\rho = 0.01$.

For these set of parameters and (13) we have a stable tumor free equilibrium at $\tilde{N} = 1.0$, $\tilde{T} = 0$, $\tilde{I} = 1.65$, and the matrix A has the form,

$$A = \begin{bmatrix} 0 & -1 & 0 \\ 0 & -0.325 & 0 \\ 0 & -1.65 & -0.2 \end{bmatrix}$$

Choosing,

$$Q = \begin{bmatrix} 1 & 0 & 0 \\ 0 & 1 & 0 \\ 0 & 0 & 0,01 \end{bmatrix} \quad ; R = 1$$

we obtain

$$P = \begin{bmatrix} 0{,}4985 & -0{,}3489 & -0{,}0006 \\ -0{,}3489 & 2{,}2879 & -0{,}0600 \\ -0{,}0006 & -0{,}0600 & 0{,}0246 \end{bmatrix}.$$

By solving the Riccati equation (11) using the LQR function in MATLAB, it follows that

$$u = -0{,}0550(N-1) + 0{,}6395T - 0{,}0132(I-1{,}65) \tag{18}$$

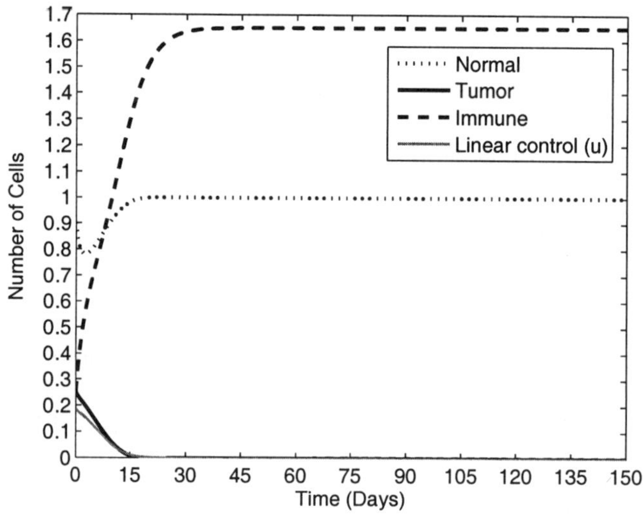

Figure 2. The time trajectories of the system (12) controlled by optimal control (18).

The time trajectories of the system (12) controlled by this optimal control are shown in Figure 2.

It is difficult to analyze the matrix $\tilde{Q}$ analytically in this case. Numerical simulations show that the function $L(t) = y^T \tilde{Q}\, y$ is positive definite (Figure 3). It implies that the matrix $\tilde{Q}$ is positive definite. Consequently, equation (18) represents an optimal control.

From the results of our numerical simulations we remark that the control functions have to be calculated with values of R greater than 1. For $R < 1$ the results show a tumor cell introduction into the system at some time of treatment, instead of tumor cell elimination, which have no biological interpretation.

4. Conclusions

In the present work, the problem of the tumor treatment was formulated in terms of the optimal control theory as the state regulator problem, aiming at the reduction of the tumor cells population. The linear feedback regulator, that stabilizes the nonlinear system with tumor around the globally stable tumor-free equilibrium point, was found. The protocol suggested by the optimal control algorithm dictates that the treatment be administered continuously over relatively long periods of time – on the order of days. The

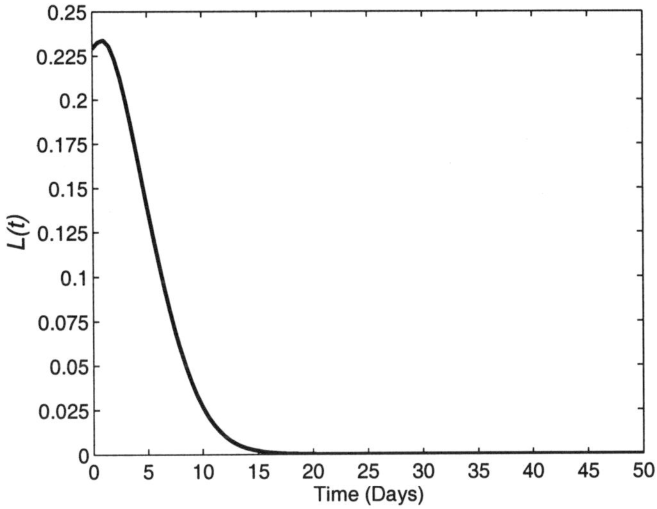

Figure 3. Time trace of the positive definite function $L(t) = y^T \tilde{Q} y$.

proposed optimal control guarantees the stability of the tumor-free equilibrium. The analysis of the simulation results show that optimal control could be used to describe treatment protocols which have the potential to be more efficient than standard periodic protocols now in use.

References

1. R.P. Araújo and L.S. McElwain, A History of the Study of Solid Tumour Growth: The Contribution of Mathematical Modelling. *Bulletin of Mathematical Biology* **66, pp.** 1039-1091 (2004).
2. L. de Pillis e A. Radunskaya, The Dynamics of an Optimally Controlled Tumor Model: A Case Study. *Mathematical and Computer Modelling*, vol. **37**, pp. 1221-1244 (2003).
3. L. Moonen and H. Bartelink, Antitumor Treatment - Fractionation in radiotherapy. *Cancer Treatment Reviews,* vol. **20**, pp. 365-378. (1994).
4. Xiangkui Mu, Per-Olov Löfroth, Mikael Karlsson, Bjöm Zackrisson, The effect of fraction time in intensity modulated radiotherapy: theoretical and experimental evaluation of an optimisation problem. *Radiotherapy and Oncology*, vol. **68**, pp. 181-187. (2003).
5. J. Stein. Prey-Predator-Protector model for cancer. *IEEE Transactions on Biomedical Engineering* 28 (5) pp 544-551 (1981).
6. V. Kuznetsov, I. Makalkin, M. Taylor and A. Perelson, Nonlinear dynamics of immunogenic tumors: Parameter estimation and global bifurcation analysis,

Bulletin of Mathematical Biology. 56 (2), pp. 295-321 (1994)
7. D. S. Bernstein, Nonquadratic Cost and Nonlinear Feedback Control, *Int. J. Robust Nonlinear Control*, Vol. 3, pp. 211-229 (1993)
8. M. Rafikov, J.M. Balthazar, Optimal linear and nonlinear control design for chaotic systems. *Proceedings of ASME International Design Engineering Technical Conferences and Computers and Information in Engineering Conference*, Long Beach, California, USA, September 24–28, 2005.

HEAT KERNEL BASED 3D RECONSTRUCTION OF OBJECTS FROM 2D PARALLEL CONTOURS

CÉLESTIN WAFO SOH

Department of Mathematics
College of Science, Engineering and Technology
Jackson State University, JSU Box 1710
1400 J R Lynch St, Jackson, MS 39217
E-mail: wafosoh@yahoo.com, celestin.wafo@jsums.edu

Imaging modalities such as Magnetic Resonance Imaging (MRI) and Computed Tomography (CT) produce sequences of planar parallel cross-sectional images taken at regular or irregular intervals of 3D objects (brain, coronary arteries, etc). In order to reconstruct a 3D object which has been imaged using MRI or CT there is a need for a fast, accurate and robust algorithm. Reconstruction usually starts with an appropriate segmentation algorithm that detects the trace of the 3D object boundary on each 2D image. Segmentation can be either manual, semi-automatic or ideally automatic. After segmentation, one needs to "glue" together the contours in order to reconstruct the 3D object. Both the segmentation technique and the method used to glue individual contours affect the quality of the reconstructed object. In this paper we introduce a novel initialization that speeds up a semi-automatic segmentation technique recently developed by Chan and Vese (IEE Trans. Im. Proc., vol. 10, No. 2, February 2001, pp 266-277). We implement Vese and Chan (Int. J. Comp. Vis. vol 50, N0. 2, 2002, pp 271-293) multi-phase segmentation using relaxed Gauss-Seidel method. As a result we obtain an algorithm faster than the one proposed by Vese and Chan . We employ the heat kernel to piece together 2D parallel contours in order to obtain 3D reconstruction. This approach is preferred because of the slowness of 3D segmentation using level sets. In case of large volumes we indicate how to use fast Gauss transform to achieve fast reconstruction. Numerical experiments are provided to support our methodology.

1. Introduction

Efforts to reconstruct 3D surfaces from 2D contours started in the early 70's[1]. The advent of new imaging tools such MRI and CT scan which produce high quality 2D cross-sectional images of 3D objects put a new spin on the need to create efficient reconstruction algorithms. Accurate diagnosis and treatment of diseases such as atherosclerosis (obstruction of arteries), aneurysm (irreversible dilation of an artery) and cancer will benefit from high quality 3D reconstruction of lesions. In the case of coronary arteries diseases which account for most of the death in the United States, 3D images of diseased arteries allow accurate quantification (obstruction level, stress distribution, ...) and monitoring of stenosis.

In any attempt to reconstruct a 3D object from images of its cross-sections, the first problem one has to tackle is that of determining the trace of the 3D object on each image. This is a nontrivial task that can be carried out manually, semi-automatically and ideally automatically. Manual segmentation is time-consuming, error-prone and depends on the skills and dexterity of the segmenter. To the best of our knowledge the problem of finding a fully automatic and accurate segmentation algorithm is still open. In this paper we shall use a semi-automatic partial differential equation (PDE) based segmentation algorithm due to Chan and Vese[2] and Vese and Chan[3].

After contours identification on each 2D slice, the next logical step is to find a way to stitch together these contours in order to produce a 3D smooth surface. We face at this stage a major problem: we have to account for missed information between two consecutive contours. The method used by several authors (e.g[1]) replaces missed information by the solution of an interpolation problem involving neighboring contours. Linear interpolation of neighboring contours is computational inexpensive but produces at most continuous surfaces. Linear interpolation results in discontinuous 3D models . Other interpolation methods such as cubic $B-$spline produce smooth surface but are computational costly and results in other types of discontinuity as described in[4]. In this work we propose and justify a reconstruction method that does not require the explicit solution of an interpolation problem and yet produces a reasonably smooth surface that necessitates minimal smoothing at most.

We have organized this paper as follows. There are five sections of which this section is the first. For the sake of self-containedness, Section 2 deals with a succinct introduction to PDE based segmentation with an emphasis on models based on Mumford and Shah functional. In particular, we present Chan-Vese active contour without edge. Section 3 deals with the numerical solution of Chan-vese segmentation models. Here we propose an initialization that speeds up 2-phase segmentation and for multi-phase segmentation, we suggest relaxed Gauss-Seidel method as a way of speeding up convergence. Numerical experiments on real images are used to back up our methodology. Section 4 uses the heat kernel to reconstruct 3D objects from their 2D cross-sectional images. The last section i.e. Section

5 summarizes the paper.

2. Partial differential equations based segmentation

A key step in the reconstruction process is image segmentation i.e. the identification of all objects present on the image. In this work, we employ a PDE based segmentation technique.

The main idea of PDE based segmentation is to deform a curve or a surface under certain constraints until it conforms to objects boundaries. In the classical snake model of Kass, Witkin and Terzopoulos[5] a curve/surface is deformed by minimizing its energy which is made of two terms: the internal energy which controls the smoothness of the curve /surface and the external energy which attracts the contour/surface towards objects boundaries. The main issue of the classical Snake method is that the choice of the external energy is problem-dependent. Thus several researchers[6,7,8] have devoted their works to the design of appropriate external energies. The other problem associated with the classical snake model is the issue of parametrization that makes the detection of topologically complicated boundaries difficult. The geometric snake model of Casselles, Kimmel and Shapiro[9] is parametrization-free but is unable to detect some topologically complicated boundaries. The level set method of Osher and Sethian[10] is the appropriate technique for representing topologically complicated curves or surfaces: it allows cusps, corners and automatic topological changes. For this reason several researchers[11,12,13] have reformulated geometric snake models in terms of level sets.

Classical and geometric snake models rely on an edge-detector that depends on the image gradient and uses the vanishing of the gradient at edges as stopping criterium. Thus they can only detect edges defined by gradient. Several imaging modalities like MRI and CT scan produce image with objects boundaries that cannot be defined by the gradient.For images produced using these modalities or noisy images, classical or geometric snake models might miss the boundary. Therefore there is a need for snake models which do not involve edge-function in the stopping process and is robust to noises. Such a model was recently introduced by Chan and Vese[2] and Vese and Chan[3]. Chan and Vese models are based on Munford and Shah[14] functional.

2.1. *Mumford and Shah functional*

In their probabilistic approach to image restoration, Geman and Geman[15] interpreted images as statistical mechanics system in which intensity levels become states of atoms or molecules. They assigned an energy function to the resulting system using Gibbs distribution. As a result, they obtained a Markov random field image model. They established that in a Bayesian framework, image restoration amounts to maximize the posterior distribution or equivalently minimize the discrete functional

$$F^{GG}(u,h,v) = \sum_{i,j} \left(|u_{i,j} - u_{0,i,j}|^2 + \nu(|u_{i+1,j} - u_{i,j}|^2(1 - v_{i,j}) \right.$$
$$\left. + |u_{i,j+1} - u_{i,j}|^2(1 - h_{i,j})) + \mu(h_{i,j} + v_{i,j}) \right), \tag{1}$$

where u is the unknown restored image, u_0 is the observed image, h and v are binary representations of unknown horizontal and vertical objects boundary respectively,and μ and ν are positive constants.

Mumford and Shah[14] considered a continuous analog of the functional (1) by treating u and u_0 as real-value functions defined on a bounded open subset of $\mathbb{R}^2$, Ω say. Let $\mathcal{C}$ be a closed subset of Ω formed by a finite number of smooth curves. Denote the length of $\mathcal{C}$ by $|\mathcal{C}|$. Using the following equivalence

$$\sum_{i,j} |u_{i,j} - u_{0,i,j}|^2 \equiv \int_\Omega |u - u_0|^2 dxdy,$$

$$\sum_{i,j} (|u_{i+1,j} - u_{i,j}|^2(1 - v_{i,j}) + |u_{i,j+1} - u_{i,j}|^2(1 - h_{i,j})) \equiv \int_{\Omega\setminus\mathcal{C}} |\nabla u|^2 dxdy$$

$$\sum_{i,j} (h_{i,j} + v_{i,j}) \equiv |\mathcal{C}|,$$

we obtain the celebrated Mumford-Shah functional given by

$$F^{MS}(u,\mathcal{C}) = \int_\Omega |u - u_0|^2 dxdy + \nu \int_{\Omega\setminus\mathcal{C}} |\nabla u|^2 dxdy + \mu|\mathcal{C}|. \tag{2}$$

Image restoration problem is thus reduced to the minimization problem:

$$\inf_{u,\mathcal{C}} F^{MS}(u,\mathcal{C}). \tag{3}$$

The restored image u may be sought amongst piecewise constant functions. In this case the Mumford-Shah functional assumes the simpler form

$$E^{MS}(u,\mathcal{C}) = \sum_i \int_{\Omega_i} (u_0 - c_i)^2 dxdy + \mu|\mathcal{C}|, \tag{4}$$

where the Ω_i's are connected components of Ω and the c_i's are the constant values of u on connected components.

The solution of the problem (3) is hindered by the non convexity of the functional F^{MS}. Besides, $\mathcal{C}$ is unknown. Thus existence, uniqueness and regularity of solutions of the minimization problem (3) is not guaranteed. Early results on the existence and regularity of solutions of the problem (3) can be found in[14,16,17].

In the next subsection, we review Chan and Vese recent approach to the solution of the minimization problem associated with the functional (4).

2.2. *Chan-Vese active contour without edge*

One of the challenges one face when attempting to solve the minimization problem associated with the energy (4) is the representation of $\mathcal{C}$ which may be topologically complicated. The level set method of Osher and Sethian[10] was designed to handle such topological complications.

A level set function is a scalar Lipschitz continuous function $\phi : \Omega \to \mathbb{R}$. Assume that the level set representation of objects boundary, $\mathcal{C}$, is

$$\mathcal{C} = \{(x, y) \in \Omega |\, \phi(x, y) = 0\}$$

where ϕ is a level set function. Thus $\mathcal{C}$ is simply the zero level set of ϕ. Now let

$$\begin{aligned} \Omega_1 &= \{(x, y) \in \Omega |\, \phi(x, y) \geq 0\}, \\ \Omega_2 &= \{(x, y) \in \Omega |\, \phi(x, y) < 0\}. \end{aligned}$$

It is easy to see that $\{\Omega_1, \Omega_2\}$ is a partition of Ω. Let H be the Heaviside function defined by

$$H(x) = \begin{cases} 1 & \text{if } x \geq 0, \\ 0 & \text{otherwise.} \end{cases}$$

The restored image u may be expressed as

$$u = c_1 H(\phi) + c_2(1 - H(\phi)).$$

Note that the restoration and segmentation problem will be solved if we are able to find ϕ, c_1 and c_2. In the sequel we shall find the relationship between these unknown.

It can be shown[2,18] that

$$|\mathcal{C}| = \int_{\Omega} \nabla H(\phi)\, dxdy = \int_{\Omega} \delta(\phi) \nabla\phi\, dxdy, \tag{5}$$

where δ is Dirac measure.

Using the representation (5), we can rewrite the functional (4) as

$$E^{MS}(\phi, c_1, c_2) =$$
$$\int_\Omega \left[H(\phi)(u_0 - c_1)^2 + (1 - H(\phi))(u_0 - c_2)^2 + \mu\, \delta(\phi)\nabla\phi\right] dxdy. \quad (6)$$

The functional E^{MS} is extremal provided

$$\frac{\partial E^{MS}}{\partial \phi} = 0, \quad \frac{\partial E^{MS}}{\partial c_1} = 0, \quad \frac{\partial E^{MS}}{\partial c_2} = 0, \quad (7)$$

where the first partial derivative is the Gateaux derivative of E^{MS} with respect to ϕ. Expanding the system (7) leads to the equations

$$\begin{cases} \delta(\phi)\left(-\mu\nabla\cdot\left(\frac{\nabla\phi}{|\nabla\phi|}\right) + (u_0 - c_1)^2 - (u_0 - c_2)^2\right) = 0 \text{ in } \Omega, \\ \delta(\phi)\frac{\partial\phi}{\partial\vec{n}} = 0 \text{ on } \partial\Omega, \end{cases} \quad (8)$$

$$c_1(\phi) = \frac{\int_\Omega u_0 H(\phi) dxdy}{\int_\Omega H(\phi) dxdy}, \quad (9)$$

$$c_2(\phi) = \frac{\int_\Omega u_0 (1 - H(\phi)) dxdy}{\int_\Omega (1 - H(\phi)) dxdy}. \quad (10)$$

In order to solve the system (8) an artificial time is introduced so that finding a solution of Eqs.(8) is equivalent to finding the steady state solution of the transient problem

$$\frac{\partial\phi}{\partial t} = \delta(\phi)\left[\mu\nabla\cdot\left(\frac{\nabla\phi}{|\nabla\phi|}\right) - (u_0 - c_1)^2 + (u_0 - c_2)^2\right],$$
$$\text{in } (0,\infty)\times\Omega, \quad (11)$$

$$\phi(0, x, y) = \phi_0(x, y) \quad \text{in } (0,\infty)\times\Omega, \quad (12)$$

$$\frac{\delta(\phi)}{|\nabla\phi|}\frac{\partial\phi}{\partial\vec{n}} = 0, \quad \text{on } \partial\Omega. \quad (13)$$

Equations (11)-(13) form the basis of Chan-Vese active contour without edge[2]. If objects boundaries is represented by several level set functions, these equations can be naturally extended[3]. For instance if $\mathcal{C}$ is represented using two level set functions ϕ_1 and ϕ_2 i.e.

$$\mathcal{C} = \{(x, y) \in \Omega |\, \phi_1(x, y) = 0 \text{ or } \phi_2(x, y) = 0\},$$

then a partition of Ω is given by the subsets

$$\Omega_{11} = \{(x,y) \in \Omega |\, \phi_1(x,y) \geq 0 \text{ and } \phi_2(x,y) \geq 0\},$$

$$\Omega_{10} = \{(x,y) \in \Omega |\, \phi_1(x,y) \geq 0 \text{ and } \phi_2(x,y) < 0\},$$

$$\Omega_{01} = \{(x,y) \in \Omega |\, \phi_1(x,y) < 0 \text{ and } \phi_2(x,y) \geq 0\},$$

$$\Omega_{00} = \{(x,y) \in \Omega |\, \phi_1(x,y) < 0 \text{ and } \phi_2(x,y) < 0\},$$

and the associated piecewise constant representation of the restored image is

$$\begin{aligned} u = {} & c_{11} H(\phi_1) H(\phi_2) + c_{10} H(\phi_1)(1 - H(\phi_2)) + \\ & + c_{01}(1 - H(\phi_1)) + c_{00}(1 - H(\phi_1))(1 - H(\phi_2)). \end{aligned}$$

Proceeding as in the derivation of (11)-(13), we arrive at the equations governing the evolution of ϕ_1 and ϕ_2:

$$\begin{aligned} \frac{\partial \phi_1}{\partial t} = {} & \delta(\phi_1) \left\{ \mu \nabla \cdot \left(\frac{\nabla \phi_1}{|\nabla \phi_1|} \right) - [(u_0 - c_{11})^2 - (u_0 - c_{01})^2] H(\phi_2) \right. \\ & \left. -[(u_0 - c_{10})^2 - (u_0 - c_{00})^2][1 - H(\phi_2)] \right\}, \end{aligned} \tag{14}$$

$$\begin{aligned} \frac{\partial \phi_2}{\partial t} = {} & \delta(\phi_2) \left\{ \mu \nabla \cdot \left(\frac{\nabla \phi_2}{|\nabla \phi_2|} \right) - [(u_0 - c_{11})^2 - (u_0 - c_{10})^2] H(\phi_1) \right. \\ & \left. -[(u_0 - c_{01})^2 - (u_0 - c_{00})^2][1 - H(\phi_1)] \right\}, \quad \text{in } (0,\infty) \times \Omega \end{aligned} \tag{15}$$

$$\phi_i(0,x,y) = \phi_i^0(x,y), \quad \text{in } (0,\infty) \times \Omega, \quad i = 1,2, \tag{16}$$

$$\frac{\delta(\phi_i)}{|\nabla \phi_i|} \frac{\partial \phi_i}{\partial \vec{n}} = 0, \quad \text{on } \partial\Omega, \tag{17}$$

where

$$c_{11} = \frac{\int_\Omega u_0 H(\phi_1) H(\phi_2) dxdy}{\int_\Omega H(\phi_1) H(\phi_2) dxdy},$$

$$c_{10} = \frac{\int_\Omega u_0 H(\phi_1)(1 - H(\phi_2)) dxdy}{\int_\Omega H(\phi_1)(1 - H(\phi_2)) dxdy},$$

$$c_{01} = \frac{\int_\Omega u_0 (1 - H(\phi_1)) H(\phi_2) dxdy}{\int_\Omega (1 - H(\phi_1)) H(\phi_2) dxdy},$$

$$c_{00} = \frac{\int_\Omega u_0 (1 - H(\phi_1))(1 - H(\phi_2)) dxdy}{\int_\Omega (1 - H(\phi_1))(1 - H(\phi_2)) dxdy}.$$

At this point, few remarks are in order. The introduction of the artificial time leads to the necessity of initial conditions. We shall see subsequently that the numerical solution of the problems (11)-(13) and (14)-(17) is extremely sensible to initialization and the choice of the parameter μ. Due to the non convexity of the functional (4), there is no guarantee that we will obtain a global solution of the associated minimization problem. Initialization together with appropriate approximation of the Dirac measure is fundamental in obtaining global minimizers.

3. Numerical implementation of active contour without edge

Here we are concerned with the numerical solution of the problems (11)-(13) and (14)-(17). We have implemented all the algorithms in Matlab.

3.1. *Numerical solution of (11)-(13)*

We employ an implicit finite difference scheme as described in[2,19,20]. However we use a completely new initialization that speeds up convergence to global minimizers.

Let represent the observed image u_0 as an $M \times N$ matrix with entries $u_{0i,j}$. Denote the time steps by Δt and introduce the notation $\phi_{i,j}^n = \phi(n\Delta t, i, j), \quad 1 \le i \le M;\ 1 \le j \le N$. Note that we have chosen space step equal 1. Let δ_ϵ be the smooth approximation of the Dirac measure given by

$$\delta_\epsilon(x) = \frac{\epsilon}{\pi(\epsilon^2 + x^2)}, \qquad 0 < \epsilon \le 1.$$

Equation (11) is approximated by

$$\begin{aligned} \frac{\phi_{i,j}^{n+1} - \phi_{i,j}^n}{\Delta t} = \delta_\epsilon(\phi_{i,j}^n) \Bigg[& \frac{\mu D_-^x D_+^x \phi_{i,j}^{n+1}}{\sqrt{(D_+^x \phi_{i,j}^n)^2 + 0.25(\phi_{i,j+1}^n - \phi_{i,j-1}^n)^2 + \epsilon_0}} \\ & + \frac{\mu D_-^y D_+^y \phi_{i,j}^{n+1}}{\sqrt{(D_+^y \phi_{i,j}^n)^2 + 0.25(\phi_{i+1,j}^n - \phi_{i-1,j}^n)^2 + \epsilon_0}} \\ & -(u_{0,i,j} - c_1(\phi^n))^2 + (u_{0,i,j} - c_2(\phi^n))^2 \Bigg], \end{aligned} \tag{18}$$

where ϵ_0 is a small positive number introduced to avoid division by zero,

$$D_\mp^x \phi_{i,j} = \pm(\phi_{i,j} - \phi_{i\mp1,j}), \quad D_\mp^y \phi_{i,j} = \pm(\phi_{i,j} - \phi_{i,j\mp1}). \tag{19}$$

Solving Eq. (18) for $\phi_{i,j}^{n+1}$ yields

$$\begin{aligned}\phi_{i,j}^{n+1} = & \frac{1}{C(\phi_{i,j})}[\phi_{i,j}^n + m(\phi_{i,j}^n)(C_1(\phi_{i,j}^n)\phi_{i+1,j}^n + C_2(\phi_{i,j}^n)\phi_{i-1,j}^{n+1} + \\ & +C_3(\phi_{i,j}^n)\phi_{i,j+1}^n + C_4(\phi_{i,j}^n)\phi_{i,j-1}^{n+1}) + \\ & +\Delta t \delta_\epsilon(\phi_{i,j}^n)(-(u_{0,i,j} - c_1(\phi^n))^2 + (u_{0,i,j} - c_2(\phi^n))^2)], \qquad (20)\end{aligned}$$

where

$$C_1(\phi_{i,j}) = \left((\phi_{i+1,j} - \phi_{i,j})^2 + 0.25(\phi_{i,j+1} - \phi_{i,j-1})^2\right)^{-1/2}, \qquad (21)$$

$$C_2(\phi_{i,j}) = \left((\phi_{i,j} - \phi_{i-1,j})^2 + 0.25(\phi_{i-1,j+1} - \phi_{i-1,j-1})^2\right)^{-1/2}, \qquad (22)$$

$$C_3(\phi_{i,j}) = \left((\phi_{i,j+1} - \phi_{i,j})^2 + 0.25(\phi_{i+1,j} - \phi_{i-1,j})^2\right)^{-1/2}, \qquad (23)$$

$$C_4(\phi_{i,j}) = \left((\phi_{i,j} - \phi_{i,j-1})^2 + 0.25(\phi_{i+1,j-1} - \phi_{i-1,j-1})^2\right)^{-1/2}, \qquad (24)$$

$$m(\phi_{i,j}) = \Delta t \delta_\epsilon(\phi_{i,j}), \quad C(\phi_{i,j}) = m(\phi_{i,j}) \sum_{k=1}^{4} C_k(\phi_{i,j}) . \qquad (25)$$

Equation (20) is Gauss-Seidel method for the linear system (18). If in Eq. (20) we replace $\phi_{i-1,j}^{n+1}$ and $\phi_{i,j-1}^{n+1}$ by $\phi_{i-1,j}^n$ and $\phi_{i,j-1}^n$ respectively, we obtain Jacobi method which has a slower convergence speed.

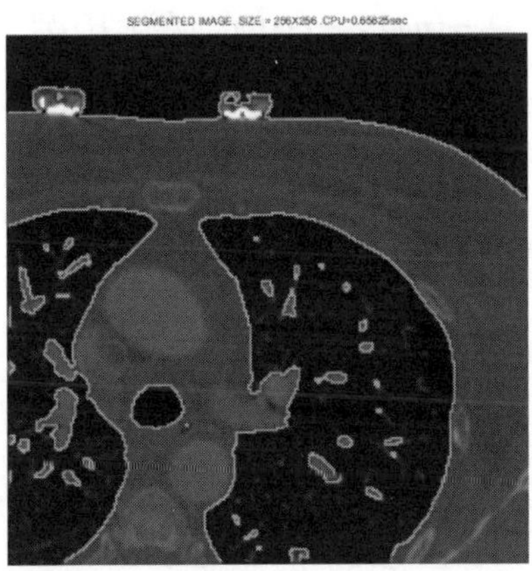

Figure 1. Segmentation of lungs MRI using a single level set. Size=256x256, CPU=0.65s

Starting from an initial level set $\phi_{i,j}^0$, we have to iterate (20) and enforce

Neuman boundary condition (7) until the quantity $\max_{i,j}|\phi_{i,j}^{n+1}-\phi_{i,j}^{n}|$ is smaller than a certain threshold. The initialization plays a fundamental role in this process. In[2], the authors chose the initial level set to be the distance function to a circle. Here we use the special initialization

$$\phi_{i,j}^{0}=\delta_{1i}\delta_{1j}, \tag{26}$$

which produces a faster convergence to the steady-state solution.

We have implemented the above algorithm in Matlab on a Laptop computer with the following characteristics: 1.66 GHz Dell Inspiron 9400 Laptop computer with 0.99 GB of RAM. Figure 1 shows the segmentation of lungs MRI. We have used the following parameters: $\epsilon = 10^{-3}$ and $\mu = 0.1 \cdot 255^2$.

3.2. *Numerical solution of (14)-(17)*

Equations (14) and (15) are discretized as in the single level set case. However we use here a relaxed Gauss-Seidel scheme:

$$\begin{aligned}\phi_{1,i,j}^{n+1} = &\frac{\omega}{C(\phi_{1,i,j})}[\phi_{1,i,j}^{n}+m(\phi_{1,i,j}^{n})(C_1(\phi_{1,i,j}^{n})\phi_{1,i+1,j}^{n}+C_2(\phi_{1,i,j}^{n})\phi_{1,i-1,j}^{n+1}\\&+C_3(\phi_{1,i,j}^{n})\phi_{1,i,j+1}^{n}+C_4(\phi_{1,i,j}^{n})\phi_{1,i,j-1}^{n+1})\\&+\Delta t\delta_\epsilon(\phi_{1,i,j}^{n})(-(u_{0,i,j}-c_{11}(\phi_1^n,\phi_2^n))^2H(\phi_{2,i,j}^{n})\\&-(u_{0,i,j}-c_{10}(\phi_1^n,\phi_2^n))^2)(1-H(\phi_{2,i,j}^{n}))\\&+(u_{0,i,j}-c_{01}(\phi_1^n,\phi_2^n))^2H(\phi_{2,i,j}^{n})\\&+(u_{0,i,j}-c_{00}(\phi_1^n,\phi_2^n))^2)(1-H(\phi_{2,i,j}^{n})))]\\&+(1-\omega)\phi_{1,i,j}^{n},\end{aligned} \tag{27}$$

$$\begin{aligned}\phi_{2,i,j}^{n+1} = &\frac{\omega}{C(\phi_{2,i,j})}[\phi_{2,i,j}^{n}+m(\phi_{2,i,j}^{n})(C_1(\phi_{2,i,j}^{n})\phi_{2,i+1,j}^{n}+C_2(\phi_{2,i,j}^{n})\phi_{2,i-1,j}^{n+1}\\&+C_3(\phi_{1,i,j}^{n})\phi_{1,i,j+1}^{n}+C_4(\phi_{1,i,j}^{n})\phi_{1,i,j-1}^{n+1})\\&+\Delta t\delta_\epsilon(\phi_{2,i,j}^{n})(-(u_{0,i,j}-c_{11}(\phi_1^n,\phi_2^n))^2H(\phi_{1,i,j}^{n})\\&-(u_{0,i,j}-c_{10}(\phi_1^n,\phi_2^n))^2)H(\phi_{1,i,j}^{n})\\&-(u_{0,i,j}-c_{01}(\phi_1^n,\phi_2^n))^2(1-H(\phi_{1,i,j}^{n}))\\&+(u_{0,i,j}-c_{00}(\phi_1^n,\phi_2^n))^2)(1-H(\phi_{1,i,j}^{n})))]\\&+(1-\omega)\phi_{2,i,j}^{n},\end{aligned} \tag{28}$$

where ω is the relaxation parameter. Note that the algorithm given in the appendix of[3] is Jacobi method. In our implementation, we utilized the so-called seed initialization as suggested in[3]. We computed the averages c_{11}, c_{10}, c_{01} and c_{00} without using the regularized Heaviside function. Also we noticed that there is no need to regularize the Heaviside functions in Eqs.(27)-(28). Overall the underrelaxed Gauss-Seidel method converges faster to the solution than the Jacobi method.

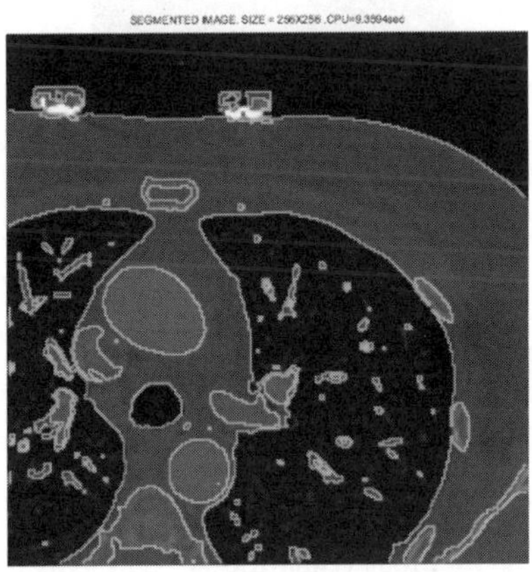

Figure 2. Segmentation of lungs MRI using two level sets. Size=256x256, CPU=9.4s

We implemented the above algorithm in Matlab on a 1.66 GHz Dell Inspiron 9400 Laptop computer with 0.99 GB of RAM. Figures 2 and 3 shows segmentations of lungs MRI and carotid artery. We have utilized the following parameters: $\epsilon = 1$, $\omega = 0.319$, $\mu = 0.005 \cdot 255^2$.

4. 3D Reconstruction from parallel contours

In many scientific and technical activities (medicine, archeology, geology, biology, etc), 3D objects must be constructed from transversal sections in order to understand their structure or facilitate their manipulation and analysis.There are roughly two types of reconstruction that can be performed on cross-sectional data: slice-based reconstruction and volume-based reconstruction. We focus only on the first type of reconstruction because of its relevancy to this work.

Most of the available research (see for instance[1,21,22,23]) on sliced-based reconstruction stitch two successive contours by using an appropriate mesh.

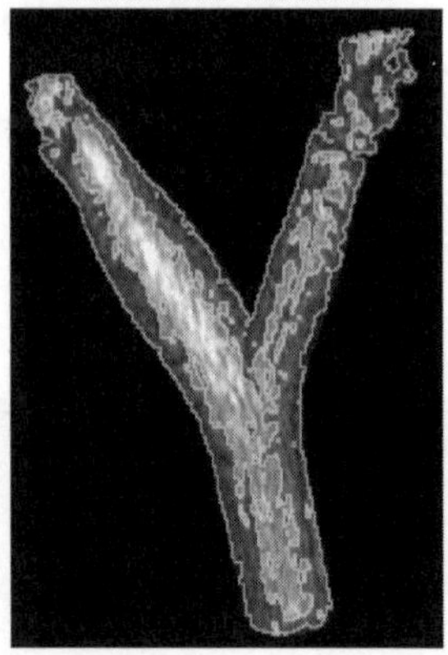

Figure 3. Segmentation of carotid using two level sets. Size=183x265, CPU=6.2s

The resulting surface is at most continuous. One can mitigate this problem by smoothing the meshed continuous surface. However extensive smoothing using available algorithms[24] tends to shrink data.Note that mesh-based stitching of contours is equivalent to a linear interpolations of contours.

Another prominent approach to sliced-based reconstruction is the point set approach. In this approach each contour is represented by points in $\mathbb{R}^3$. This representation results in the 3D object being given by a set of points. The reconstruction problem is thus reduced to the following one: find an implicit surface that contains a set of points. The latter is an interpolation problem that has been solved using radial basis functions method (see for example[25,26]). Reconstruction using using a point-based algorithm is computational expensive and sometimes requires the knowledge of the normal vector at each point. Beside, one is confronted as a rule by the problem of poor conditioning of the interpolation matrix.

In this section, we propose to use a reconstruction method that does not require the explicit solution of an interpolation problem and yet produces smooth and accurate reconstructed 3D objects.

We segment individual cross-section images of a given 3D objects using level sets. Assume that there are N cross-sectional images. There will be N functions $\phi_1(x,y),\ldots,\phi_N(x,y)$ representing the contours. Let $z_1, z_2, \ldots z_N$ be the location of the parallel contours, we have to solve the following problem: Find a function $\phi(x, y, z)$ such that $\phi(x, y, z_i) = \phi_i(x, y)$ for $i = 1, \ldots, N$. A solution to this problem is

$$\phi(x, y, z) = \sum_{i=1}^{N} \delta(z - z_i)\phi(x, y, z_i), \tag{29}$$

where δ is Dirac distribution. The solution (29) should be understood in the sense of distributions.

Let $(\delta_h)_{h>0}$ be a sequence of smooth functions such that

$$\lim_{h \to 0} \delta_h(x) = \delta(x) \tag{30}$$

in the sense of distributions. We approximate ϕ using the equation

$$\phi(x, y, z) \approx \sum_{i=1}^{N} \delta_h(z - z_i)\phi_i(x, y), \tag{31}$$

where h is sufficiently small.

In our numerical calculations we use

$$\delta_h = \frac{1}{2\sqrt{\pi h}} \exp\left(\frac{-z^2}{4h}\right) \tag{32}$$

which is the *fundamental solution* or *kernel* of the heat equation

$$u_h - u_{zz} = 0.$$

The calculation of the 3D level set $\phi(x, y, z)$ is reduced to the evaluation of the sum appearing in (31). For very large volumes, the computation of this sum can be slow. Fortunately, there is the fast multi-pole method[28,29,30] that was designed to calculate such sums in $O(N \ln N)$.

The reconstructed 3D surface is given by the implicit equation $\phi(x, y, z) = 0$. There are several algorithm for representing this implicit surface. We have used the one implemented in Matlab.

Figures 4 to 7 demonstrate our reconstruction procedure on real images. Note that the success of reconstruction depends greatly on the quality of segmentation. For segmentation using a single level set, we use $\mu = 0.1 \cdot 255^2$ and $\epsilon = 10^{-3}$. For segmentation using two level sets, we use $\epsilon = 1$, $\omega = 0.319$ and $\mu_i = 0.00165 \cdot 255^2 \cdot (0.385)^{i-1}$, where i is the slice number. All the reconstructed 3D images use $h = 0.2$.

Figures 4 and 5 depict human head reconstruction using single and two level sets respectively. Slices are under-segmented when we use a single level set since the images are have multiple phases. As a result we obtain a partial reconstruction of the brain. With two level sets there is a clear improvement of the brain reconstruction.

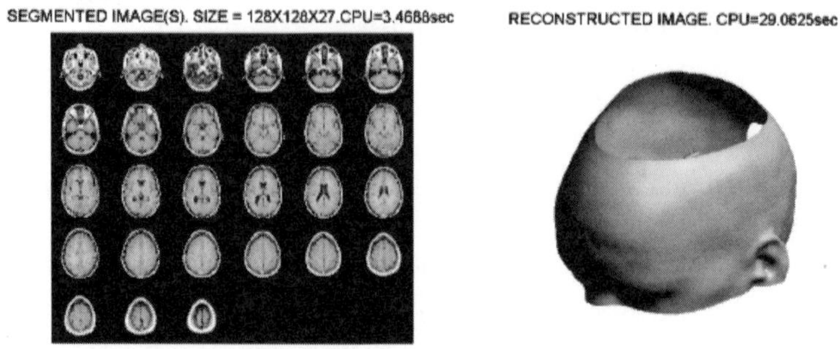

Figure 4. Single level set Segmentation and reconstruction of human head. Size=128x128x27, CPU=3.2s + 29.1s

Figures 6 and 7 deals with the reconstruction of bat cochlea and human iliac bone respectively.

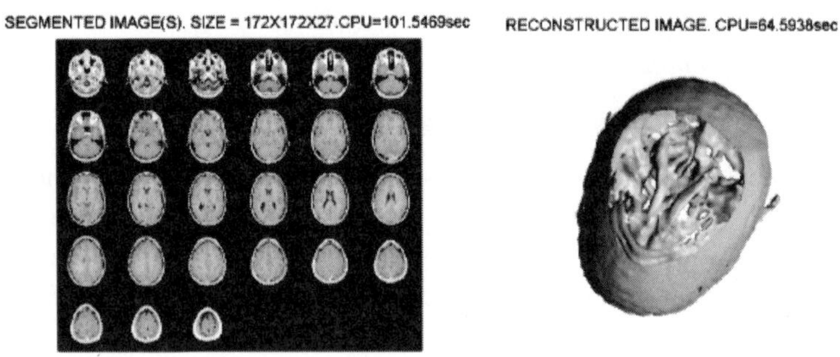

Figure 5. Two level sets Segmentation and reconstruction of human head,172x172x27, CPU=101.5s + 64.6s

5. Conclusion

In Chan-vese active contour without edge, initialization plays a crucial role. We have utilized in this paper a new initialization that speeds up the convergence of Chan-vese segmentation algorithm. For segmentation using multiple level sets we employed relaxed Gauss-Seidel method. This resulted in a fast segmentation algorithm. We illustrated our approach on medical images.

Figure 6. Reconstruction of bat cochlea (segmentation not shown), 121x151x111, CPU=16.6s+267.9s

Figure 7. Reconstruction of human iliac bone (segmentation not shown), 128x128x90, CPU=10.9s+171.5s

We have introduced a novel slice-based reconstruction methodology that relies on level set segmentation and the heat kernel. We have demonstrated the new procedure on real images.

Acknowledgement

The author acknowledges partial financial support from the Jackson State University Summer Scholars program.

References

1. H Fuchs, Z M Kedem and S P Uselton 1977 Optimal surface reconstruction from planar contours, *Communication of the ACM*, vol. 20, No. 10, 693–702.

2. F T Chan and L A Vese 2001 Active contours without edge, *IEEE Transaction on Image Processing*, vol. 10, No.2, 266–277.
3. L A Vese and T F Chan 2002 A Multiphase level set framework for image segmentation using the Mumford and Shah model, *Int. J. Comp. Vis.*, vol. 50, No. 3, 271–293.
4. A Sandholm and K Museth 2004 3D reconstruction from non-euclidian distance field, *Lincoping Electronic Conference Proceedings*, Gavle, Sweden, 55.
5. M Kass, A Witkin and D Terzopoulos 1988 Snake: active contour models, *Int. J. Comp. Vis.*, vol. 1, 321–331.
6. L D Cohen 1991 On active contours and balloons, *CVGIP: Image Understanding*, vol. 53, 211–218.
7. L D Cohen and I Cohen 1993 Finite-element method for active contour models and balloons for 2D and 3D images, *IEEE Trans. Patern Anal. Machine Intell.*, vol 15, 1131–1147.
8. C Xu and J L Prince 1998 Snakes, shapes and gradient vector flow, IEEE Transaction on Image Processing, vol. 7, No. 3, 359–369.
9. V Casselles, R Kimmel and G Shapiro 1995 Geodesic active contours,in *Proceedings of 5th Int. Conf. Computer Vision*, 694–699.
10. S Osher and J A Sethian 1988 Front propagating with curvature-dependent speed: algorithms based on Hamilton-Jacobi Formulation, *J. Comp. Phys.*, vol. 79, 12–49.
11. V Casselles, F Catté and F Dibos 1993 A geometric model for active contours in image processing, *Num. Math.* vol. 66, 1–31.
12. R Malladi, J A Sethian and B C Vemuri 1993 A topology independent shape modeling scheme,in *Proc. SPIE Conf. Geometric Methods Computer Vision II*, vol 2031, San Diego, CA, 73–78.
13. R Malladi, J A Sethian and B C Vemuri 1994 Evolutionary fronts for topology-independent shape modeling and recovery, in *Proc. 3rd Eur. Conf. Computer Vision*, Stockholm, Sweden, vol. 800, 3–13.
14. D Munford and J Shah 1989 Optimal approximation by piecewise smooth functions and associated variational problems, *Commun. Pure Appl. Math.*, vol. 42, 577–685.
15. D Geman and S Geman 1984 Stochastic relaxation, Gibbs distribution and Bayesian restoration of images, *IEEE Trans. Pattern Anal. Mach. Intell.*, vol. 6, 721–741.
16. J M Morel and S Solinini 1988 Semegmentation of images by variational methods: a constructive approach, it Revista Mathematica Universidad Complutense de Madrid vol. 1, 169–182.
17. E De Giorgi, M Carriero and A Leaci 1989 Existence theorem for a minimum problem with free discintinuity set, *Arch. Rational Mech. Anal.* 108(3), 199–210.
18. L C Evans and R F Gariepy 1992 Measure theory and fine properties of functions, CRC Press: Boca Raton, FL.
19. L Rudin, S Osher and E Fatemi 1992 Nonlinear total variation based noise removal algorithms, *Phys. D*, vol. 60, 259–268.
20. G Aubert and L Vese 1997 A variational method for image recovery, *SIAM*

J Numer. Anal., vol. 34, No. 5, 1948–1979.

21. S Gana and T G Dennehy 1982 A new general triangulation method for contours, in *SIGGRAPH'82: Proceedings of the 9th annual conference on computer graphics and interactive techniques*, 69–75.
22. G Barequet and M Sharir 1996 Piecewise-linear interpolation bewteen polygonal slices, *Computer vision and Image Understanding: CVIU*, 63(2), 116–272.
23. G Barequet, D Shapiro and A Tal 2000 Multilevel sensitive reconstruction of polyhedral surfaces from parallel slices, *The visual Computer*, 16(2), 116–133.
24. M Desbrun, M Meyer, P Shröder and A H Barr 1999 Implicit fairing of irregular meshes using diffusion and curvature flow, in *SIGGRAPH' 99: Proceedings of the 9th annual conference on computer graphics and interactive techniques*, 317–324.
25. J C Carr, W R Fright and R K Beatson 1997 Surface interpolation with radial basis functions for medical imaging, *IEEE Transactions on Medical Imaging*, Vol. 16, No. 1, 96–107.
26. Y Ohtake, A Belyaev and H Seidel 2003 A multi-scale approach to 3D scattered data interpolation with compactly sypported basis functions, in *SMI'03: Proceedings of the Shape Modeling International 2003*, Washington DC, 292.
27. C Wafo Soh 2004 *Mesh-free reconstruction of coronary arteries*, Technical report, Biomedical Systems Modeling Lab, MMAE, University of Central Florida.
28. L Greegard 1988 The rapid evaluation of potential fields in particle systems,MIT press, Cambridge.
29. L Greengrad and J Strain 1991 The fast Gauss transform, *SIAM J. Sci Stat. Comp.*, No. 12, 79–94.
30. L Greengard 1994 Fast algorithm for classical physics, *Science*, No. 235, 909–915.

MATHEMATICAL PREDICTION OF HIGH ENERGY METABOLITE GRADIENTS IN MAMMALIAN CELLS

RAYMOND MEJIA

NHLBI, National Institutes of Health
10 Center Drive, Room 4A15
Bethesda, MD 20892-1348 USA
E-Mail: ray@helix.nih.gov

RONALD M. LYNCH

Department of Physiology, University of Arizona
P.O. Box 245051
Tucson, AZ 85724 USA
E-mail: rlynch@u.arizona.edu

Described is a mathematical model that evaluates the distribution of cellular adenosine nucleotides (ATP/ADP) to test the hypothesis that local changes in the concentrations of these molecules can modulate cell function without a significant change in global cytosolic concentrations. The model incorporates knowledge of cell structure to predict the spatial concentration profiles of ATP, ADP and inorganic phosphate. The steady state was perturbed by increasing the activity of membrane bound ion transporters including the Na-K ATPase on the cell periphery or the V-type H^+ ATPase which serves to acidify intracellular compartments including endosomes/lysosomes. Both of these transporters utilize ATP and produce ADP and P_i in the process. Both models are run over a range of cytosolic diffusivities, including local low diffusivity near the pump sites. Results suggest that local changes in the concentration of ADP (not ATP) during activation of ion transport, in particular at near membrane sites, may serve to modulate ion transport, and thereby cell behavior.

1. Introduction

Previous studies have considered the role of barriers to diffusion of high energy metabolites between organelles in MDCK[4] and cardiac muscle[2,5] cells. We have used imaging techniques to identify the location of organelles within the cell and mathematical modeling to test the hypothesis that local changes can modulate function without significant changes in global cytosolic concentrations.

2. Mathematical Model

2.1. *MDCK Cell*

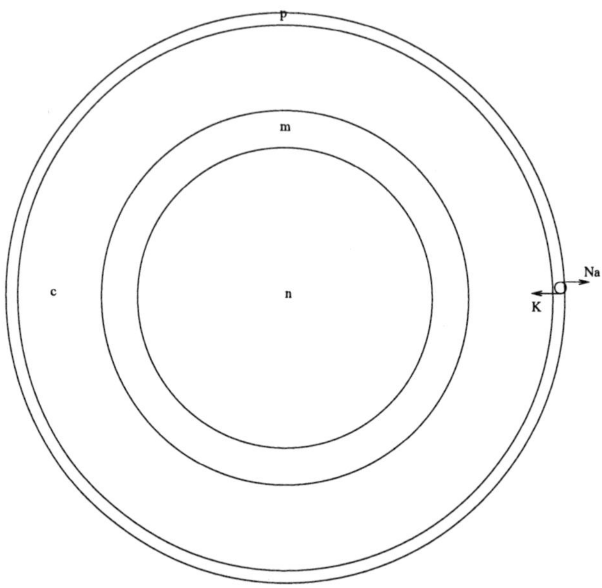

Figure 1. A schematic diagram of a spherical section through a cell is shown. Na-K pumps (labeled **p**) are shown on the periphery. Mitochondria (**m**), are shown about 55–65% from the center of the cell. The cytosol is labeled **c**, and the nucleus **n**.

A mathematical model of diffusion in an MDCK cell (Figure 1) is based on a model described by Lynch, et al.[4]. The model includes Na-K pumps, mitochondria and the cytosol. Mitochondria are restricted to approximately 10% of cell volume as shown in Fig. 2 of[4], and Na-K ATPase is distributed uniformly in the plasmalemma in the outer 1% of the cell volume. The concentration of species within the cell is described by the equation

$$\mathbf{C}' = \nabla \cdot \mathbf{D}\nabla\mathbf{C} + \mathbf{S} \ , \tag{1}$$

where $\mathbf{C} = [C_1, C_2, C_3]$ is the vector of concentrations of ATP, ADP, and P_i, respectively; $\mathbf{D} = [D_1, D_2, D_3]$ is the corresponding vector of diffusion coefficients; $\mathbf{S} = [S_1, S_2, S_3]$ describes production and consumption of species.

Sources are described by

$$
\begin{aligned}
S_1(r) = {} & \frac{k_+}{K_{eq}}\, C_2\, C_3 - k_+\, C_1 \text{ for } 0 \le r \le r_C \text{ (entire cell)} \\
& + \frac{a_m\, k_+}{K_{eq}}\, C_2\, C_3 \text{ for } r_0 \le r \le r_1 \text{ (mitochondria)} \\
& - a_c\, k_+\, C_1 \text{ for } 0 \le r \le r_2 \text{ (cytosol + nucleus)} \\
& - a_p\, k_+\, C_1 \text{ for } r_2 \le r \le r_C \text{ (plasma membrane)}\ ,
\end{aligned}
$$

where k_+ is the dephosphorylation rate; K_{eq} is the apparent equilibrium constant; a_m is the local (pointwise) phosphorylation rate; a_c is local ATP consumption in the cytosol and nucleus; and a_p is local ATP consumption at the pump sites near the plasma membrane. The remaining source terms are defined by

$$
S_2 = S_3 = -S_1\ .
$$

The rate of ATP phosphorylation at the mitochondria, A_m, is given by

$$
A_m = \frac{4}{3}\, \pi\, (r_1^3 - r_0^3)\, \frac{a_m\, k_+}{K_{eq}}\, C_2^0\, C_3^0\ .
$$

The rate of ATP dephosphorylation in the cytosol, A_c, and in the plasma membrane, A_p, is given by

$$
\begin{aligned}
A_c &= \frac{4}{3}\, \pi\, r_2^3\, a_c\, k_+\, C_1^0\ , \\
A_p &= \frac{4}{3}\, \pi\, (r_c^3 - r_2^3)\, a_p\, k_+\, C_1^0\ .
\end{aligned}
$$

Initial and boundary conditions for all species are

$$
\begin{aligned}
C_k(r,0) &= C_k^0, \quad \forall k \\
\frac{\partial C_k}{\partial r} &= 0 \quad \forall k \text{ and } r = r_C
\end{aligned}
$$

2.2. *Muscle Cell*

A smooth muscle cell is represented in the model by a cylinder with a nucleus at the center of the cell (Figure 2). Mitochondria are distributed uniformly throughout the cell with a central nucleus (Figure 3). Endosomes and lysosomes are distributed as shown in (Figure 4). V-type channels that hydrolyze ATP are located at the endosomes/lysosomes.

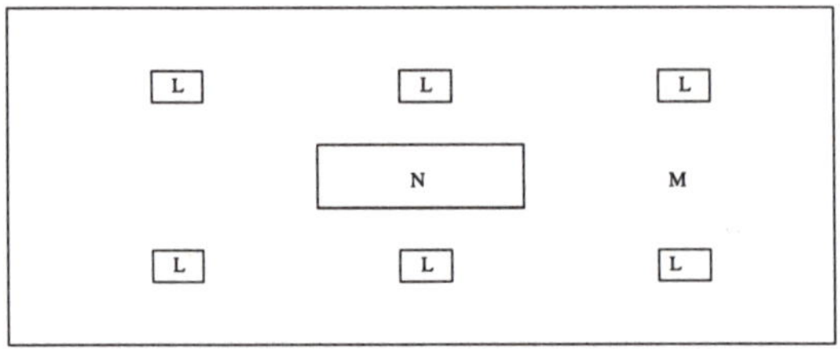

Figure 2. A schematic diagram of a rectangular section through a cylindrical smooth muscle cell shows mitochondria distributed uniformly throughout the cell (**M**), lysosomes labeled **L** at six locations on the section, and a nucleus (**N**).

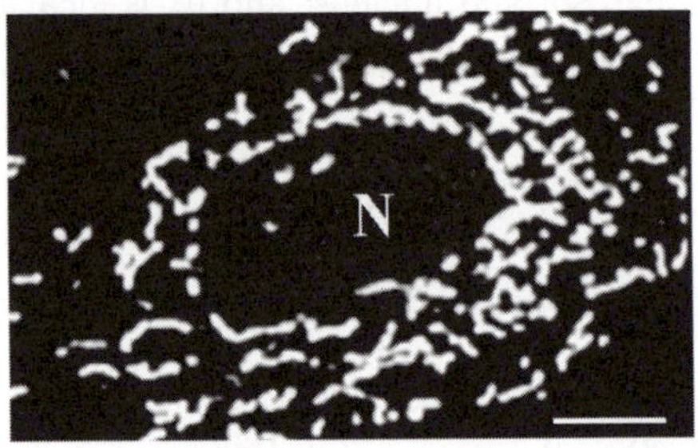

Figure 3. Mitochondria were labeled in an A7R5 smooth muscle cell using an inner membrane specific antibody. The nucleus is labeled (**N**), and the scale bar is 20 μm in length.

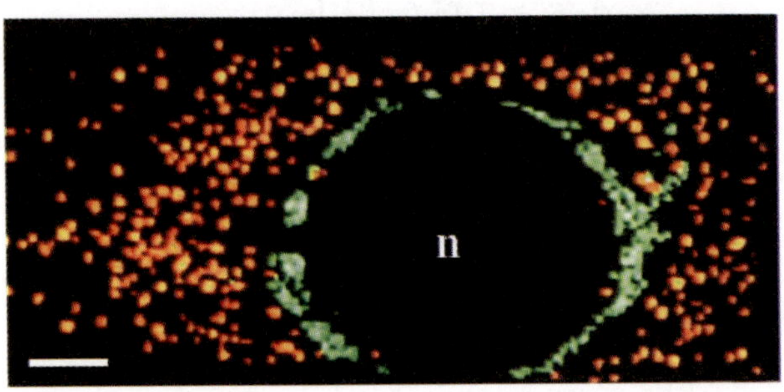

Figure 4. The endosomal/lysosomal compartments in an A7R5 cell were loaded with Texas Red labeled dextran, and the cell was transfected with a construct encoding EGFP (enhanced green fluorescent protein) coupled to a target sequence that retains the EGFP in the Golgi apparatus (not included in this model). The cell nucleus is labeled **n**, and the scale bar is 10 μm.

Equation 1 is solved with the ATP source term defined by

$$\begin{aligned} S_1(h,r) = & \frac{k_+}{K_{eq}}\, C_2\, C_3 - k_+\, C_1 \text{ for } (h,r) \in M - N \text{ (cell excluding nucleus)} \\ & + \frac{a_m\, k_+}{K_{eq}}\, C_2\, C_3 \text{ for } (h,r) \in M - N - L \text{ (mitochondria)} \\ & - a_c\, k_+\, C_1 \text{ for } (h,r) \in M - N \text{ (cytosol)} \\ & - a_p\, k_+\, C_1 \text{ for } (h,r) \in L \text{ (endosomes/lysosomes)}\ , \end{aligned}$$

where h is the major coordinate (height of the cylinder) and r the minor coordinate (radius) of point (h,r). M labels the cell; N the cell nucleus, and L a lysosome.

For muscle, the phosphorylation and dephosphorylation terms are

$$\begin{aligned} A_m &= \pi\left(r_M^2 h_M - r_N^2 h_N - 4\ell(r_L^2 + \delta r_L)h_L\right) a_m\, \frac{k_+}{K_{eq}}\, C_2^0\, C_3^0\ , \\ A_c &= \pi(r_M^2 h_M - r_N^2 h_N)\, a_c\, k_+\, C_1^0\ , \\ A_p &= 4\pi\ell\, (r_L^2 + \delta r_L)\, h_L\, a_p\, k_+\, C_1^0\ , \end{aligned}$$

where r_M is the cell radius, and r_N is the radius of the nucleus. h_M, h_N and h_L are the width of the cell, nucleus and a lysosome, respectively. ℓ is the number of lysosomes on a cylindrical plane, and δ is the distance of a lysosome from the major axis of the cell.

Initial and boundary conditions for all species are

$$\begin{aligned} C_k(r,h,0) &= C_k^0 \ \ \forall k, (h,r) \in M - N\ , \\ \frac{\partial C_k}{\partial r} &= 0 \ \ \forall k \text{ and } (r,\cdot) \in M \cap N \textit{ or } (r,\cdot) \in \partial M\ , \\ \frac{\partial C_k}{\partial h} &= 0 \ \ \forall k \text{ and } (\cdot,h) \in M \cap N \textit{ or } (\cdot,h) \in \partial M\ . \end{aligned}$$

3. Model Results

In each model a system of partial differential equations (Equation 1) is solved using FEMLAB[1]. The time-dependent equations are solved until a steady-state solution is obtained.

3.1. *Results of MDCK Model*

Parameters used to describe the model are shown in Table 1.

Table 1. Parameters used by MDCK Model.

Symbol	Value	Description
r_C	6.4×10^{-4}	cell radius (cm)
r_0	$55\% r_C$	inside radius for distribution of mitochondria (cm)
r_1	$65\% r_C$	outside radius for distribution of mitochondria (cm)
r_2	$r_C - 1\% r_C$	inside radius for distribution of pumps (cm)
C_1^0	$1 - 4$	initial [ATP] (mM)
C_2^0	$28 - 40$	initial [ADP] (μM)
C_3^0	1	initial $[P_i]$ (mM)
K_{eq}	$C_2^0 C_3^0 / C_1^0$	apparent equilibrium constant (M)
k_+	1×10^{-5}	dephosphorylation rate at equilibrium (s^{-1})
D_1	$10^{-8} - 10^{-6}$	ATP diffusion (cm^2/s)
D_2	$10^{-8} - 10^{-6}$	ADP diffusion (cm^2/s)
D_3	$10^{-8} - 10^{-6}$	P_i diffusion (cm^2/s)
A_m	7.8×10^{-14}	initial rate of phosphorylation at mitochondria ($moles/min$)
A_c	7.0×10^{-14}	ATP consumption rate in cytosol and nucleus ($moles/min$)
A_p	7.8×10^{-15}	initial rate of ATP consumption by Na-K ATPase at plasma membrane ($moles/min$)

Concentrations reported in the literature, [ATP] from 1 – 4 mM and for [ADP] from 28 – 40 μM, were used as initial values without any qualitative difference in the solution. Hence, results for one set of initial values are described here. Figure 5 shows [ATP] for pump rate $A_p = 7.8 \times 10^{-15}$ $(moles/min)^3$. The concentration of ATP is a maximum at the mitochondria, decreases in the cytosol with distance from the mitochondria, and reaches a minimum at the periphery of the cell.

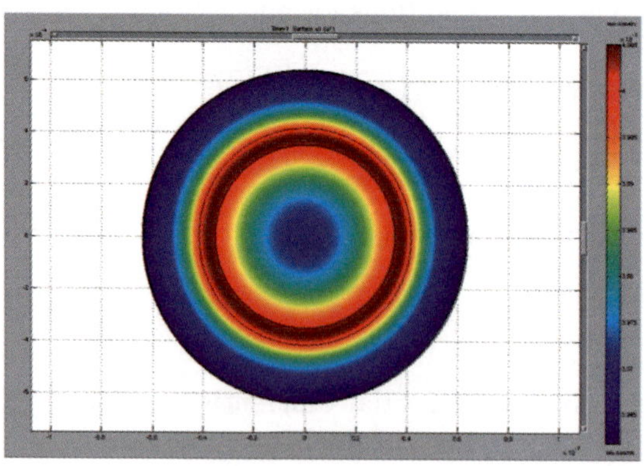

Figure 5. ATP concentration is shown for a circular section of an MDCK cell. The nucleus is in the center, and the mitochondria are restricted (as shown in Figure 1) to the band indicated as dark red. Since ATP is produced at this location, ATP concentrations are highest. Analogously, ADP utilization (to form ATP) is the highest in this location causing ADP to reach a minimum (not shown).

When diffusion is restricted near the pump sites, even as the pump rate increases, ATP concentration is reduced only moderately - from a maximum of 4 mM at the mitochondria to a minimum of 3.8 mM at the pumps under the most extreme conditions shown. However, [ADP] increases significantly as A_p increases and as diffusion is restricted as shown in Figure 6). $[P_i]$ changes moderately in a manner analogous to [ATP].

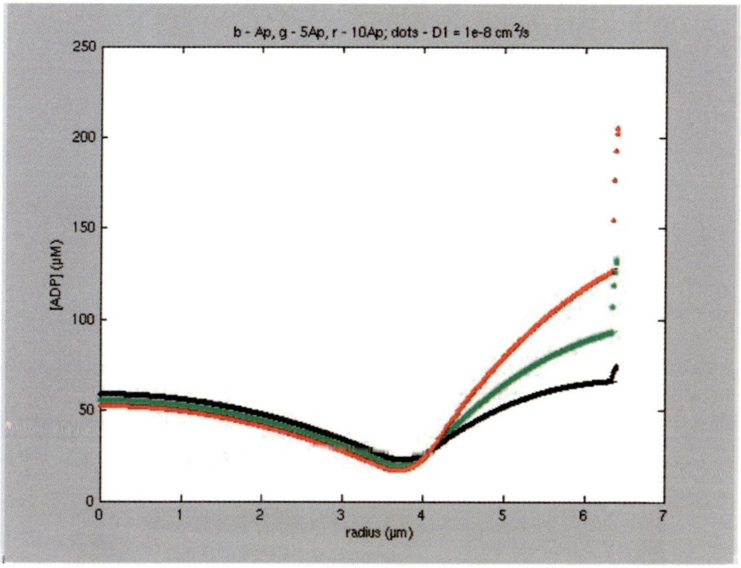

Figure 6. [ADP] is shown as a function of cell radius as A_p (ATP utilization rate at the plasma membrane) is increased from the initial low rate (*black line) to five times this rate (green) and to ten times the initial rate (red line). The dotted lines show the concentration when diffusion at the periphery (the outer 1% of the cell) is reduced by two orders of magnitude to* 10^{-8} cm^2/s.

3.2. *Results of Muscle Model*

Parameters used to describe this model are shown in Table 2. The number of endosomes/lysosomes per plane perpendicular to the centerline of the cell, ℓ, was varied to obtain a representative distribution (Figure 4). Figure 7 shows ADP concentration for $A_p = 10\%$ A_c and uniform diffusion throughout the cell of 10^{-6} cm^2/s to vary from a minimum of 25.8 μM in the cytosol to a maximum of 29.4 μM at the lysosomes. However, with diffusion reduced to 10^{-8} cm^2/s at the lysosomes, [ADP] increases to 72.5 μM (Table 3).

Table 2. Parameters used by Muscle Model.

Symbol	Value	Description
h_M	50×10^{-4}	cell width (cm)
h_N	12.5×10^{-4}	width of the cell nucleus (cm)
h_L	2×10^{-4}	width of endosome/lysosome (cm)
r_M	10×10^{-4}	cell radius (cm)
r_N	2.5×10^{-4}	radius of cell nucleus (cm)
r_L	0.5×10^{-4}	radius of endosome/lysosome (cm)
ℓ	$2 - 8$	number of endosomes/lysosomes per cell plane
δ	$0 - 4.5 \times 10^{-4}$	distance of lysosomes from centerline of a plane (cm)
C_1^0	4	initial [ATP] (mM)
C_2^0	28	initial [ADP] (μM)
C_3^0	1	initial [P_i] (mM)
K_{eq}	$C_2^0 C_3^0 / C_1^0$	apparent equilibrium constant (M)
k_+	1×10^{-5}	dephosphorylation rate at equilibrium (s^{-1})
D_1	$10^{-8} - 10^{-6}$	ATP diffusion (cm^2/s)
D_2	$10^{-8} - 10^{-6}$	ADP diffusion (cm^2/s)
D_3	$10^{-8} - 10^{-6}$	P_i diffusion (cm^2/s)
A_m	7.7×10^{-14}	rate of phosphorylation at mitochondria for $A_p = 10\% A_c$ ($moles/min$
A_c	7.0×10^{-14}	ATP consumption rate in cytosol ($moles/min$)
A_p	$1\% - 10\% A_c$	initial rate of ATP consumption by H^+ ATPase ($moles/min$)

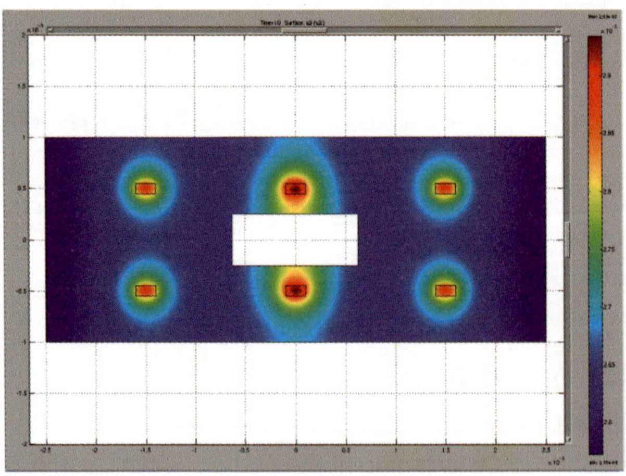

Figure 7. ADP concentration is shown with the assumption of uniform diffusion throughout the cell, and the geometry as shown in Figure 2.

Table 3. ADP and ATP concentration as function of A_p and diffusivity at L.

A_p $\%A_c$	Diffusivity cm^2/s	[ADP] μM		[ATP] mM	
		max	min	max	min
5	10^{-6}	28.8	26.6	4.00	4.00
5	10^{-7}	31.2	26.6	4.00	4.00
5	10^{-8}	54.7	26.6	4.00	3.97
10	10^{-6}	29.4	25.8	4.00	4.00
10	10^{-7}	33.5	25.8	4.00	3.99
10	10^{-8}	72.5	25.8	4.00	3.96
20	10^{-6}	30.6	24.2	4.00	4.00
20	10^{-7}	38.0	24.2	4.00	3.99
20	10^{-8}	108.0	24.2	4.00	3.92

4. Discussion

By setting $A_p = 0$, the equivalent of blocking all transport of Na-K ATPase, we observe a quiescent MDCK cell with no concentration gradients. [ADP] increases at the pump sites as A_p is increased from the control pump rate[4] (black line of Figure 7) to five (green line), and to ten times the control rate (red line). Concomitantly, [ADP] decreases at the mitochondria (Figure 7) as A_p is increased, while [ATP] and $[P_i]$ vary inversely. As diffusivity is reduced at the plasma membrane, [ADP] increases (dots on Figure 7). This shows an increase by a factor of two and a factor of four, respectively, in local [ADP] at the plasmalemma, while the global concentration of ATP, ADP and P_i does not change much.

The model of a muscle cell shows gradients analogous to those for MDCK cells as a consequence of increased ATPase activity or restricted local diffusion. An increase in acidification rate in lysosomes shows a moderate increase in [ADP] locally (Table 3) that increases more sharply when diffusion is restricted locally. Since the activities of the transporters themselves are sensitive to the relative ATP and ADP concentrations via mass action, our findings suggest that local changes in pump activity (increased local ADP) may negatively influence turnover rate. In addition, ion channels that are sensitive to ATP/ADP, such as the KATP channels could be regulated by changes in local pump activity. Based on these considerations, activity based changes in local ADP, and thereby the ATP/ADP, may provide a mechanism for altering cell function. Moreover, mechanisms for buffering of these local changes in adenine nucleotides may be important for regulating this signaling activity.

References

1. FEMLAB Version 2.3.0.1-48, COMSOL, Inc.
2. O. Kongas, and J. H. G. M. Van Beek, *Mol. Biol. Rep.* ***29****, 141 (2002).*
3. R. M. Lynch, and R. S. Balaban, *Am. J. Physiol.* ***253*** *(Cell Physiol. 22), C269 (1987).*
4. R. M. Lynch, R. Mejia, and R. S. Balaban, *Comments Mol. Cell Biophys.* ***5****, 151 (1988).*
5. M. Vendelin, M. Eimre, E. Seppet, N. Peet, T. Andrienko, M. Lemba, J. Engelbrecht, E. K. Seppet, and V. A. Saks, *Mol. Cell. Biochem.* ***256-257****, 229 (2004).*

IDEAL PROTEIN FORMS AND THEIR APPLICATION TO DE-NOVO STRUCTURE PREDICTION*

WILLIE TAYLOR, V. CHELLIAH, DAN KLOSE, TOM SHELDON, G. J. BARTLETT†

Division of Mathematical Biology, National Institute for Medical Research, The Ridgeway, Mill Hill, London NW7 1AA, U.K.
E-mail: wtaylor@nimr.mrc.ac.uk

INGE JONASSEN

Computational Biology Unit and Department of Informatics, University of Bergen, Norway.
E-mail: inge@ii.uib.no

An abstract representation, similar to a Periodic Table, was used to generate a large number of idealised protein folds. Each of these was taken as a framework onto which a variety of predicted secondary structures were mapped and the resulting models constructed at a more detailed level. The best of these were refined and rescored. On a set of five proteins, the correct fold was scored highly in each with the top models having a low root-mean-square deviation from the known structure.

1. Introduction

For over thirty years it has been accepted that the amino acid sequence of a protein is sufficient to specify the three dimensional structure, or fold, of the polypeptide chain. Despite the apparent simplicity of this relationship, over the same period, there has been limited progress in explaining how the protein sequence dictates its fold. The definitive test of our ability to understand this relationship is to predict the 3D structure of the protein given just its sequence. Specifically, our formulation of the problem should be encoded in a computer program that, given a protein sequence as input,

*This work is supported by the medical research council (uk) and the research council of norway (functional genomics programme FUGE)

†Work partially supported by DARPA PDP grant, University of Washington Seattle, USA.

will generate the 3D coordinates of the structure as output.

Two distinct computational approaches to this protein structure prediction problem have been developed. One is to imitate Nature by allowing a flexible chain to fold under the direction of purely physical and chemical factors (referred to as the *ab-initio* approach)[1,2]. Where the *ab-initio* approach allows a single structure to evolve over time, an alternative approach is to generate many static 'snap-shots' of possible structures for the protein and try and pick the correct one (referred to as the *combinatorial* approach)[3,4].

Both approaches have the common problem that they need an evaluation function that can recognise a good structure when it has been constructed. However, to evaluate each model in sufficient detail requires considerable calculation and, in the past, progress has been limited by insufficient computational capacity to simulate the folding process for long enough or to generate and evaluate enough models.

1.1. *Ab-initio and de-novo prediction*

Several years ago, the most successful *ab-initio* method was able to approximate the structure of a very small protein (36 residues) with 4.5Å root-mean-square-deviation (RMSd) from the native[5]. More recently, augmenting the purely physical approach with information of the known protein structures (in the form of fragments) has allowed progress to larger structures, resulting in accurate predictions for several small proteins under 90 residues in length[6]. (This approach is referred to as *de-novo* since it can no longer be considered to build strictly from first principles).

Both these methods operate close to the limits of reasonably powerful computer resources and to extend either of them to larger proteins would require computation that is likely to increase well in excess of a linear extrapolation with chain length. For proteins under 100 residues in length, especially those composed mainly of α-helices (including most of the results mentioned above), the fold can be constructed largely from sequentially local packing which is the ideal situation for prediction methods that simulate folding. However, protein chains of over 100 residues give scope for increasingly non-local interactions and the search time required to explore this additional complexity is expected to be considerable.

1.2. *Reevaluating the combinatorial approach*

Early studies indicated that the combinatorial approach had some potential to produce useful results for large proteins. Its ability to tackle these larger problems derived from its analysis of structure at the higher level of secondary structure elements. However, this capacity came at a price as the method required both an accurate knowledge of the secondary structure elements and a suitable framework (or secondary structure lattice) on which to place them. Given this information, reasonable structures could be generated for proteins around 200 residues in length[4,7]. In the intervening time, our knowledge of the variety of protein structure has improved greatly as have the methods for secondary structure prediction, suggesting it may be timely to reevaluate the combinatorial approach to structure prediction.

Over the years, the accuracy of secondary structure prediction methods has slowly increased and current secondary structure prediction methods, which typically employ an artificial neural net applied to an aligned sequence family, can attain up to 80% accuracy[8,9]. Progress towards systematically representing the variety of protein structure has also been made with the recent automatic classification of protein folds into a 'Periodic Table'-like structure[10] which used idealised secondary structure frameworks (Forms) similar to those used previously for prediction.

These two sources of data were combined into a method to generate protein models that were then evaluated by relatively conventional methods. The predicted secondary structures were combined with the Periodic Table frameworks by placing every secondary structure variation over every framework. This leads to an enormous number of possible structures and we have restricted our attention to the more limited problem of addressing only the $\beta\alpha$-class of protein and, specifically, only those that have an open β-sheet (excluding β-barrels) below 200 residues in length. These proteins typically adopt a three-layer arrangement (or architecture) consisting of α-helices packing on either side of a twisted β-sheet.

2. Computational Methods

2.1. *Ideal Protein Forms*

The ideal Forms for proteins were derived from stick models[11] in which the hydrogen-bonded links across a β-sheet impose a layer structure onto the arrangement of secondary structures in a protein domain[12,13]. These layers can consist of either α-structure (packed α-helices) or β-structure (hydrogen-bonded β-strands). There are seldom more than four layers in any one domain and each layer tends to be exclusively composed of one of these two types of secondary structure. The spacing between the axes of packed α-helices is typically 10Å, as is the spacing between β-sheets and between helices and sheets[4,7], while the spacing between the hydrogen-bonded strands in a sheet is close to 5Å. This makes 10 Ångstroms a convenient unit with which to 'digitise' protein structure.

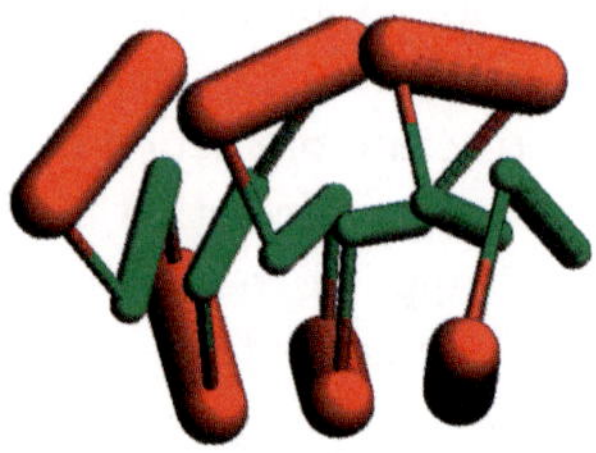

(a) $\alpha\beta\alpha$ layers

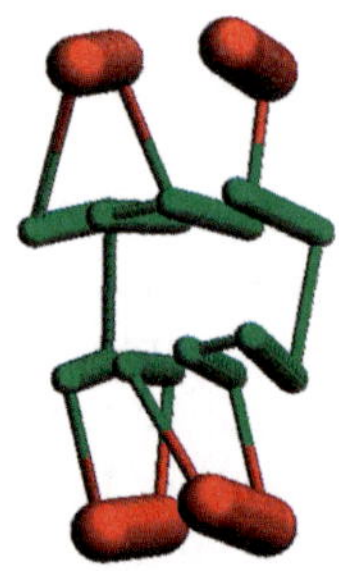

(b) $\alpha\beta\beta\alpha$ layers

Figure 1. **Stick-figure representations of the basic Forms**. Each of the basic generating Forms are represented by 'stick' models in which α-helices are red and drawn thicker than the green β-strands. (*a*) $\alpha\beta\alpha$ layers. Six strands are shown but the sheet can extend indefinitely. (*b*) $\alpha\beta\beta\alpha$ layers. As in *a*, the sheets can be extended. (Removal of the α layers leaves the common β-'sandwich').

If the layers of secondary structure are systematically filled with alpha and beta elements, the resulting arrangement is not dissimilar to the Periodic Table of elements. In this loose analogy, the layers are equivalent to electron orbital valance shells that become progressively filled with electrons (secondary structure elements): firstly, the inner β layer (S orbitals) followed by the outer α layers (P orbitals) then repeating with a second β layer (D orbitals). Extending the model to incorporate the permutations arising from additional α layers would even be reminiscent of the interjection of the rare-earth series. (Even de-localised helices in the outer shell might be imagined acting like metallic electrons).

2.2. *Building and evaluating models*

2.2.1. *Mapping predicted structure onto ideal Forms*

Those who develop secondary structure prediction methods have aimed always to increase the average number of residues predicted in the correct state (α, β, other). However, even if the average accuracy is 80%, it is difficult to rely on any individual prediction knowing the standard deviation is at least $\pm 15\%$. To avoid this we generated variations in the secondary structure predictions by biasing each prediction towards one member of the sequence family and by using two different methods: PsiPred[8] and YASPIN[9]. This degree of variation was sufficient that usually at least one of the variants was a reasonable approximation to the true secondary structure.

For every secondary structure prediction generated from a protein family (typically, 50), the number of predicted α-helices and β-strands restrict the range of ideal Forms onto which they can be mapped. For example, specifying the secondary structure elements in each layer of the lattice in the order α-β-α, then, if there were five predicted β-strand and five α-helices, the possible Forms would be 0-5-5, 1-5-4 and 2-5-3. Five helices are too big to fit along a 5-stranded sheet so the first can be excluded and the more balanced distribution of helices (2-5-3) would be preferred.

Secondary structure elements can easily be over-predicted and further variation was introduced by omitting the weakest α and β prediction incorporating the additional Forms 2-4-3 (0-4-5 and 1-4-4 can be neglected) and 0-5-4, 1-5-3 and 2-5-2.

Occasionally, some secondary structure predictions are ambiguous, being part α and part β. Rather than force a choice, all possible combinations were considered allowing each ambiguous element to be first α then β.

Finally, as some secondary structures are occasionally broken incorrectly, any pairs of adjacent helices were reconsidered as a single helix across a gap of up to four residues and similarly for pairs of β-strand separated by one or two residues.

2.2.2. *Combinatoric fold generation*

Every secondary structure variation was mapped onto all its compatible Forms, specifying only the overall architecture of the protein structure but not the order in which the secondary structure elements are connected. For a given sequence of secondary structure elements there are very many ways in which they can be connected and to speed this enumeration, basic features of structure organisation were used. These included the observation that connections between strands in the same sheet are almost invariably made as a right-handed connection and that two surface loops seldom cross and that protein chains are seldom knotted[14].

In addition to these constraints, a bias was included to give preference to architectures with an even balance of helices on either side of the sheet, a preference for an antiparallel connection between sequential secondary structure elements and slight bias for folds with the shortest chain path.

Even with the above restrictions, many tens of thousands of possible folds remain for a moderate sized protein (between 100 and 200 residues) and it is not possible to evaluate all of these at a fine level of resolution. A rough selection was imposed by considering the hydrophobicity of each secondary structure element. From the ideal Form, an estimation can be made of the extent of solvent exposure for each secondary structure and a simple correlation score was used to compare this to the hydrophobicity of each element, giving double weight to β-strands to reflect their relative importance in specifying the protein fold. On the basis of this score (s), the folds were ranked and the top scoring selection were constructed as α-carbon models as described previously[15].

2.2.3. *Threading over ideal α-carbon models*

The α-carbon models constructed in this way still retain idealised features, such as the way in which each predicted secondary structure is centred on a lattice position. To allow the model to deviate from this ideal, each α-carbon model was used as a template over which the sequence was threaded, using a simple evaluation function that optimised predicted secondary structure match to the model structure together with hydrophobic positions onto buried positions[16]. This threading method generates many variants from each template, each of which has a score that can be used to select the best. But before making the effort to do this, the scoring function was applied to the template itself. This check was originally introduced simply to save on computation (as a bad template is unlikely to give rise to good models) but, as will be explored below, it proved to be one of the most critical points in the procedure.

Because of limited computational resources, only a few hundred of the best folds could be expanded at the threading stage, specifically, 100 plus the length (N) of the protein. The variation generated by the threading method expanded these, typically, by 100-fold and again each model was evaluated and ranked. Since the threading models were constructed at a finer level, more realistic measures could be used to evaluate them. These included the observed/predicted secondary structure match, the observed/predicted residue exposure and a residue packing measure calculated by the SPREK method[17].

2.2.4. *Model refinement*

After ranking, the best $100 + N$ models from the threading procedure were taken for further refinement using a novel fragment-based method derived from the SPREK program. This encodes the model as a series of linear patterns that describe the environment around each residue and this set of patterns is matched against a much larger set derived from a non-redundant copy of all known proteins.

As the proteins used below for testing have a known structure, the set of patterns was filtered using BLAST to remove any proteins that had a similarity at the sequence level to the protein being predicted or to any member of its family included in its sequence alignment. Unlike artificial neural net methods, where the structures are encoded in an abstract network of weights, the patterns used by the SPREK based method can be directly checked to ensure that no fragment from a homologous protein of known structure contributed to the final model.

The refined models were increased again in their level or representation to include main-chain atoms which were used to calculate an estimation of the number of hydrogen-bonds. The final score was a combination of predicted/observed solvent exposure, the SPREK score and the number of hydrogen bonds, with those in β-strands counting double. All the refined variants were ranked on this value and the top 100+N models were classified using the 'Periodic Table'[14].

2.3. *Development and Testing of the method*

The large numbers of structures generated for even a single protein means that the full procedure from sequence alignment to final structure takes about 12 hours for a protein in the range 100–150 residues when run in parallel on a cluster of 50 computer processors. This makes development and testing of the method time-consuming and it has not been practical to optimise parameters and cutoffs over a large number of proteins. Instead, five proteins were selected with a variety of folds and lengths in the 100–150 residue range.

2.3.1. *Selection of proteins*

The set included two proteins with the common flavodoxin fold, one was the bacterial chemotaxis Y protein (PDB code `3chy`, 128 residues; Form 2-5-3 with strand order 21345) and a flavodoxin (PDB code `1f4p`, 158 res.; 2-5-3 strand order 21345). Che-Y is a smaller and more compact protein compared to the flavodoxin which has larger loops between its secondary structures. The enzyme glycerol-3P cytidyltransferase (PDB code `1coz`, 126; 2-5-3 order 32145) has the same 2-5-3 architecture but with a different strand order in the sheet as well as some long loops. In addition it has a small C-terminal α-helix that does not pack on the sheet. The longer enzyme lumazine synthase (PDB code `1di0`, 147; 2-4-2 order 2134) has only four strands but these are long and are packed with some very long helices. Finally, the smallest structure considered was thioredoxin (PDB code `2trx`, 108; 2-5-2 order 13245). Although the smallest protein in the set, thioredoxin has the unusual feature of a helical connection between two antiparallel strands, a feature that is not ideally represented on the ideal lattice.

2.3.2. *Parameter adjustment*

A number of parameters were adjusted in an *ad hoc* manner with the only aim being to prevent the selection of models that were manifestly unprotein-like (while retaining those with the known native folds). During this preliminary phase it became clear that both the threading and refinement stages could construct a reasonable model when given the correct fold to start with and that the critical 'bottle-neck' in the procedure was in the reduction of the many possible starting folds to produce the relatively small number of a few hundred passed for further consideration.

The parameters that controlled this stage were the secondary structure level score (s), the maximum number of models (ranked on s) evaluated by the threading method as templates (n) and the threading score of these templates (t). In addition, the total surface area (S) of the model was used to identify those with unpacked secondary structure elements. When ranked on s, one third of the total number of folds were evaluated up to a maximum of $n/3$. These were then scored as: $(s + wt)/(S - a - 40N)$, where N is the number of residues in the protein and w is a weight on the threading score. The line $a + 40N$ provides a good baseline for the total surface area of native proteins with a in the range of 500—1000 (data not shown) and the difference between this estimate and the surface area of the model provided a reasonable penalty against models with over-exposed surface.

The values of a, w and n were then adjusted and the number of correct folds for the five proteins that got through 'bottle-neck' were monitored. With $a = 750(\pm 100)$, $w = 1.5(\pm 0.5)$ and $n = 7000(\pm 1000)$, correct folds were selected for each of the proteins. However, as the threading component of the score contains a stochastic contribution, the tolerance ranges associated with each value above indicates the range within which the parameter values are equivalent.

2.3.3. *Sequence alignment and analysis*

Multiple sequence alignments were constructed for a small test-set of five proteins using a standard alignment method that involved no knowledge of the structure of the proteins. The alignments contained between 10–20 sequences covering a broad range of phylogenetic spread and the selection and alignment of the sequences was completely automatic. From the alignment, secondary structures were predicted using the PsiPred and YASPIN methods.

The predicted models were evaluated as a simple Root Mean Square deviation (RMSd) over the α-carbon positions against the native. As an alternative, this simple one-to-one register was allowed to slip (as calculated by the `SAPit` program[18]), allowing a distinction to be made between accuracy in alignment and accuracy in the structural model.

3. Predicted structures for small proteins

The method was developed on a set of five small $\beta\alpha$ proteins in the length range of 100–150 residues. For each protein in the set, the method was applied four times and the results considered both individually and pooled.

3.0.4. *Thioredoxin (2trx)*

The native fold of thioredoxin was ranked top and in one of the runs another run at fourth position. The RMSd for the top scoring model was 4.8Å which is close enough to be clearly correct by visual inspection. (Figure 3(b)). Some deviation in the chain is apparent in a short helix and its adjacent loops, otherwise the register of the sequence over the structure is remarkably

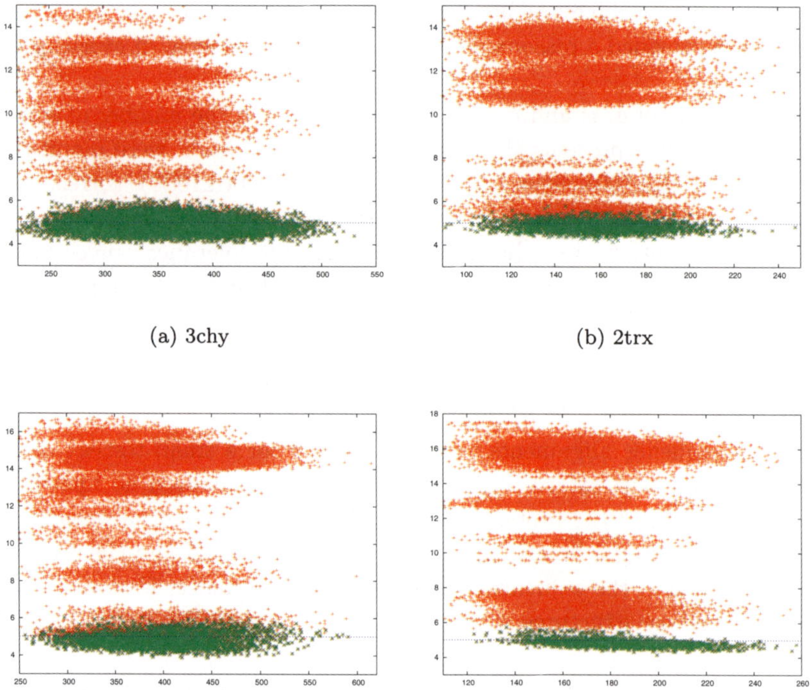

(a) 3chy (b) 2trx

(c) 1f4p (d) 1di0

Figure 2. **Score plots for the test proteins.** The RMSd against the native structure for each refined model (Y-axis) was plotted against the evaluation score (X-axis). Models with a high score are best and appear towards the lower-right corner. Models with the correct fold are plotted in green and others in red. (*a*) `3chy`, a large number of correct models dominate. (*b*) `2trx`, a number of equivalent high scoring models are found. (*c*) `1f4p`, two alternative high scoring folds occur. (*d*) `1di0`, two alternative folds score highly but the correct fold is dominant.

good with the one-to-one superposition giving an RMSd value almost the same as the structure based alignment. Two shifts of one position were made in the edge strands but little else.

Despite being the smallest protein in the set, the native thioredoxin fold faced strong competition from other folds for top position Figure 2(b)). It seems likely that this was a result of poor modelling for the helix that bridges two strands (3 and 4) across the edge of the sheet rather than in the more usual $\beta\alpha\beta$ connection along the length of the β-strands (mak-

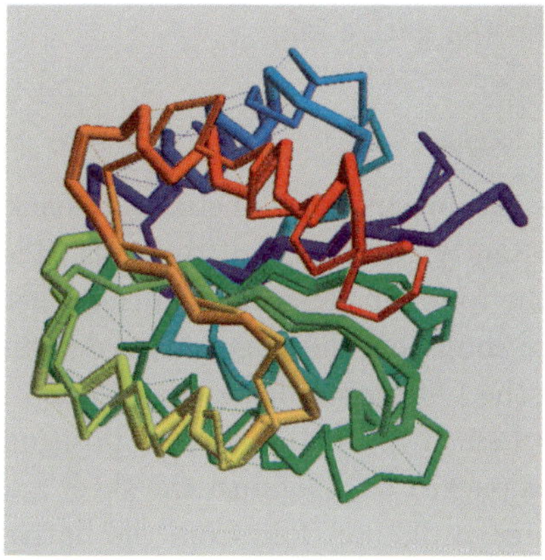

(a) 3chy

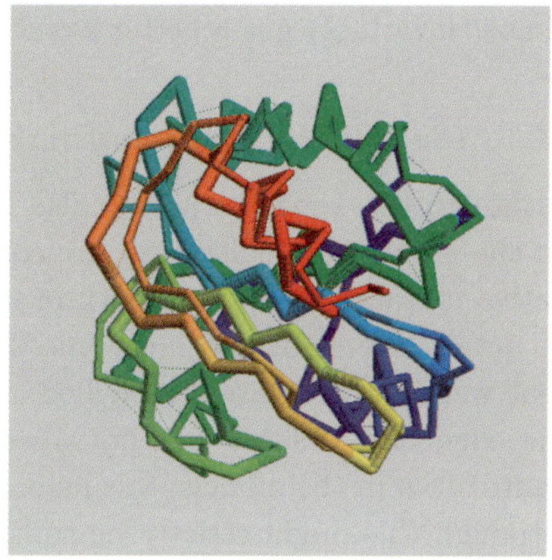

(b) 2trx

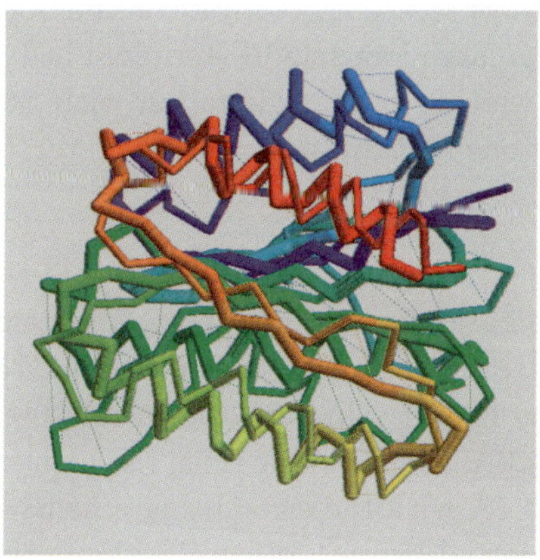

(c) 1f4p

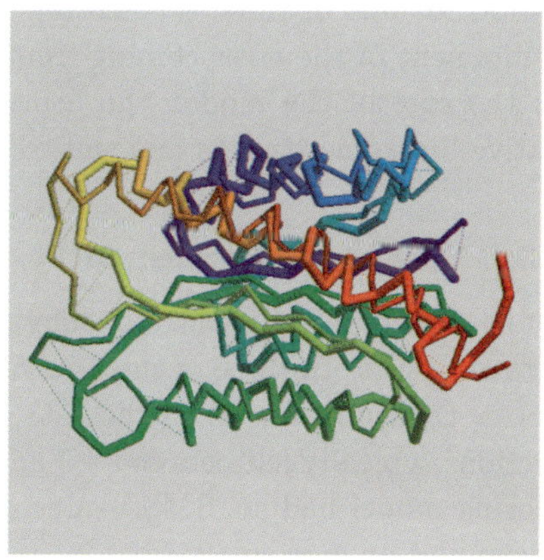

(d) 1di0

Figure 3. **Highest scoring models compared to the native structure**. The highest scoring correct model for each of the test proteins was compared to the corresponding native structure. For each, this was the highest scoring model overall except for `1f4pA` for which the top correct fold was ranked fourth. All superpositions are shown as the α-carbon trace with the native structure drawn slightly thinner. Traces are coloured from blue to red following amino to carboxy terminus. Each view is roughly equivalent, viewed along the edge of the β-sheet, with the amino terminus towards the back right.

ing consecutive antiparallel connections). Models with this configuration almost invariably got a better score than the native fold.

3.0.5. *Glycerol-3P-cytidyltransferase (1coz)*

Although the `1cozA` structure also has an unusual short helix that packs off the sheet, its predictions were consistently good over all the runs. The native fold was ranked top in three of the four runs and third in the other. In detail, the structure has some large loops on the edge of its domain that were not accurately modelled and the C-terminal helix that packs off the edge of the sheet was not always helical or in the correct position. Quite often in the models, this helix was packed back against the sheet and although this might strictly be considered a distinct fold from the native, it was considered close enough to be counted as correct.

In the four sets of results for `1coz`, two performed very well with all-bar-one of the top scoring 20 models having the correct fold. The best RMSd for one of these top scoring models was 5.1Å, with a considerable component of the error coming from the large loops and C-terminal helix. In the core of the model, the helices were in good registration with the native but some strands had shifted by one and two positions.

3.0.6. *Chemotaxis Y protein (3chy)*

Of the five proteins in the set, `3chy` produced the best and most consistent sets of results. In three of the four runs, only one incorrect fold was ranked better than position 25 in the list. The RMSd values for the correct predictions typically fell between 4.5 and 5.0Å. Over all four runs, the highest scoring model had an RMSd value of 4.4Å with the error being a general accumulation over small shifts in strands and helices. When the alignment was allowed to shift (giving the slightly improved RMSd of 3.8Å), it was more clearly revealed that three of the β-strands had slipped by one residue position. (Figure 3(a)).

The only fold to present any 'challenge' for top position had the strand order 23415, maintaining the first strand in a similar environment to that found in the correct topology of 21345.

3.0.7. *Flavodoxin (1f4p)*

Flavodoxin has the same fold as the chemotaxis Y protein but is larger by 20 residues. The extra chain length is accommodated both in longer secondary structures and more extensive loop regions connecting them. The longer sequence allows for a greater variety of secondary structure to be predicted leading to an increased number of folds to be evaluated. For this reason it was not unexpected that the flavodoxin models were not dominated by the correct fold and in almost all runs, the top position was taken by folds that placed the first strand in the middle of the sheet. Despite this, the correct fold remained a close contender, being placed twice in second position and once in third position.

Unlike the `3chy` structure, flavodoxins tend to have a loop instead of a helix connecting strands 2 and 3 in the sheet. In most of the top scoring models with the correct fold, this had become attached to the edge of the sheet forming a sixth β-strand. Although strictly a different topology, the shift required for this change is not large (Figure 3(c)) but did contribute to the elevated RMSd values that were slightly over 5Å for this protein.

3.0.8. *Lumazine synthase (1di0)*

Despite being the largest protein of the five, `1di0` has only four strands in its β-sheet with the extra length, like flavodoxin, being incorporated in long secondary structure elements and loops. The correct fold came in top position in each of the four runs and, although these did not always dominate the top positions (like in CheY), there was typically three correct folds in the top 10 positions.

The highest scoring fold over all the runs had an RMSd of 4.7Å, which is reasonable considering the larger size of the protein and the longer less structured loop regions. (Figure 3(d)). In two of the top scoring models, one of these loops has been 'bent-back' to form a small addition to the edge of the β-sheet. Although strictly a different topology, the overall fold of the protein is unchanged but the RMSd compared to the native structure increased by around 2Å whenever this feature was present.

4. Discussion

4.1. *Summary of the results*

Despite very little in the way of extensive parameter optimisation, the results over the five proteins were remarkably good. Taking the highest scoring model over all the runs placed the native fold in top position for four of the proteins and fourth place for one of them (`1f4p`). The RMSd values for all these models were around 5Å. (Figure 2).

From the point of view of the prediction methodology, perhaps the more unexpected aspect of this result is that the top models were based on secondary structure predictions selected from a range of predictions that were mostly incorrect. For example, the reliable prediction of `3chy` required the prediction of five β-strands, yet less than a third of the secondary structure predictions had this number. For the more problematic prediction of the thioredoxin structure, only an eighth of the predictions had the correct order of secondary structure elements and for `1di0`, this fraction was halved again.

Minor errors in the correct folds were found in the displacement of α-helices and β-strands, typically by $\pm$one turn in the former and $\pm$one or two residues in the latter. The folds that were found in strongest competition to the native almost always had differences that corresponded to a swap of strand positions in the sheet, often between adjacent strands but sometimes between more distant positions that had an equivalent environment. These small changes in the chain position make an absolute difference in topology but are often negligible at the level of RMSd compared to the other non-topological shifts described above. In any analysis it is important to distinguish these two types of error and to do this we used an automatic fold classification method[10]

4.2. *Conclusions*

We have demonstrated that a return to the combinatorial approach to protein structure prediction has allowed progress towards the prediction of larger proteins. The top models, whether right or wrong, all contained features that are typical of globular proteins, demonstrating the approach was sampling fold-space in a realistic manner.

The main limitation of the method remains the poor quality of secondary structure prediction. This was substantially overcome by trying many variants and the one of the most encouraging aspects of the method is the power it displayed in selecting the correct prediction from a wealth of alternatives. A greater variety of sequences in the alignment could be used to further increase the range of predictions or extending the permutation of the weaker predictions.

Despite these limitations, the method was able to predict the tertiary structures of some proteins that were almost twice as large as anything achieved before. In others, the errors often involved only the swapping of two elements between equivalent environments (usually buried). The current evaluation method, being based only on α-carbon positions has little power to distinguish these alternatives and it may be necessary to consider side-chain packing to be able to do this.

Acknowledgments

Some of the calculations for this work were made on the TSUBAME (10,000 CPU) computer at the Tokyo Institute of Technology and Motonori Ota is thanked for his help in facilitating this.

References

1. M. Levitt. A simplified representation of protein conformations for rapid simulation of protein folding. *J. Molec. Biol.*, 104:59–107, 1976.
2. I. D. Kuntz, G. M. Crippen, P. A. Kollman, and D. Kimelman. Calculation of protein tertiary structure. *J. Molec. Biol.*, 106:983–994, 1976.
3. F. E. Cohen, T. J. Richmond, and F. M. Richards. Protein folding: Evaluation of some simple rules for the assembly of helices into tertiary structures with myoglobin as as example. *J. Molec. Biol.*, 132, 275–288 1979.
4. F. E. Cohen, M. J. E. Sternberg, and W. R. Taylor. Analysis and prediction of protein β-sheet structures by a combinatorial approach. *Nature*, 285:378–382, 1980.
5. Y. Duan and P. A. Kollman. Pathways to a protein folding intermediate observed in a 1-microsecond simulation in aqueous solution. *Science*, 282:740–744, 1998.
6. P. Bradley, K. M. S. Misura, and D. Baker. Toward high-resolution de novo structure prediction for small proteins. *Science*, 309:1868–1871, 2005.
7. F. E. Cohen, M. J. E. Sternberg, and W. R. Taylor. Analysis and prediction of the packing of α-helices against a β-sheet in the tertiary structure of globular proteins. *J. Molec. Biol.*, 156:821–862, 1982.
8. D. T. Jones. Protein secondary structure prediction based on position-specific scoring matrices. *J. Molec. Biol.*, 292:195–202, 1999.
9. K. Lin, V. A. Simossis, W. R. Taylor, and J. Heringa. A simple and fast secondary structure prediction method using hidden neural networks. *Bioinformatics*, 21:152–159, 2005.
10. W. R. Taylor. A periodic table for protein structure. *Nature*, 416:657–660, 2002.
11. W. R. Taylor. Defining linear segments in protein structure. *J. Molec. Biol.*, 310:1135–1150, 2001.
12. C. Chothia and A. V. Finkelstein. The classification and origins of protein folding patterns. *Ann. Rev. Biochem.*, 59:1007–1039, 1990.
13. A. V. Finkelstein and O. B. Ptitsyn. Why do globular proteins fit the limited set of folding patterns? *Prog. Biophys. Molec. Biol.*, 50:171–190, 1987.
14. W. R. Taylor. A deeply knotted protein and how it might fold. *Nature*, 406:916–919, 2000.

15. W. R. Taylor. Protein fold refinement: building models from idealised folds using motif constraints and multiple sequence data. *Prot. Engng.*, 6:593–604, 1993.
16. W. R. Taylor. Multiple sequence threading: an analysis of alignment quality and stability. *J. Molec. Biol.*, 269:902–943, 1997.
17. W. R. Taylor and I. Jonassen. A structural pattern-based method for protein fold recognition. *Proteins: struc. funct. bioinf.*, 56:222–234, 2004.
18. W. R. Taylor. Protein structure alignment using iterated double dynamic programming. *Prot. Sci*, 8:654–665, 1999.

EUCLIDEAN FULL STEINER TREES AND THE MODELLING OF BIOMOLECULAR STRUCTURES

RUBEM P. MONDAINI

Federal University of Rio de Janeiro - UFRJ - CT - COPPE
BIOMAT Institute for Advanced Studies of Biosystems
21941-972, P.O. Box 68511, Rio de Janeiro, RJ, Brazil
E-mail: mondaini@cos.ufrj.br, rpmondaini@gmail.com

Some considerations about the modelling of the structure of biological macromolecules are studied in the present work. It is emphasized the usefulness of the concept of Steiner trees and some derived parameters like the Steiner Ratio and chirality functions for characterizing the potential energy minimization process of these structures.

1. Introduction and Motivation

The phenomenon of molecular chirality is essential for understanding the interaction of biomolecules and their stability[1] since the folding of protein structures is expected to be driven by the influence of chirality in a thermodynamic approach of molecular evolution. In spite of the best scientific efforts, a successful theory is still missing in the scientific literature, but we believe that the modelling of molecular structures by Steiner networks can span the bridge across some preliminary difficulties by providing a robust approach to the foundations of a consistent theory of chirality and folding. Some recent works[2,3] have been very useful at convincing their readers of these methods since they are able to derive the actual relative placement of the atoms of a biomolecule in terms of the connection with the problem of energy minimization. In order to give an insight of the application of Steiner trees, we start with the energy problem as disguised in a generalized Fermat problem.

2. The Fermat Problem and Maxwell's Theorem

Let $\vec{r}_j$, $j = 0, \ldots, n-1$ the position vectors of n atoms. Let us examine the stability of the molecular cluster formed by their interaction with an atom placed at $\vec{S}$. By assuming that the potential energy of this configuration is given by

$$U_S = \mu_S \sum_{j=0}^{n-1} \frac{\mu_j}{\|\vec{r}_j - \vec{S}\|^d} , \tag{1}$$

where $\|\cdot\|$ is the euclidean norm and μ_S, μ_j are characteristic constants of the force law, the equilibrium conditions are then given by

$$0 = \frac{\partial U_S}{\partial \vec{S}} = -\mu_S d \sum_{j=0}^{n-1} \frac{\mu_j}{\|\vec{r}_j - \vec{S}\|^{d+1}} \frac{\partial}{\partial \vec{S}} \|\vec{r}_j - \vec{S}\| . \tag{2}$$

A plausible assumption of equality of the interaction strengths[4] of each atom at $\vec{r}_j$ with the atom at $\vec{S}$ leads to

$$0 = -\mu_S d \frac{\mu_0}{\|\vec{r}_0 - \vec{S}\|^{d+1}} \frac{\partial}{\partial \vec{S}} \sum_{j=0}^{n-1} \|\vec{r}_j - \vec{S}\| . \tag{3}$$

But this is actually the generalized Fermat problem of minimizing the sum of distances $\|\vec{r}_j - \vec{S}\|$. Eq.(3) can be also written as

$$\sum_{j=0}^{n-1} \hat{r}_{jS} = 0 \tag{4}$$

where

$$\hat{r}_{jS} = \frac{\vec{r}_j - \vec{S}}{\|\vec{r}_j - \vec{S}\|}$$

is the unit vector in the direction of the j-edge of the molecular cluster.

For $n = 2, 3$, eq.(4) has only the solution corresponding to equal angles among the edges

$$n = 2 : \hat{r}_{1S} \cdot \hat{r}_{2S} = -1 ; \tag{5}$$

$$n = 3 : \hat{r}_{1S} \cdot \hat{r}_{2S} = \hat{r}_{1S} \cdot \hat{r}_{3S} = \hat{r}_{2S} \cdot \hat{r}_{3S} = -\frac{1}{2} . \tag{6}$$

Eq.(6) corresponds to the usual situation on each equilibrium node (Steiner vertex) of a Steiner tree. There are analogous solutions for $n > 3$ with $\hat{r}_{iS} \cdot \hat{r}_{jS} = -1/(n-1)$, $i \neq j$, but the unit vectors of our modelling could not satisfy this requirement. We notice that the cases $p = 3$ and $p = 4$ are feasible for the kind of modelling which we present in the following pages. It is also interesting to report that the position of the atoms in biomolecules satisfy the equilibrium requirement for $p = 3$ (carbon and nytrogen of the amide planes[2,3]) and $p = 4$ (α-carbon[5]).

Let us now assume that the structure is not in equilibrium. We have to apply forces $\vec{f}_j$, $\vec{f}_S$ on each vertex $\vec{r}_j$ and $\vec{S}$ in order to re-establish the equilibrium. These forces will be decomposed in a component orthogonal to the edge connecting $\vec{r}_j$ to $\vec{S}$ and other parallel to this edge, or

$$\vec{f}_{jS} = a_{jS}\hat{f}_{jS^{\parallel}} + b_{jS}\hat{f}_{jS^{\perp}} \,, \quad \hat{f}_{jS^{\parallel}} = \frac{\vec{r}_j - \vec{S}}{\|\vec{r}_j - \vec{S}\|} = \hat{r}_{jS} \,. \tag{7}$$

We have:

$$\vec{f}_{jS} \cdot (\vec{r}_j - \vec{S}) = a_{jS}\|\vec{r}_j - \vec{S}\| \tag{8}$$

The total length of edges for this out-of-equilibrium stage is:

$$l_F = \sum_{j=0}^{n-1} \|\vec{r}_j - \vec{S}\| = \sum_{j=0}^{n-1} \vec{r}_j \cdot \frac{\vec{f}_{jS}}{a_{jS}} - \vec{S} \cdot \sum_{j=0}^{n-1} \frac{\vec{f}_{jS}}{a_{jS}} \tag{9}$$

If we now suppose the forces $\vec{f}_j$ as collinear with the edges or $\vec{f}_{jS} = a_{jS}\hat{f}_{jS^{\parallel}} = a_{jS}\hat{r}_{jS}$, we then have for the equilibrium length:

$$l_F^{\text{eq}} = \sum_{j=0}^{n-1} \vec{r}_j \cdot \hat{f}_{jS^{\parallel}} = \sum_{j=0}^{n-1} \vec{r}_j \cdot \hat{r}_{jS} \tag{10}$$

where $\hat{r}_{jS}$ is given by eq.(7). Eq.(10) is the content of the Maxwell's theorem on the forces to be applied to keep the equilibrium of a truss[6]. The application of the theorem to this work is done with unit forces.

There are $(n+1)$ vertices and $(n+1-1) = n$ edges. A number of $2n$ forces will be necessary to re-establish the equilibrium. The first variation of the equilibrium length under variation of the point $\vec{S}$ is

$$\delta l_F^{\text{eq}} = \delta\vec{S} \cdot \sum_{j=0}^{n-1} \vec{T}_{jS} \tag{11}$$

where

$$\vec{T}_{jS} = \frac{(\vec{r}_j \times \hat{r}_{jS}) \times \hat{r}_{jS}}{\|\vec{r}_j - \vec{S}\|} \,. \tag{12}$$

We also have for the finite length difference between the generic and the equilibrium configurations:

$$\Delta l = l - l^{\text{eq}} = -\sum_{j=0}^{n-1} \hat{r}_{jS} \cdot \vec{S} \,. \tag{13}$$

3. The Search for a New Equilibrium Configuration

If the equilibrium configuration cannot be maintained due to a special perturbation in the point $\vec{S}$, the configuration will search for a new equilibrium. We now assume that $\vec{S}$ is the average position of a set of points S_k, $k = 1, \ldots, q$ which we can imagine to be produced from a physical instability of a large unstable ion originally at position $\vec{S}$ (an explosion process). Furthermore, we assume that the system of n fixed points $\vec{r}_j$ and q variable points $\vec{S}_k$ will reach this new equilibrium configuration when this system is a tree formed by connecting the points $\vec{S}_1$ to $\vec{S}_q$ consecutively and by connecting $\vec{S}_1$ to the first $(p-1)$ fixed points and $\vec{S}_q$ to the last $(p-1)$ fixed points. Each intermediary S_k point $(k = 2, \ldots, q-1)$ will be connected to $(p-2)$ fixed points. We notice that[1] $q = (n-2)/(p-2)$ and the tree can be formed only when q is an integer number.

In order to generalize the process of section 1 to the present configuration, we have to stress that there is now $(n+q)$ vertices and $(n+q-1)$ edges[6]. We have to apply $2(n+q-1)$ forces to re-establish the equilibrium. The usual decomposition of these forces in a parallel and an orthogonal component to the edge can be written as

$$\vec{f}_{jk} = a_{jk}\hat{f}_{jk^{\parallel}} + b_{jk}\hat{f}_{jk^{\perp}} \, , \quad \hat{f}_{jk^{\parallel}} = \frac{\vec{r}_j - \vec{S}_k}{\|\vec{r}_j - \vec{S}_k\|} = \hat{r}_{jk} \tag{14}$$

$$\begin{aligned} \vec{f}_{S_{k+1},S_k} &= a_{S_{k+1},S_k}\hat{f}_{S_{k+1},S_k^{\parallel}} + b_{S_{k+1},S_k}\hat{f}_{S_{k+1},S_k^{\perp}} \, , \\ \hat{f}_{S_{k+1},S_k} &= \frac{\vec{S}_{k+1} - \vec{S}_k}{\|\vec{S}_{k+1} - \vec{S}_k\|} = \hat{r}_{S_{k+1},S_k} \, . \end{aligned} \tag{15}$$

We then have,

$$\vec{f}_{jk} \cdot (\vec{r}_j - \vec{S}_k) = a_{jk}\|\vec{r}_j - \vec{S}_k\| \, , \tag{16}$$

$$\vec{f}_{S_{k+1},S_k} \cdot (\vec{S}_{k+1} - \vec{S}_k) = a_{S_{k+1},S_k}\|\vec{S}_{k+1} - \vec{S}_k\| \, . \tag{17}$$

The total length of the edges of this out-of-equilibrium configuration is now

$$\begin{aligned} l_S = & \sum_{j=0}^{p-2}\|\vec{r}_j - \vec{S}_1\| + \sum_{k=2}^{q-1}\sum_{j=k(p-2)-p+3}^{k(p-2)}\|\vec{r}_j - \vec{S}_k\| \\ & + \sum_{j=n-p+1}^{n-1}\|\vec{r}_j - \vec{S}_q\| + \sum_{k=1}^{q-1}\|\vec{S}_{k+1} - \vec{S}_k\| \end{aligned} \tag{18}$$

Analogously to eq.(9), after using eqs.(16) and (17), we can also write the last equation as

$$l_S = \sum_{j=0}^{p-2}(\vec{r}_j - \vec{S}_1)\cdot\frac{\vec{f}_{j1}}{a_{j1}} + \sum_{k=2}^{q-1}\sum_{j=k(p-2)-p+3}^{k(p-2)}(\vec{r}_j - \vec{S}_k)\cdot\frac{\vec{f}_{jk}}{a_{jk}}$$
$$+ \sum_{j=n-p+1}^{n-1}(\vec{r}_j - \vec{S}_q)\cdot\frac{\vec{f}_{jq}}{a_{jq}} + \sum_{k=1}^{q-1}(\vec{S}_{k+1} - \vec{S}_k)\cdot\frac{\vec{f}_{S_{k+1},S_k}}{a_{S_{k+1},S_K}} \qquad (19)$$

If the forces $\vec{f}_{jk}$, $\vec{f}_{S_{k+1},S_k}$, $j = 0, 1, \ldots, n-1$, $k = 1, 2, \ldots, q$ are collinear with the edges, this can be also written as

$$\Delta l_S = l_S - l_S^{\text{eq}} = -\vec{S}_1 \cdot \left(\hat{r}_{S_2,S_1} + \sum_{j=0}^{p-2}\hat{r}_{j1}\right)$$
$$-\sum_{k=2}^{q-1}\vec{S}_k \cdot \left[\hat{r}_{S_{k+1},S_k} + \left(-\hat{r}_{S_k,S_{k-1}}\right) + \sum_{j=k(p-2)-2k+3}^{k(p-2)}\hat{r}_{jk}\right]$$
$$-\vec{S}_q \cdot \left[\left(-\hat{r}_{S_q,S_{q-1}}\right) + \sum_{j=n-p+1}^{n-1}\hat{r}_{jq}\right] \qquad (20)$$

An elementary check will convince us of the number of forces[6] present in the structure introduced above. We have

$$n + p + (q-2)p + p = n + qp = 2(n + q - 1)$$

The quantity l_S^{eq} is the length of the tree for the equilibrium configuration, or

$$l_S^{\text{eq}} = \sum_{j=0}^{p-2}\vec{r}_j \cdot \hat{r}_{j1} + \sum_{k=2}^{q-1}\sum_{j=k(p-2)-p+3}^{k(p-2)}\vec{r}_j \cdot \hat{r}_{jk} + \sum_{j=n-p+1}^{n-1}\vec{r}_j \cdot \hat{r}_{jq} \qquad (21)$$

This means that the sum of the p terms on each square brackets in eq.(20) is equal to zero, which is analogous to eq.(4) for the present case. The existence of solutions will be related to a good Ansatz for the coordinates of the vectors $\vec{r}_j$, $\vec{S}_k$ and these should be compatible with the values $p = 3$ and $p = 4$ since these correspond to biomolecular structures which can be easily seen from the examination of 3D-structure proteins in data banks[5].

The vectors $\hat{r}_{jk}$ and $\hat{r}_{S_{k\pm1},S_k}$ could be required to satisfy

- $\hat{r}_{jk} \cdot \hat{r}_{ik} = -\dfrac{1}{p-1}, \quad i \neq j,$

$$\begin{cases} k = 1: \; i,j = 0,\ldots,(p-2) \\ 1 < k < n-2: \; i,j = (k(p-2)-p+3),\ldots,k(p-2) \\ k = n-2: \; i,j = (n-p+1),\ldots,(n-1) \end{cases}$$

- $\hat{r}_{S_{k-1},S_k} \cdot \hat{r}_{S_{k+1},S_k} = -\dfrac{1}{p-1}, \quad k = 1,2,\ldots,q \qquad (22)$

The perturbation of the configuration around the equilibrium position is formed by the combination of the independent variations of position of the points S_k. We can write in analogy with eq.(11), after using eq.(21),

$$\delta l_S^{\text{eq}} = \sum_{j=0}^{p-2} \vec{T}_{j1} \cdot \delta\vec{S}_1 + \sum_{k=2}^{q-1} \sum_{j=k(p-2)-p+3}^{k(p-2)} \vec{T}_{jk} \cdot \delta\vec{S}_k + \sum_{j=n-p+1}^{n-1} \vec{T}_{jq} \cdot \delta\vec{S}_q \qquad (23)$$

where

$$\vec{T}_{jk} = \frac{(\vec{r}_j \times \hat{r}_{jk}) \times \hat{r}_{jk}}{\|\vec{r}_j - \vec{S}_k\|} \qquad (24)$$

The concepts introduced in this section and the quantities derived there are enough for modelling the stability of the structure of biomolecules by networks of points and connecting edges.

4. The Sets of Evenly Spaced Points

We now assume that the points $\vec{r}_j$ belong to a continuous and differentiable curve are evenly spaced in terms of the euclidean norm or

$$\|\vec{r}_{j+2} - \vec{r}_{j+1}\| = \|\vec{r}_{j+1} - \vec{r}_j\|, \quad j = 0,1,\ldots,n-1 . \qquad (25)$$

Eq.(14) can be satisfied by

$$\vec{r}_j = (r(\omega)\cos j\omega, r(\omega)\sin j\omega, jh(\omega)), \quad j = 0,1,\ldots,n-1 , \qquad (26)$$

where $r(\omega)$ and $h(\omega)$ are continuous and differentiable functions.

We do the same assumption for the points $\vec{S}_k$. The sequence of evenly spaced points is given by

$$\vec{S}_k = (r_S(\omega)\cos k\omega, r_S(\omega)\sin k\omega, kh_S(\omega)), \quad k = 1,2,\ldots,q \qquad (27)$$

The functions $r_S(\omega)$ and $h_S(\omega)$ are also continuous and differentiable and we have

$$\|\vec{S}_{k+2} - \vec{S}_{k+1}\| = \|\vec{S}_{k+1} - \vec{S}_k\|, \quad k = 1,2,\ldots,q \qquad (28)$$

We leave to the reader to check that the restrictions imposed on the Ansatz of the eqs.(26), (27) by the equilibrium conditions of eq.(22) are satisfied only for $p = 3, 4$. The $p = 3$ case (Steiner Trees) correspond to a unique solution. The solution for $p = 4$ is not unique as an equilibrium configuration is concerned but it should corresponds to the minimal length equilibrium configuration among the other solutions with $p = 4$ which do not satisfy eqs.(22). It seems that we have a good Ansatz for working in the modelling of biomolecular structures. As a final remark, we notice another property of the Ansatz: the position vectors of points r_j and S_k will satisfy the equations[7]

$$\vec{r}_{j+4} = (3-A)\vec{r}_{j+3} - 2(2-A)\vec{r}_{j+2} + (3-A)\vec{r}_{j+1} - \vec{r}_j\,, \quad j = 0, 1, \ldots, n-1 \quad (29)$$

$$\vec{S}_{k+4} = (3-A)\vec{S}_{k+3} - 2(2-A)\vec{S}_{k+2} + (3-A)\vec{S}_{k+1} - \vec{S}_k\,, \quad k = 1, 2, \ldots, q \quad (30)$$

where $A = 1 - 2\cos\omega$.

5. The Case of Steiner Trees and a Proposal for a Chirality Measure

We examine the behaviour of the quantities introduced in section 3 for the case $p = 3$ (Steiner Trees). First of all, we notice that eqs.(22) lead to

$$h = h_S\,; \quad h_S^2 = r_S^2 A(A+1) \quad (31)$$

If we make the restriction of working with full Steiner Trees, we have to require that the angles between consecutive edges should be lesser than $2\pi/3$. This means that

$$(\vec{r}_{j-1} - \vec{r}_j) \cdot (\vec{r}_{j+1} - \vec{r}_j) > -\frac{1}{2} \cdot \|\vec{r}_{j-1} - \vec{r}_j\|\, \|\vec{r}_{j+1} - \vec{r}_j\| \quad (32)$$

After using eqs.(26) and (31), we get

$$r > r_S \quad (33)$$

The length of the out-of-equilibrium configuration, eq.(18), will be given now by

$$\begin{aligned} l_S = {} & [(r - r_S)^2 + h_S^2 + (A+1) r r_S]^{1/2} + [(r - r_S)^2 + (h_S - h)^2]^{1/2} \\ & + \sum_{j=2}^{n-3} [(r - r_S)^2 + k^2 (h_S - h)^2]^{1/2} \\ & + [(r - r_S)^2 + (n-2)^2 (h_S - h)^2]^{1/2} + (n-3)[h_S^2 + (A+1) r_S^2]^{1/2} \\ & + [(r - r_S)^2 + ((n-2) h_S - (n-1) h)^2 + (A+1) r r_S]^{1/2} \end{aligned} \quad (34)$$

The first and last terms correspond to contributions of the two ends of this tree. We should take a limiting process $r_S \to r$ on them in order to satisfy the conditions of eq.(22) in these regions. From these considerations and after using eqs.(31) and (33), we get

$$l_S = l_S^{\text{eq}} = (n-2)r + [A(n-1)+1]r_S \tag{35}$$

The variation of the equilibrium length due to a variation of the curve given by eq.(27) will be calculated with the usual prescription which we have used to derive eq.(35). We can write

$$\delta l_S^{\text{eq}} = \sum_{j=0}^{1} \vec{T}_{j1} \cdot \delta\vec{S}_1 + \sum_{k=2}^{n-3} \vec{T}_{kk} \cdot \delta\vec{S}_k + \sum_{j=n-2}^{n-1} \vec{T}_{j\,n-2} \cdot \delta\vec{S}_{n-2} \tag{36}$$

where $\vec{T}_{jk}$ is given by eq.(24).

From eqs.(26) and (27) we then have,

$$\begin{aligned}
\|\vec{r}_j - \vec{S}_k\|^3 \vec{T}_{jk} \cdot \delta\vec{S}_k = {} & [(jhr_S - krh_S)(r - r_S) \\
& - rr_S(jh + kh_S)(1 - \cos(j-k)\omega)]k\,\delta h_S \\
& - [(jhr_S - krh_S)(jh - kh_S) \\
& + rr_S(r + r_S)(1 - \cos(j-k)\omega)]\cos((j-k)\omega)\delta r_S
\end{aligned} \tag{37}$$

From eq.(37), we can now write:

$$\begin{aligned}
\vec{T}_{01} \cdot \delta\vec{S}_1 &= -\frac{1}{2}\frac{\delta h_S}{\sqrt{(A+1)^3 A}} \\
\vec{T}_{11} \cdot \delta\vec{S}_1 &= -\frac{h_S\,\delta h_S}{r - r_S} \\
\vec{T}_{kk} \cdot \delta\vec{S}_k &= -\frac{k^2}{r - r_S} h_S\,\delta h_S\,, \quad 2 \le k \le n-3 \\
\vec{T}_{n-2,n-2} \cdot \delta\vec{S}_{n-2} &= -\frac{(n-2)^2}{r - r_S} h_S\,\delta h_S \\
\vec{T}_{n-1,n-2} \cdot \delta\vec{S}_{n-1} &= -\frac{1}{2}\frac{[1 + A(n-1)(2n-5)]}{\sqrt{(A+1)^3 A}}\delta h_S
\end{aligned} \tag{38}$$

From eqs.(36) and (38), we get the first and second variations of equilibrium length of the Steiner tree as

$$\frac{\delta l_S^{\mathrm{eq}}}{\delta h_S} = -\frac{\left[1+\frac{1}{2}A(n-1)(2n-5)\right]}{\sqrt{(A+1)^3 A}} - \frac{1}{6}(n-2)(n-1)(2n-3)\frac{h_S}{r-r_S} \quad (39)$$

and

$$\delta\left(\frac{\delta l_S^{\mathrm{eq}}}{\delta h_S}\right) = -\frac{1}{6}(n-2)(n-1)(2n-3)\frac{r\,\delta h_S}{(r-r_S)^2}\,. \quad (40)$$

This can be seen as a consequence of the molecular search for minimum energy configurations. In the process of looking for new minima, the biomolecule increases its linear extension in order to minimize its energy and the total length of its associated Steiner Tree.

We can obtain an elementary consistency check of the calculations of eq.(35) by working with eq.(21) with the same procedure of the two calculations above. We have,

$$l_S^{\mathrm{eq}} = \sum_{j=0}^{1} \vec{r}_j \cdot \hat{r}_{j1} + \sum_{k=2}^{n-3} \vec{r}_k \cdot \hat{r}_{kk} + \sum_{j=n-2}^{n-1} \vec{r}_j \cdot \hat{r}_{j\,n-2} \quad (41)$$

and

$$\|\vec{r}_j - \vec{S}_k\| \vec{r}_j \cdot \hat{r}_{jk} = r(r-r_S) + r r_S(1-\cos(j-k)\omega) + jh(jh-kS) \quad (42)$$

After using eqs.(31) and (33), we have

$$\begin{aligned} \vec{r}_0 \cdot \hat{r}_{01} &= \frac{1}{2} r_S \\ \vec{r}_1 \cdot \hat{r}_{11} &= r \\ \vec{r}_k \cdot \hat{r}_{kk} &= r\,, \quad 2 \le k \le n-3 \\ \vec{r}_{n-2} \cdot \hat{r}_{n-2,n-2} &= r \\ \vec{r}_{n-1} \cdot \hat{r}_{n-1,n-2} &= \frac{1}{2} r_S + (n-1)A r_S \end{aligned} \quad (43)$$

From eq.(41), these results will lead to the same result of eq.(35).

The proposals for chirality functions should be based on the possibility of forming pseudoscalar quantities[8] from the geometrical structures of points and trees. Our first proposal will be given by

$$\chi_t = \sum_{k=1}^{n-2} \sum_{j=0}^{n-1} \chi_{jk} \quad (44)$$

where

$$\chi_{jk} = \frac{1}{6} \cdot (\vec{r}_j - \vec{S}_k) \times (\vec{r}_{j+1} - \vec{S}_k) \cdot (\vec{S}_{k+1} - \vec{S}_k) \quad (45)$$

and $\chi_{jk} = 0$ for $j \neq k-1$.

The definition of eq.(45) corresponds to the volume of tetrahedra formed by the three vectors of the parentheses. These are vectors along the edges of the Steiner tree.

The last tetrahedra which value is $n-3$, $n-2$ should not be taken into account, since it does not correspond to the prescription of eq.(45). The first one should be considered as having a volume $\chi_{01} = 0$ for the validity of eqs.(31) at this end of the tree.

We then have,

$$\chi_t = \sum_{k=2}^{n-3} \chi_{k-1,k} = \frac{1}{6} \cdot (n-4)(r-r_S)^2 h_S \sin\omega \,. \tag{46}$$

There are also other elementary examples of proposals for chirality function as derived from the volume of sequences of tetrahedra. From the sequences of vectors $\vec{r}_j$ and $\vec{S}_k$ we can have, respectively

$$\chi_r = \frac{1}{6} \sum_{j=0}^{n-4} (\vec{r}_{j+1} - \vec{r}_j) \times (\vec{r}_{j+2} - \vec{r}_j) \cdot (\vec{r}_{j+3} - \vec{r}_j) \tag{47}$$

$$\chi_s = \frac{1}{6} \sum_{k=1}^{n-5} (\vec{S}_{k+1} - \vec{S}_k) \times (\vec{S}_{k+2} - \vec{S}_k) \cdot (\vec{S}_{k+3} - \vec{S}_k) \tag{48}$$

The terms of the sums above do not depend on j or k, as can be seen from substitution of the identities given by eqs.(29) and (30), respectively. From eq.(31) we can write,

$$\chi_r = \frac{1}{6}(n-3) r^2 h_S (A+1)^2 \sin\omega \tag{49}$$

$$\chi_s = \frac{1}{6}(n-5) r_S^2 h_S (A+1)^2 \sin\omega \tag{50}$$

The function $(\chi_r - \chi_s)$ for $n \gg 1$ has been used as a constraint of a constrained minimization process of the Steiner Ratio function in ref.[3].

The examples of the last section, eqs.(46), (49) and (50) should be discarded as good proposals for chirality measure. The chirality should be zero for configurations of $n = 3, 4, 5$ points if the definitions χ_r, χ_t, χ_s were adopted as proposals of chirality measure. The chirality of a configuration $n = 1, 2$ is obviously zero and also for $n = 3$ since a triangle is always non-chiral in 3D-dimensional Euclidean space, according Kelvin's definition of coincidence of a configuration with its mirror image through translations and rotations. The chirality will be zero for $n = 4, 5$ if the configurations are formed by regular tetrahedra glued together at common faces. This is due to the existence of three-fold rotation axes in the structure for $n = 4, 5$ points. The ω−value which corresponds to this configuration is $\omega_T = \pi \pm \arccos(2/3)$ and it can be derived from the vector $\vec{r}_j$

$$\|\vec{r}_{j+l}(\omega_T) - \vec{r}_j(\omega_T)\| = \|\vec{r}_{j+1}(\omega_T) - \vec{r}_j(\omega_T)\|\,, \quad l = 2, 3\,. \tag{51}$$

There is not an analogous relation for the vectors $\vec{S}_k$, or

$$\|\vec{S}_{k+l}(\omega) - \vec{S}_k(\omega)\| \neq \|\vec{S}_{k+1}(\omega) - \vec{S}_k(\omega)\|\,, \quad l = 2, 3\,. \tag{52}$$

From these considerations, we now make the following proposal for a function of geometric chirality:

$$\chi = (n-1)(n-2)\chi_r(\omega)[\chi_t(\omega)\chi_s(\omega) + \chi_r^2(\omega) - \chi_r^2(\omega_T)]\,. \tag{53}$$

From eq.(53) and eqs.(46), (49) and (50), we can write,

$$\begin{aligned}\chi(\omega) = {} & \frac{r^9}{216}(n-1)(n-2)(n-3)\left(\frac{h_S}{r}\right)^3 (A+1)^4 \sin\omega \\ & \left[(n-4)(n-5)\left(1-\frac{r_S}{r}\right)^2\left(\frac{r_S}{r}\right)^2 \sin^2\omega \right. \\ & \left. + (n-3)^2(A+1)^2(\sin^2\omega - \sin^2\omega_T)\right]\end{aligned} \tag{54}$$

This definition satisfies the obvious requirements for a chirality measure, since only for $n \geq 4$ we can have a chiral object. Some additional information about the special structure of regular tetrahedra is also included in the definition.

6. Concluding Remarks

Some fundamental steps for a full geometric formulation of the organization of molecular structures are shown in the present contribution. The concepts of a Steiner Ratio Function and of a Chirality Function are essential in the theoretical scheme introduced here. These functions could be used in a constrained optimization problem[1] in which the chirality function is playing the role of a constraint. The problem to be solved is a translation in geometrical language of the Free Energy minimization problem of formation and stability of molecular structures. It is a well-posed problem and efficiency of this approach has been proved by comparing some of its predictions with the observed behaviour of biomolecular configurations. In spite of many sound results, some essential pieces of this theoretical formulation are still missing. A precise formulation of molecular chirality and its expression into a mathematical formula should be free from phenomenological choice. The present work is only a putative description of the difficulties to be circumvented in the construction of a fundamental theory of biomolecular formation and the search for equilibrium stages which are necessary to life maintenance.

References

1. Mondaini, R. P. (2004). The Geometry of Macromolecular Structure: Steiner's Points and Trees, *Proc. BIOMAT Symp.* 4: 347–356.
2. MacGregor Smith, J. (2006). Steiner Minimal Trees, Twist Angles, and the Protein Folding Problem, *in BIOMAT 2005, International Symposium on Mathematical and Computational Biology, World Scientific Co. Pte. Ltd.*: 299–326.
3. Mondaini, R. P. (2006). Steiner Trees as Intramolecular Networks of the Biomacromolecular Structures, *in BIOMAT 2005, International Symposium on Mathematical and Computational Biology, World Scientific Co. Pte. Ltd.*: 327–342.
4. Mondaini, R. P. (2004). The Euclidean Steiner Ratio and the Measure of Chirality of Biomacromolecules, *Gen. Mol. Biol.* 27(4): 658–664.
5. Protein Data Bank (2004). Education Section. http://www.rothamsted. bbsrc.ac.uk/notebook/courses/guide/aa.htm.
6. Gilbert, E. N. and Pollak, H. O. (1968). Steiner Minimal Trees, *SIAM J. Appl. Math.* 16(1).
7. Mondaini, R. P. (2007). Steiner Ratio of Biomolecular Structures, *C. A. Floudas, P. M. Pardalos (eds), Encyclopedia of Optimization, 2nd ed., Springer, in press.*
8. de Gennes, P.-G. (1992). Simple Views on Condensed Matter, *Series in Modern Condensed Matter Physics, World Scientific Co. Pte. Ltd.* 4: 391–393.

DEFINING REDUCED AMINO ACID SETS WITH A NEW SUBSTITUTION MATRIX BASED SOLELY ON BINDING AFFINITIES

ARMIN A. WEISER

Institute for Theoretical Biology, Humboldt University Berlin
Invalidenstr. 43,10115 Berlin, Germany

RUDOLF VOLKMER

Institute of Medical Immunology, Charité – University Hospital Berlin
Hessische Str. 3., 10115 Berlin, Germany

MICHAL OR-GUIL

Institute for Theoretical Biology, Humboldt University Berlin
Invalidenstr. 43,10115 Berlin, Germany
E-mail: m.orguil@biologie.hu-berlin.de

One main characteristic of proteins is their ability to bind other molecules with high binding affinity and specificity, enabling them to realize their function. The structural and functional diversity of proteins, however, is much smaller than the enormous combinatorial diversity of amino acid sequences. One reason for this loss of complexity is that some naturally occurring amino acids have very similar physico-chemical properties. This paper discusses an empirical method to determine groups of amino acids similar with respect to their binding properties. It is founded on binding affinity data for 68 peptide-antibodies pairs, including measurements of binding strength for all peptides with a single amino acid substitution. Tho frequency with which a substitution of an amino acid by another preserves the original high binding affinity is determined, resulting in a similarity measure which is used to define a substitution matrix and group amino acids with similar binding properties. Each group can be represented by an amino acid, thus defining a reduced alphabet of maximally dissimilar amino acids. Restraining investigation of peptide sequences to this reduced alphabet can diminish the combinatorial diversity in such a way that it renders experimental assessment of binding affinity landscapes for maximally dissimilar substitutions possible. The same applies when restraining substitutions to amino acids within similarity groups. Our results suggest that a reduced set of amino acids can coarsely cover sequence space and thus be used to find antibody epitopes rapidly and economically.

1. Introduction

Proteins are the elementary blocks which execute biological functions in living organisms. There are many types of proteins in nature that carry out various complicated activities. Proteins are composed of 20 types of naturally occurring amino acids, and the majority of proteins are encoded by complex patterns of these 20 types of amino acids. That is, 20 types of amino acids introduce not only diversity and complexity into proteins, but also some specific propensities. For example, some amino acids are similar in physico-chemical properties and mutations of amino acids can be tolerated in many regions of a sequence [1]. It has been discovered experimentally that some designed proteins with fewer than 20 types of residues can have stable native structures and contain nearly as much information as natural proteins[2-4]. Recently, a 57 residue Src SH3 domain with a ß-barrel-like structure was studied [4], and 38 out of 40 targeted residues in the domain could be replaced with five types of residues (Ile, Ala, Glu, Lys, Gly = IAEGK). From a physics viewpoint, this may imply that a 20 letter alphabet can be reduced into an N letter alphabet by partition of the similar amino acids into N groups, and then N letters can be chosen as the representative residues of the groups [5, 6]. Obviously, the simplest reduction is the so-called HP model [7, 8], where 20 types of amino acids are divided into two groups: H group and P group (H, hydrophobic residues; P, polar residues). Interestingly, such a type of simple two-letter HP model or the HP-like patterns could reproduce, to some extent, the kinetics and thermodynamics of protein folding and could be used to study the mechanism of folding [2, 3]. Previously, a five-letter alphabet based on the statistical potential matrix by Miyazawa and Jernigan (MJ) (a pairwise interaction potential between amino acids [9]) was studied [5, 6]. In that reduction, the same five representative residues as above were given (IAEGK). One of the advantages of such a reduction is that it reduces greatly the complexity of the protein sequences. Some other simplified alphabets were also proposed [10-14].

All previous work relies on protein folding. There was no reduced set or a substitution matrix relying on protein ligand binding affinity until now. This is discussed in this paper based on peptide-protein interactions.

2. Results

2.1. *Creation of a substitution matrix based on binding affinity measurements*

Our data base consists of mutation analyses of 68 epitopes and their corresponding monoclonal antibodies obtained in binding experiments of the SPOT synthesis method [15] (Figure 1).

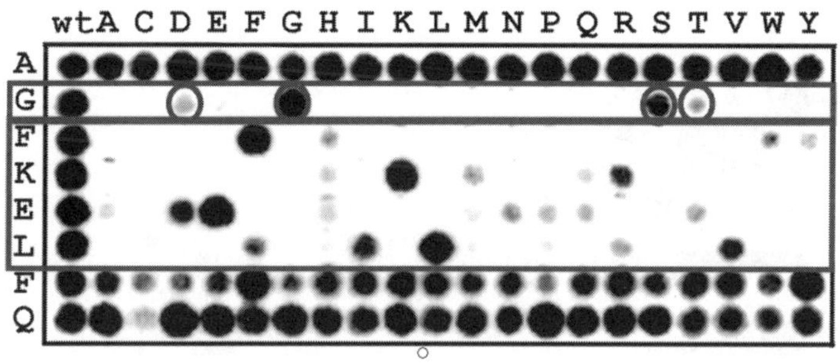

Figure 1. Mutation analysis of a peptide (AGFKELFQ) against the monoclonal antibody MiB5. The spots in the first column contain the original peptide (wildtype), which binds strongly to the antibody. In the second column, each spot contains a sequence in which the corresponding positions was substituted by an Alanine (A), and so on. Dark spots indicate strong binding affinity. Rows with more than 15 significant signals, as the first and last ones, have not been considered for further analysis, as the corresponding position probably does not contribute substantially to binding.

It was previously shown that successful binding and loss of binding between peptide and antibody can be distinguished with high significance with this method [16]. The threshold of this differentiation is about 10^{-6}M for the dissociation constant. Therefore we have binary data: binding — non-binding peptides with identical sequence except for one exchanged amino acid. We call mutations that do not affect the binding affinity of the epitope *conserving* mutations and *harmful* mutations otherwise.

Furthermore, we call a residue a *key residue* if there are harmful mutations in this position. More precisely, amino acids with at most 15 conserving mutations are key residues (15 "black spots" in a row or less in Figure 1 indicate a key residue). Obviously, a threshold of 20 would include amino acid mutations that have at most backbone effects to the binding. With an undersized threshold we would exclude mutations that cause binding damage — only very closely related amino acids would be emphasized.

Taking only these key residues into account, we determine the frequency with which an amino acid substitution leads to a change in binding behavior for all amino acid pairs. This defines, after symmetrization, the substitution matrix *AFFI* (Figure 2). Each entry in this matrix describes the probability that binding is conserved when substituting within the corresponding amino acid pair. The mean value of all entries is 0.21, meaning that approximately each fifth substitution is a conserving one.

	A	C	D	E	F	G	H	I	K	L	M	N	P	Q	R	S	T	V	W	Y
A	**1.00**																			
C	0.33	**1.00**																		
D	0.24	0.24	**1.00**																	
E	0.38	0.14	0.52	**1.00**																
F	0.21	0.04	0.17	0.22	**1.00**															
G	0.45	0.08	0.27	0.25	0.23	**1.00**														
H	0.29	0.00	0.28	0.25	0.39	0.18	**1.00**													
I	0.46	0.19	0.01	0.07	0.39	0.06	0.26	**1.00**												
K	0.23	0.13	0.02	0.09	0.11	0.13	0.50	0.21	**1.00**											
L	0.16	0.22	0.09	0.09	0.39	0.13	0.12	0.69	0.22	**1.00**										
M	0.25	0.25	0.16	0.26	0.30	0.09	0.36	0.36	0.33	0.33	**1.00**									
N	0.15	0.05	0.19	0.20	0.09	0.12	0.36	0.16	0.22	0.19	0.18	**1.00**								
P	0.45	0.13	0.11	0.17	0.16	0.16	0.27	0.22	0.16	0.16	0.29	0.21	**1.00**							
Q	0.22	0.13	0.29	0.38	0.13	0.13	0.28	0.24	0.29	0.07	0.54	0.29	0.29	**1.00**						
R	0.32	0.00	0.06	0.08	0.14	0.14	0.28	0.20	0.62	0.19	0.30	0.24	0.12	0.32	**1.00**					
S	0.64	0.35	0.25	0.29	0.09	0.34	0.15	0.22	0.13	0.17	0.27	0.15	0.27	0.32	0.22	**1.00**				
T	0.38	0.19	0.16	0.16	0.11	0.25	0.21	0.48	0.15	0.18	0.16	0.27	0.22	0.23	0.25	0.75	**1.00**			
V	0.45	0.14	0.01	0.13	0.17	0.11	0.21	0.78	0.21	0.38	0.21	0.13	0.12	0.15	0.20	0.21	0.50	**1.00**		
W	0.08	0.00	0.03	0.19	0.42	0.10	0.16	0.18	0.01	0.17	0.14	0.12	0.11	0.01	0.09	0.01	0.07	0.15	**1.00**	
Y	0.14	0.05	0.05	0.06	0.63	0.11	0.26	0.23	0.15	0.15	0.22	0.10	0.08	0.12	0.08	0.11	0.06	0.16	0.23	**1.00**

Figure 2. The matrix AFFI gives the probability with which the substitution of an amino acid in a peptide sequence will not destroy binding. Gray colored entries denote probabilities larger than 0.40, the dark ones those smaller than 0.10.

2.2. *Reliability of the substitution matrix*

An approach to measure the predictive power of a method is the receiver-operator curve (ROC-curve). We divided the data set into a test data set (25%) and a training set (75%) for cross-validation. *AFFI* was determined using the training set. A threshold between 0 and 1 was fixed. If the matrix entry was larger than the threshold, a substitution of the test data set was predicted to conserve binding affinity. The accuracy of prediction was determined for the whole test set, for different values of the threshold. The resulting ROC-curve is shown in Figure 3a. The prediction quality is measured by the area under the curve (AuC); the determined value of 0.79 implies a good prediction quality.

A comparison of this AuC-value with values obtained using other substitution matrices is shown in Figure 3b. The AuC-value of the *AFFI* matrix is the largest, followed by PAM250 (0.75), Blosum62 (0.68) and the identity matrix (0.54), showing that *AFFI* is better suitable for binding affinity predictions.

The sample sizes in the data base vary for different amino acid pairs. The precision of the matrix was determined by bootstrapping as a resampling method for each entry. Figure 3c gives an overview on the coefficient of variation for each matrix entry.

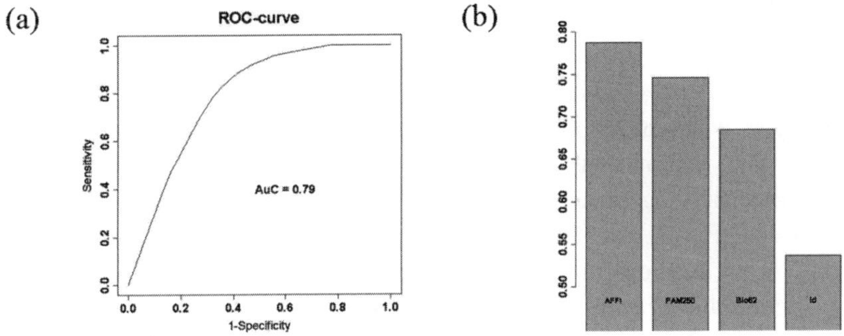

(c)

	A	C	D	E	F	G	H	I	K	L	M	N	P	Q	R	S	T	V	W	Y
A	0%																			
C	15%	0%																		
D	11%	13%	0%																	
E	10%	25%	7%	0%																
F	15%	42%	14%	11%	0%															
G	7%	35%	11%	14%	12%	0%														
H	16%	0%	9%	13%	9%	19%	0%													
I	9%	20%	0%	29%	10%	31%	21%	0%												
K	14%	21%	43%	23%	19%	15%	9%	14%	0%											
L	20%	16%	19%	22%	8%	19%	23%	5%	13%	0%										
M	21%	24%	14%	13%	13%	32%	16%	17%	11%	12%	0%									
N	18%	40%	11%	15%	20%	22%	12%	18%	12%	15%	20%	0%								
P	10%	19%	18%	14%	12%	15%	11%	14%	15%	13%	11%	13%	0%							
Q	17%	25%	10%	9%	17%	20%	13%	16%	11%	27%	10%	11%	10%	0%						
R	10%	0%	28%	22%	17%	16%	10%	15%	5%	13%	14%	11%	16%	10%	0%					
S	6%	14%	11%	12%	19%	10%	22%	16%	19%	14%	15%	18%	12%	14%	13%	0%				
T	11%	17%	13%	17%	17%	13%	19%	8%	16%	15%	23%	13%	13%	15%	12%	3%	0%			
V	7%	20%	0%	24%	14%	18%	13%	5%	13%	9%	15%	19%	21%	20%	12%	15%	6%	0%		
W	29%	0%	49%	15%	9%	24%	22%	19%	0%	14%	30%	25%	19%	0%	22%	0%	27%	18%	0%	
Y	20%	41%	25%	32%	5%	17%	13%	13%	14%	17%	14%	20%	27%	22%	18%	20%	25%	16%	13%	0%

Figure 3. (a) ROC-curve generated by cross-validation, showing the prediction quality of the substitution matrix *AFFI*. The AuC-value of 0.79 implies good prediction quality. We used the conserving mutation probability as sweeping parameter between 0 and 1 as threshold for binding. (b) Comparison of AuC values with other well known substitution matrices. (c) Coefficient of variation for each matrix entry determined via bootstrapping. Gray colored entries denote standard deviations that are larger than 25%, and the dark ones are larger than 35%. Their corresponding mutation probabilities might be less confidential.

2.3. *Similarity grouping of amino acids*

Conserving mutations occur with higher probability between similar amino acids. Our aim is to define groups of amino acids that combine similar properties optimally. We thus search for groups of amino acids such that the average probability that a mutation will conserve binding is maximized within each group. Therefore we divide the 20 amino acids into k groups and require that there have to be at least l amino acids in each group. Eq. (1) computes the average probability p of the defined cluster.

$$p = \frac{2 \sum_{i=1}^{k} \frac{\sum_{j=1}^{|C_i|} \sum_{m=j+1}^{|C_i|} p_{jm}}{|C_i|(|C_i|-1)}}{k} \tag{1}$$

Here, C denotes a cluster of amino acids and $|C_i|$ is the number of elements in C_i. p_{jm} is the probability that a mutation between the two amino acids indexed by j and m conserves binding. Firstly, this equation computes the average conserving probability within each group, and then averages over all groups (Figure 5). In the following we will call the maximal p for a cluster with k groups and at least l elements in each group $p_{\max}^{k,l}$.

The number of ways of partitioning a set of n elements into k nonempty sets is described by the Stirling numbers of the second kind $S(n,k)$ [17]. For example, for $n = 20$ and $k = 5$ we obtain $S(20,5) = 749.206.090.500$. An extension of the formula [18] considers the number of elements each group should contain at least as new parameter l. If we demand that $l = 3$, the number of possible combinations would be $S(20,5,3) = 172.096.749.825$.

Generating a random sample of about 100.000.000 partition implementations renders a distribution of the average p of all groups for $k = 5$ and $l = 3$ shown in Figure 4 Only 0.1% of the sample are partitions with $p > 0.30$.

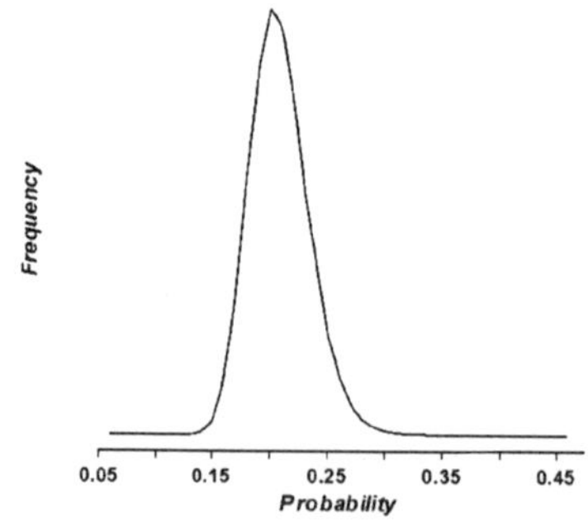

Figure 4. Distribution of the average conserving mutation probabilities within groups found in random cluster implementations. Around 100.000.000 different random clusters were realized. The clusters have a minimal group size $l = 3$ and $k = 5$ groups. Only 0.1% of the samples are clusters with an average probability larger than 0.30. The cluster found via simulated annealing, however, has an average probability of 0.46.

Applying simulated annealing using Eq. (1) as objective function, we are able to identify the optimal parameters k and l. The best performing partition according to Eq. (1) is reached for $k = 5$ and $l = 2$: $p^{5,2}_{\max} = 0.60$ (Figure 6). So it is nearly 3 times more probable to find a conserving mutation for a walk inside the same group within this partition. The optimal partition has groups that contain only few amino acids. If one wants to have larger group sizes, the parameter l has to be increased (see also Figure 5 and Figure 6).

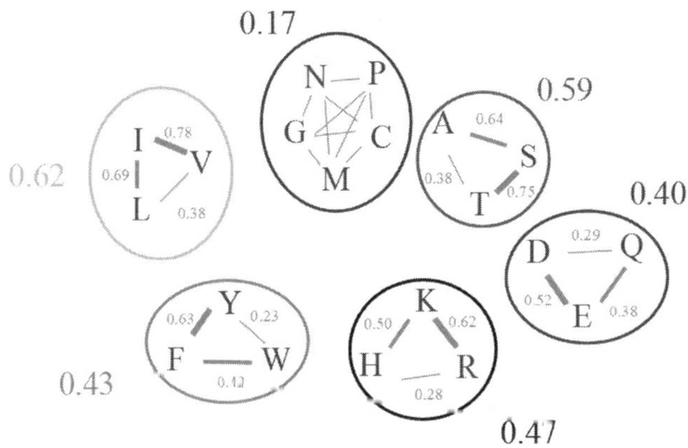

Figure 5. Grouping of amino acids obtained for the parameters $l = 3$ (minimal group size) and $k = 6$ (number of groups) using Eq. (1) as objective function. The numbers represent the averaged probabilities for a conserving binding behaviour in case of a mutation.

Each resulting group can be interpreted as a main component of sequence space. For $l = 3$ and $k = 5$ we obtain the aliphatic [ILV], hydrophobic/aromatic [FWY], basic [HKR], small/hydroxyl [AST] and the reminder [CDEGMNPQ] group.

Each group can be represented by an amino acid, yielding a reduced alphabet. The representatives chosen here have the greatest sum of probabilities to leave binding conserved compared to their group mates.

3. Discussion

We compiled a substitution matrix that is based on binding affinity only. Furthermore the entries of the matrix have a simple interpretation: they are the probabilities that a mutation conserves the binding behavior.

l	k	p-max	group1	group2	group3	group4	group5	group6	group7	group8	group9	group10	Representatives
2	2	0,49	ACDEFGHKLMNPQRSTWY	IV									[A-I]
2	3	0,58	ACDEFGHKLMNPQRWY	IV	ST								[M-I-S]
2	4	0,59	ACDEGHKLMNPQRW	IV	ST	FY							[A-I-S-Y]
2	5	0,60	ACDEGHLMNPQW	IV	ST	FY	KR						[A-I-S-Y-R]
2	6	0,58	ACDEGHLNPW	IV	ST	FY	KR	MQ					[A-I-S-Y-R-Q]
2	7	0,57	ACGHLNPW	IV	ST	FY	KR	MQ	DE				[A-I-S-Y-R-Q-E]
2	8	0,56	CHLNPW	IV	ST	FY	KR	MQ	DE	AG			[N-I-S-Y-R-Q-E-G]
2	9	0,53	CLPW	IV	ST	FY	KR	MQ	DE	AG	HN		[L-I-S-Y-R-Q-E-G-N]
2	10	0,50	CL	IV	ST	FY	KR	MQ	DE	AG	HN	PW	[L-I-S-Y-R-Q-E-G-N-W]
3	2	0,41	ACDEFGHKMNPQRSTWY	ILV									[A-I]
3	3	0,47	CDEFGHKMNPQRWY	ILV	AST								[H-I-S]
3	4	0,46	CDEFGMNPQWY	ILV	AST	HKR							[M-I-S-K]
3	5	0,46	CDEGMNPQ	ILV	AST	HKR	FWY						[Q-I-S-K-F]
3	6	0,45	CGMNP	ILV	AST	HKR	FWY	DEQ					[G-I-S-K-F-E]
4	2	0,35	ACDEGHKMNPQRSTWY	ILTV									[A-I]
4	3	0,38	CDEHKMNPQRWY	ILTV	AGST								[M-I-S]
4	4	0,38	ACDEGNPS	ILTV	FHWY	KMQR							[A-I-F-K]
4	5	0,38	HKNR	ILTV	YFWG	DEMQ	ACPS						[K-I-F-E-A]
5	2	0,34	CDEFGHKLMNPQRWY	AISTV									[M-I]
5	3	0,35	CDEFGLNPWY	AISTV	HKMQR								[F-I-K]
5	4	0,34	CDEMQ	AGSTV	HKNPR	FILWY							[Q-S-K-F]
6	2	0,31	ACDEGHKMNPQRST	AILSTV									[H-I]
6	3	0,32	ACDEGPST	FILVWY	HKMNQR								[S-I-K]

Figure 6. Amino acid groups and their representatives estimated via simulated annealing for several group sizes (l) and numbers of groups (k) according to Eq. (1).

The reduced sets of amino acids and also the clusters presented here are similar to previous findings [5, 6, 10-14]. Note, that the data base used here consists solely of binary measurements of peptide binding affinities. In contrast to data used in the previous works, protein function, protein structure, and overall amino acid frequency do not play a role here.

The reduced set of amino acids can be used to capture properties of peptide affinity landscapes, as demanded by Kaufmann [19]. For experimental sequence space search the combinatorial diversity of peptides is too large. Even a peptide with length seven — which covers most of the investigated epitopes (Dong, Kramer, unpublished) — has more than 10^9 possible amino acid combinations. By using a reduced set of only five or six amino acids the size of the sequence space would decrease tremendously to less than 10^6. Further investigation of this space could be simplified by allowing substitutions only within the defined similarity groups — also multiple substitutions at once are feasible with a good chance of preserving binding affinity.

Altogether this offers the opportunity to measure highly correlated (allowing mutations only within the groups) and less correlated (allowing representatives only) affinity landscapes. It might give a nearly complete overview of the landscapes, although the correlations are separately measured.

Acknowledgment

This work was in part supported by the Deutsche Forschungsgemeinschaft (SFB 618 and SFB 449). AAW and MOG thank the Volkswagen Foundation for funding. We thank L. Dong and A. Kramer (Institute of Medical Immunology, Charité, Berlin, Germany) for providing the mutation analyses of the epitopes.

References

1. Sinha, N. and R. Nussinov, *Point mutations and sequence variability in proteins. Redistributions of preexisting populations.* Proceedings of the National Academy of Sciences of the United States of America, 2001. **98**(6): p. 3139-3144.
2. Davidson, A.R., K.J. Lumb, and R.T. Sauer, *Cooperatively Folded Proteins in Random Sequence Libraries.* Nature Structural Biology, 1995. **2**(10): p. 856-864.
3. Regan, L. and W.F. Degrado, *Characterization of a Helical Protein Designed from 1St Principles.* Science, 1988. **241**(4868): p. 976-978.
4. Riddle, D.S., et al., *Functional rapidly folding proteins from simplified amino acid sequences.* Nature Structural Biology, 1997. **4**(10): p. 805-809.
5. Chan, H.S., *Folding alphabets.* Nature Structural Biology, 1999. **6**(11): p. 994-996.
6. Wang, J. and W. Wang, *A computational approach to simplifying the protein folding alphabet.* Nature Structural Biology, 1999. **6**(11): p. 1033-1038.
7. Chan, H.S. and K.A. Dill, *Intrachain Loops in Polymers - Effects of Excluded Volume.* Journal of Chemical Physics, 1989. **90**(1): p. 492-509.
8. Lau, K.F. and K.A. Dill, *A Lattice Statistical-Mechanics Model of the Conformational and Sequence-Spaces of Proteins.* Macromolecules, 1989. **22**(10): p. 3986-3997.
9. Miyazawa, S. and R.L. Jernigan, *Residue-residue potentials with a favorable contact pair term and an unfavorable high packing density term, for simulation and threading.* Journal of Molecular Biology, 1996. **256**(3): p. 623-644.
10. Cieplak, M., et al., *Amino acid classes and the protein folding problem.* Journal of Chemical Physics, 2001. **114**(3): p. 1420-1423.
11. Li, T.P., et al., *Reduction of protein sequence complexity by residue grouping.* Protein Engineering, 2003. **16**(5): p. 323-330.
12. Liu, X., et al., *Simplified amino acid alphabets based on deviation of conditional probability from random background.* Physical Review E, 2002. **66**(2): p. -.
13. Murphy, L.R., A. Wallqvist, and R.M. Levy, *Simplified amino acid alphabets for protein fold recognition and implications for folding.* Protein Engineering, 2000. **13**(3): p. 149-152.
14. Solis, A.D. and S. Rackovsky, *Optimized representations and maximal information in proteins.* Proteins-Structure Function and Genetics, 2000. **38**(2): p. 149-164.
15. Frank, R., *The SPOT-synthesis technique. Synthetic peptide arrays on membrane supports–principles and applications.* J Immunol Methods, 2002. **267**(1): p. 13-26.

16. Weiser, A.A., et al., *SPOT synthesis: Reliability of array-based measurement of peptide binding affinity.* Analytical Biochemistry, 2005. **342**(2): p. 300-311.
17. Anderson, I., *A First Course in Combinatorial Mathematics.* Oxford University Press, 1974.
18. Cernuschi-Frías, B., *A Combinatorial Generalization of the Stirling Numbers of the Second Kind.* Proceedings of the 8th IEEE International Conference on Electronics, Circuits and Systems (ICECS), 2001. **2**: p. 593-596.
19. Kauffman, S.A. and W.G. Macready, *Search strategies for applied molecular evolution.* J Theor Biol, 1995. **173**(4): p. 427-40.

MULTI-OBJECTIVE EVOLUTIONARY APPROACH TO *AB INITIO* PROTEIN TERTIARY STRUCTURE PREDICTION

TELMA W. DE LIMA, PAULO GABRIEL, ALEXANDRE DELBEM
Institute of Mathematic and Computer Sciences,
400 Trabalhador Sao Carlense Avenue, Sao Carlos, Brazil
E-mail: telma@icmc.usp.br, phrg@grad.icmc.usp.br, acbd@icmc.usp.br

RODRIGO FACCIOLI, IVAN SILVA
Sao Carlos Engineering School,
400 Trabalhador Sao Carlense Avenue, Sao Carlos, Brazil
E-mail: faccioli@sel.eesc.usp.br, insilva@sel.eesc.usp.br

The Protein Structure Prediction (PSP) problem aims at determining protein tertiary structure from its amino acids sequence. PSP is a computationally open problem. Several methodologies have been investigated to solve it. Two main strategies have been employed to work with PSP: homology and *ab initio* prediction. This paper presents a Multi-Objective Evolutionary Algorithm (MOEA) to PSP problem using an Ab initio approach. The proposed MOEA uses dihedral angles and main angles of the lateral chains to model a protein structure. This article investigates advantages of multi-objective evolutionary approach and discusses about methods and other approaches to the PSP problem.

1. Introduction

The post-genomic era have been characterized by two different scenarios: on one hand, the huge amount of available biological data sets all over the world requires suitable tools and methods both for modelling biological process and analyzing biological sequences: on the other hand, many new computational models and paradigms inspired and developed as metaphors of biological system are ready to be applied to the context of computer science[16].

The interest in discovering a methodology for solving the Protein Structure Prediction (PSP) extends into many fields of research, including biochemistry, medicine, biology, engineering and scientific disciplines. Native protein structures have been determined using x-ray crystallography and magnetic nuclear resonance methods[4]. The latter has its application restricted to proteins with small size, while the former needs a great amount of laboratory processing requiring high cost. On the other hand, approaches for PSP range from empirical researches to mathematical modelling for protein potential energy.

One strategy to solve PSP uses information from protein homology to guide the search process. Despite of relevant results obtained using such strategies, algorithms based on protein homology are highly dependent on the set of proteins with native known structure. This set is extremely smaller than the universe of proteins.

On the other hand, Ab initio PSP does not depend on previous knowledged of protein structures. This problem is one of the most important unsolved problems in molecular biophysics[2]. At first glance, this problem may seem not complex since knowing the exact formulation of the physical environment within a cell, where proteins fold, it is possible to mimic the folding process in nature by computing the molecular dynamics based on our knowledge about physical laws[15]. Nevertheless, we do not completely understand the driving forces involved in protein folding. Perturbations in the potential energy landscape may result in a different folding pathway, generating a different folded structure.

Due to insufficient results obtained for the tertiary protein structure from amino acid sequences, many different computational algorithms have been investigated. Among these algorithms, Evolutionary Algorithms (EA) have presented relevant results[11,12,9]. EAs are powerful tools of optimization inspired in natural evolution and they has been applied to many complex problems in the most different areas of human knowledge[17].

Development of effective computational tools for PSP is fundamental in order to deal with PSP computational complexity. Moreover, the development of these tools may guide part of the future scientific effort in molecular biology. Nevertheless, there has been no computational breakthrough for accurately forecasting the final folded state of a protein. The growth of interest in "non conventional computer algorithms" applied to biological domains relies on the so named "data revolution". Among these algorithms, Evolutionary Algorithms (EA) has presented relevant results[29,10].

2. Evolutionary computation and evolutionary algorithms

Evolutionary computation research algorithms are inspired in the Evolutionary Theory[18]. Genetic algorithms (GAs) is the most known EAs. The approach proposed in this paper is based on GAs.

GAs are search techniques based on the mechanism of natural selection[18], which simulate the processes of nature evolution aiming at solving optimization problems. Genetic algorithms are robust and efficient for optimization problems with a very large search space. Another important characteristic of the GAs is that they provide a set of solutions, named population, instead of not single solution. A GA is an optimization algorithm capable of solving single objective or multiple objective problems. Multi-Objective Optimization Problems (MOOP) have a solution set that represents a trade off among the objectives[27]. Next section discusses multi-objective EAs.

2.1. *Multiobjective EAs*

Several aspects can be relevant in order to evaluate a solution for a problem. If no aspect can be measurable better than other, a solution will dominate the other only if it is better for all aspects. The entire set of solutions that are not dominated can be represented by a curve in the Cartesian space, named Pareto Front[27].

The first implementation of a Multi-Objective Evolutive Algorithm (MOEA) was proposed by Schaffer in 1985[24]. This implementation was a modification from the conventional GA for the purpose of estimating each objective independently. However, this approach can not obtain an adequate diversity in Pareto Fronts solution.

The main difference between MOEAs and traditional AEs is the selection operator since the comparison among these solutions must be performed according to Pareto's dominance. The most efficient algorithms are SPEA[14], PAES[14] and NSGA-II. The proposed approach for PSP employs NSGA-II (see Section 2.2).

2.2. *NSGA-II*

The basic idea behind NSGA-II is the ranking process performed before the selection operation. This process identifies non-dominated solutions in the population (P_i) to compose non-dominated fronts (Figure 1) approximating the Pareto Front. Afterward usual GA operators (selection, crossover, mutation) are applied to generate new solutions (offsprings - Q_i). P_i and Q_i are grouped and named R_i. Then, R_i is ranked by a non-dominated ranking procedure[14].

In the ranking procedure, the non-dominated individuals in the current population are first identified. Then, these individuals are assumed to

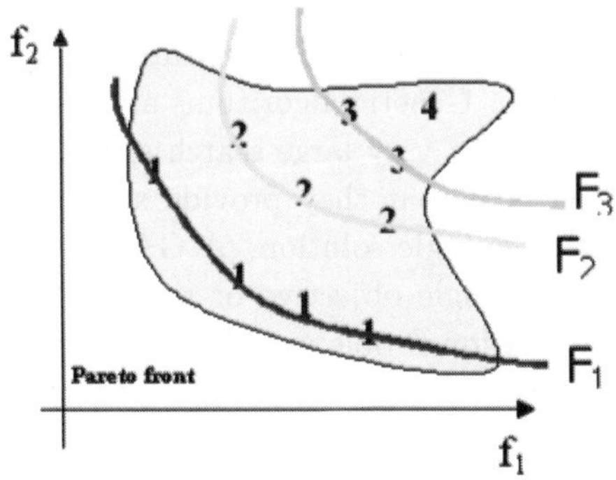

Figure 1. Illustration of Pareto Fronts (F1, F2 and F3) generation process for minimization of objectives f1 and f2.

constitute the first non-dominated front (F_1 from Figure 1) with a small dummy fitness value[25]. Afterwards, the individuals of the first front are ignored and the resultant population is processed in the same way in order to identify individuals for the second non-dominated front (F_2 from Figure 1). This process continues until the whole population is classified into non-dominated fronts. A new population (P_{i+1}) is obtained with the N first non-dominated individuals of R_i; the remaining of the individuals are rejected. This process is represented in Figure 2[14]. The individuals in the first front represent the Pareto Front solution.

The NSGA was first proposed with a stochastic remained proportional selection (SRS) procedure. However, it is possible to use any other selection technique as roulette wheel or tournament[18]. As individuals in the first front have smaller fitness value, they always get more copies than the remaining individuals. This process emphases exploration of non-dominated regions of the search space[25].

3. Protein Tertiary Structure Prediction

Proteins are macro molecules built from 20 basic units, named amino acids. All amino acids have a same generic chemical structure. There is a central carbon atom (Cα) angle attached to an hydrogen atom, an amino group (NH_2), a carboxyl group (COOH) and a lateral chain (R), which distinguishes one amino acid from the others. Every residue is assigned a 3-letter or an 1-letter code. During DNA transcription-translation phases, proteins are composed from peptide bonds, where the carboxyl group of one amino acid is joined with the amino group of another to release water. In this

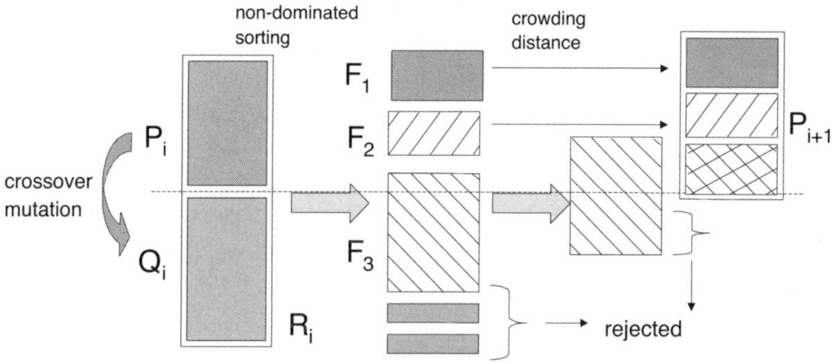

Figure 2. NSGA-II Design.

way, we can talk of a protein as a polypeptide formed by a backbone (the sequence of peptide bonds) and a side chain (the sequence of residues)[29].

The structure of a protein is hierarchically represented with three structural description levels. The primary structure is the sequence of amino acids in the polypeptide chain, which can be described as a string from a finite alphabet. The secondary structure refers to the set of local conformation motifs of the protein and schematizes the path followed by backbone in the space. The most important description level and main objective of experimental and prediction efforts is to obtain protein tertiary structure. It describes the three-dimensional organization of polypeptide chain atoms (both backbone and side chain atoms).

The formation process of tertiary structure is named folding. There are some physical properties that define this process[20] (see Section 3.1):

- Hardness of the backbone of the sequence;
- Interactions between amino acids, including electrostatic interactions;
- van der Waals forces;
- Volume constraints;
- Hydrogen or disulfate bounds;
- Interactions of amino acids with water.

Due to difficulty of understanding the protein folding process, this problem has been modelled as an optimization process. Different computational strategies have been investigated to solve this problem: homology[4], threading[4] and *Ab initio*[5] and Semi *Ab initio*[19]. We will focus on the *Ab initio* approach (see Section 3.2).

The dihedral angles ϕ and ψ determine the protein fold. Unfortunately, there is a large number of free degrees, i.e. ϕ and ψ, making the *Ab initio* approach computationally very complex.

3.1. *The folding problem*

Protein chains are subject to a folding process. Starting from the primary structure, they are capable of organize themselves into an unique three-dimensional stable (native) conformation which is responsible of their biological functions[29]. The task of searching through all the possible conformations of a polypeptide chain to find those with low energy is very complex. It requires enormous amounts of computing time. Moreover, the energy difference between a stable folded molecule and its unfolded state is very small[7].

The folding problem is one of the most difficult challenging open problem in structural genomic. The number of proteins for which the sequences are known is about a half million[1]. On the other hand, the Protein Data Bank (PDB) has about 40 thousands[16]. Excluding experimental difficulties, the reason for this impressive difference is largely due to a lack of a comprehensive theory of the folding process[29].

3.2. *Ab initio*

The *Ab initio* structure prediction aims at predicting a protein structure from its amino acid sequence. It is generally assumed that a protein sequence folds to a native conformation or an ensemble of conformations that is near the global free-energy-minimum[5].

In *Ab initio* approach, no homology between proteins are employed. *Ab initio* prediction is more challenging than homology modelling or threading. Moreover, it is the only way to derive a prediction, when no similar test fold is known.

4. Proposed Approach

An important task when proposed a search procedure for the PSP is defined good representation of the conformations and cost function for evaluating conformations. These aspects are discussed in the sequel.

4.1. *Representation of the conformation*

Few conformation-representations are commonly used:

(1) all-atom three-dimensional coordinates;
(2) all-heavy-atom coordinates;
(3) backbone atom coordinates and side-chain centroid;
(4) C_α coordinates;
(5) backbone and side-chain torsion angles and
(6) lattice models.

In our approach we choice the backbone and side-chain torsion angles representation, based on the fact that each residue type requires a fixed number of torsion angles to determine the three-dimensional coordinates for all atoms. The bond lengths and angles are considered at their ideal values. The dihedral angle ω is fixed at 180°. Thus, in order to represent a solution (chromosome) we need two torsion angles of the backbone (ϕ and ψ) and the side-chain torsion angles (χ_i $i = 0, \ldots, 4$) depending on the each residue type.

4.2. *Cost Function*

In order to evaluate the molecule structure is needed to use some cost or energy function. Quantum mechanics produces the most adequate energy functions. However, they are too computationally complex to be employed in modelling larger systems. Thus, the proposed approach uses energy function obtained from classical physics. Named potential energy functions or force fields, these functions return a energy value based on the molecule conformation. They provide information on molecule conformations are better or worse. The lower energy value indicates the better conformation.

The most typical potential energy functions have the form[8]:

$$\text{Energy}(R) = \sum_{\text{bonds}} B(R) + \sum_{\text{angles}} A(R) + \sum_{\text{torsions}} T(R) + \sum_{\text{non-bonded}} N(R) \quad (1)$$

where R is the vector representing the molecule conformation, typically in cartesian coordinates or torsion angles.

The literature on cost functions is enormous[8,21]. The proposed approach, in order to evaluate the protein conformation we use the TINKER (Software Tools for Molecular Design) energy functions and the CHARMM (Chemistry at HARvard Macromolecular Mechanics) parameters v.27. It is a composite sum of several molecular mechanics functions that can be grouped in two major types: bonded (stretching, bending, torsion, Urey-Bradley and improper) and non-bonded (van der Waals and electrostatic).

The Tinker energy functions has the form:

$$
\begin{aligned}
E_{\text{TINKER}} = & \sum_{\text{bonds}} k_r(r-r_0)^2 + \sum_{\text{angles}} k_\theta(\theta-\theta_0)^2 + \sum_{\text{UB}} k_{\text{urey}}(s-s_0)^2 \\
& + \sum_{\text{torsions}} \sum_n V_n(1+\cos(n\phi-\gamma_n)) + \sum_{\text{impropers}} k_{\text{improper}}(\omega-\omega_0)^2 \\
& + \sum_{\text{non-bond}} \varepsilon_{i,j}\left[\left(\frac{R_i+R_j}{r_{i,j}}\right)^{12} - 2*\left(\frac{R_i+R_j}{r_{i,j}}\right)^{6}\right] + \frac{q_i q_j}{Dr_{i,j}}
\end{aligned}
$$

where

(*i*) r is the bond length, r_0 is the bond length equilibrium and k_r is the bond energy constant;
(*ii*) θ is the bond angle, θ_0 is the bond angle equilibrium and k_θ is the valence angle energy constant;
(*iii*) s is the distance between two atoms separated by two covalent bonds (1-3 distance), s_0 is the equilibrium distance and $kurey$ is the Urey-Bradley energy constant;
(*iv*) ϕ is the dihedral or torsion angle, V_n is the dihedral energy constant, n is the multiplicity and γ is the phase angle;
(*v*) ω is the improper angle, ω_0 is the equilibrium improper angle and k_{improper} is the improper energy constant;
(*vi*) $\varepsilon_{i,j}$ is the Leonard-Jones well depth, $r_{i,j}$ is the distance between atoms i and j, R_i is the van der Waals atom i radius, R_j is the van der Waals atom j radius, q_i and q_j are the partial atomic charges from atom i and j and D is the dielectric constant.

4.3. *Multi-Objective Formulation*

In order to reduce the size of the conformational space the backbone torsion angles are constrained in regions derived from the CADB-2.0[22] database, that contains the most torsion angles to each residue. The side-chain torsion angles are constrained in regions derived from the the backbone-independent rotamer library of Tuffery[28]. Side-chain constraint regions are of the form: $[m-\sigma, m+\sigma]$; where m and σ are the mean and the standard deviation for each side-chain torsion angle computed from the rotamer library. Under these constraints, the conformation is still highly flexible and the structure can take on various shapes that are vastly different from the native shape.

The proteins can be seem as a collection of atoms linked by a chemical bond. The Tinker energy functions are used to evaluate the protein conformation and the atoms can be divided into bond and non-bonded groups. The bond group represents the local interaction and considers all atoms chains of max length four. The non-bond group represents the non-local interaction and considered all atoms separated by at least three or more covalent bonds. This division reflects the function energy decomposition in two partial sums: bonded and non-bonded atom energies. This is the most used decomposition of the function cost in two objectives[9]. The proposed approach uses a different decomposition:

$$\text{Energy}_1 = E_{\text{angle}} + E_{\text{bond}} + E_{\text{dihe}} + E_{\text{impr}} \tag{2}$$

$$\text{Energy}_2 = E_{\text{vdw}} \tag{3}$$

$$\text{Energy}_3 = E_{\text{elec}} \tag{4}$$

The first equation grouped the potential energies bonded, relative to the bonded atoms. The second equation is relative to the van der Waals interactions and the last equation is relative the electrostatic interactions between non-bonded atoms. This objectives are relatives to different interactions among the atoms, so is more interesting minimize them separately. These three functions represent our minimization objectives, the torsion angles of the protein are the decision variables of the multi-objective problem and constraint regions are the variable bounds.

4.4. *MOEA Proposed*

The proposed MOEA is based in NSGA-II. The algorithm starts by initializing a random conformation. The torsion angles (ϕ, ψ, χ_i) are generated at random from the constrained regions. Afterward, the energy of the conformation is evaluated. First, the protein structure in internal coordinates (backbone and side-chain torsion angles) is transformed in Cartesian coordinates. Then, the energy potential is calculated using the Tinker routines.

At this point, we have the main loop of the algorithm. From the current solutions new solutions are obtained using genetic operators. We proposed three kinds of recombination operators. The first operator is the BLX-α operator especially development to floating point representation[14]. The second operator uses the uniform crossover. The last operator is two-point crossover.

Three kinds of mutation operators were proposed. When the first mutation operator acts on a peptide chain, all the values of the backbone and side-chain torsion angles of a residue chosen at random are re-selected from their corresponding constrained regions. The second and the third mutation operators applied a uniform mutation. These operators modifies all the values of the backbone and side-chain torsion angles of a selected residue are perturbed with a uniform distribution. The difference between the second and the third operators is in the uniform distribution. For the second operator, the uniform distribution is between 0 and 1. For the third operator the interval is between 0 and 0.1. The remains steps of the proposed approach are the same of the standard NSGA-II.

5. Results

This section reports the results obtained using the multi-objective proposed approach for PSP. This algorithm was applied to four protein sequences from the Protein Data Bank (PDB): 1ZDD, 1ROP, 1CRN and 1UTG. The population size is 200 chromosomes and the maximum number of generations is 1000.

Table 1. Minimum potential energy of Pareto front and DME, with dielectric constant $D = 4.0$, the maximum distance between atom i and atom j to compute van der Waals interaction 8Å and electrostatic interaction 13Å

Protein (PDB Id)	Minimum Energy (C)	DME(Å)
1*CRN*	974.6351	5.7
1*ROP*	1246.7000	4.1
1*UTG*	1888.1380	4.9
1*ZDD*	754.0605	5.2

The cost functions have dielectric constant equal 4.0. In order to compute van der Waals and electrostatic energies were defined a maximum and a minimum distance (d) between atom i and atom j were defined. Only the atoms-interactions in this interval ($d_{min} \leq d \leq d_{max}$) are calculated. When the distance (d) is smaller than the minimum distance, the energy is computed using $d = d_{min}$.

Table 1 presents the lowest energy of each protein tested and their distance matrix error DME.

Disulphide-stabilized mini protein A domain (1ZDD): 1ZDD is a two-helix peptide of 34 residues[26]. Inspecting the Pareto front of the 1ZDD protein (Figure 3) we can note that when minimizing one of the objectives the others are maximized, this is more evident in the Figures 4, 5 and 6 where the colors represent the values of the other objective do not present in the graphic, with dark colors referent to minimum values and light colors referent to maximum values of the other objective. The Figure 7 shown the 1ZDD native structure and the Figure 8 the 1ZDD predict structure by our approach, this shown one α-helice turn and a trend to form other turns.

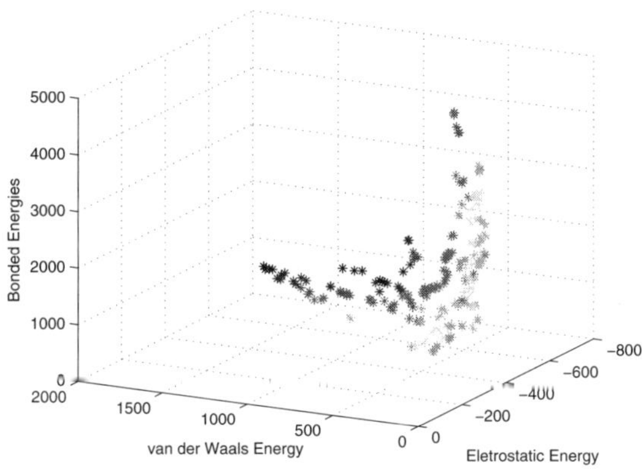

Figure 3. Pareto front for 1ZDD protein.

Repressor of primer (1ROP): is a four-helix bundle protein that is composed of two identical monomers[3]. Each monomer has 56 residues and forms α -turn- α structure (PDB id. 1ROP). The Figure 9 shown the Pareto front obtained, where the colors indicates the values of electrostatic energy, with dark colors represent minimum values and light colors the maximum values obtained for this objective. Inspecting the Pareto front we can note that the region of low values of bonded and van der Waals energies has relatively high electrostatic energy. This conflicting values show that the use of three objectives is adequate. Figure 10 presents the 1ROP native structure and Figure 11 shows the predict 1ROP structure with the lowest energy from the Pareto front, showing α-helix folding.

Uteroglobin (1UTG): is a 4-helix protein that has 70 residues[23]. Fig-

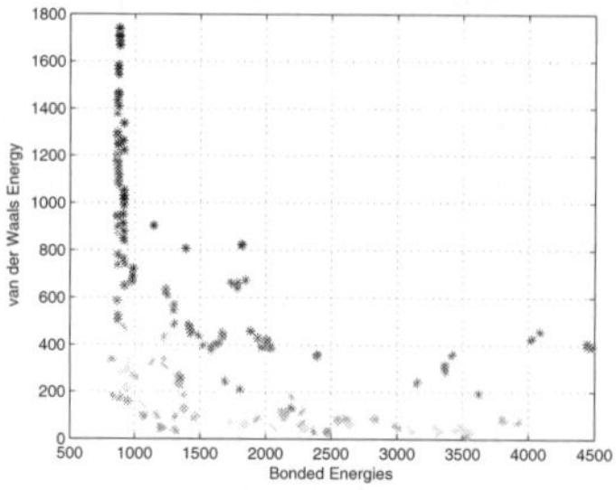

Figure 4. Pareto Front 1ZDD (Bonded x van der Waals).

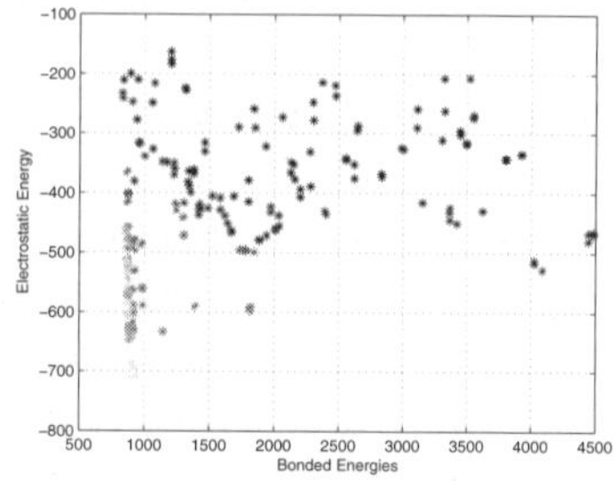

Figure 5. Pareto Front 1ZDD (Bonded x Electrostatic).

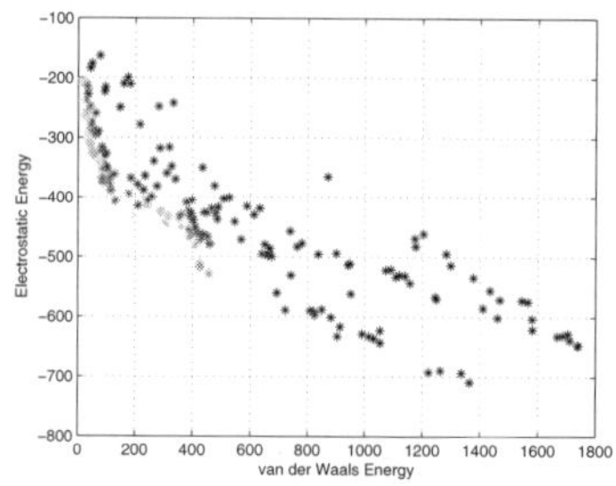

Figure 6. Pareto Front 1ZDD (van der Waals x Elestrostatic).

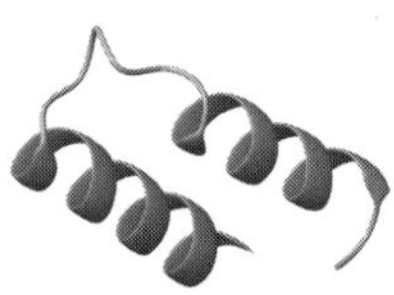

Figure 7. 1ZDD Native Structure.

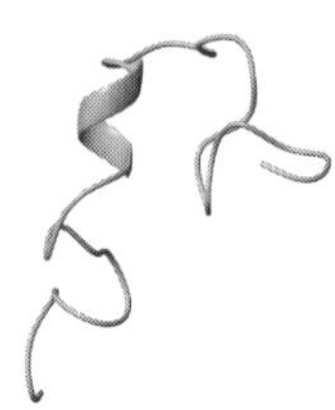

Figure 8. 1ZDD Predict Structure.

ure 12 presents the obtained Pareto front, which is similar to the front from Figure 9. Different colors correspond to different electrostatic values. Figures 13 and 14 present the native and predict 1UTG structures respectively. The obtained structure presents folding of two helices.

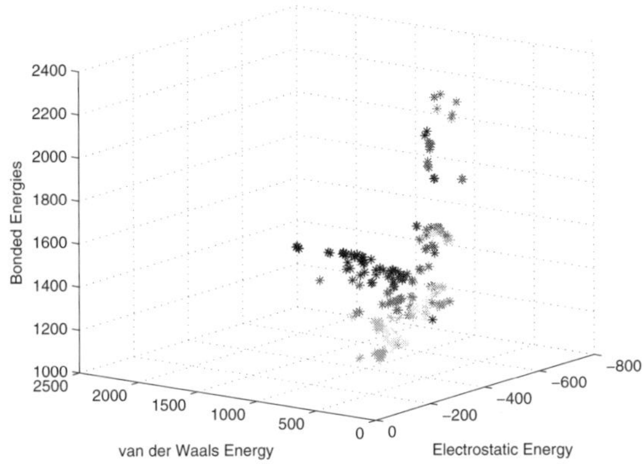

Figure 9. Pareto front for 1ROP protein.

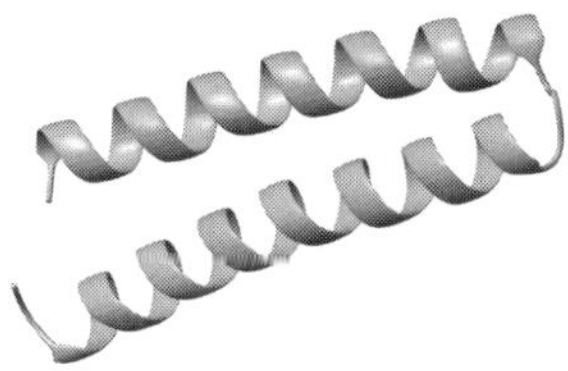

Figure 10. 1ROP Native Structure.

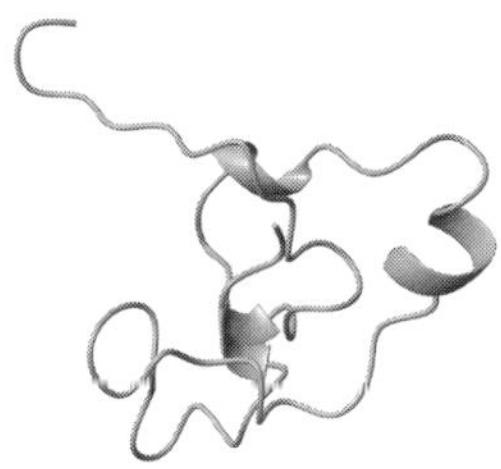

Figure 11. 1ROP Predict Structure.

Crambin (1CRN): is a 46-residue protein with two α-helix and a pair of β-strands[30]. It has three disulphide bonds. Figure 15 shows the Pareto front obtained by the proposed approach. Different colors correspond to different electrostatic energy values. Figure 16 presents the 1CRN native structure and Figure 17 shows the obtained structure, illustrating the initial folding of an α-helix and a β sheet.

5.1. *Comparisons with other Ab initio approaches*

Approaches available in the literature for PSP problem[9,6] have better results than the proposed approach for the same proteins evaluated in this paper. Our proposal is a pure Ab initio algorithm that do not use any

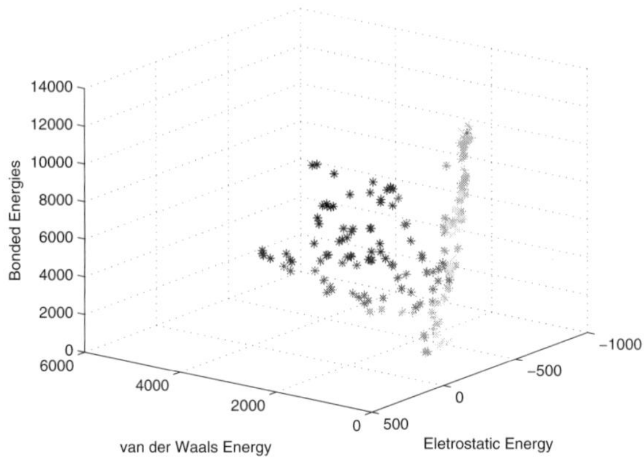

Figure 12. Pareto front for 1UTG protein.

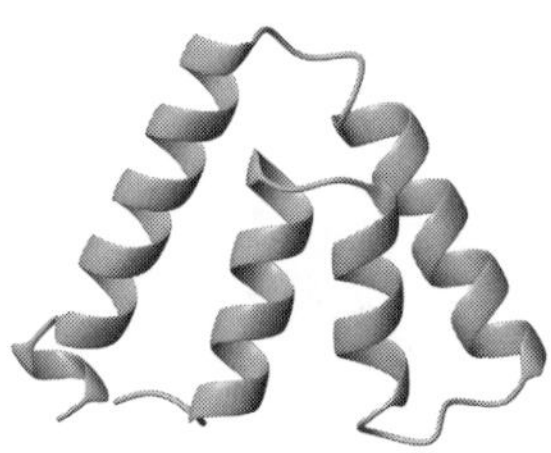

Figure 13. 1UTG Native Structure.

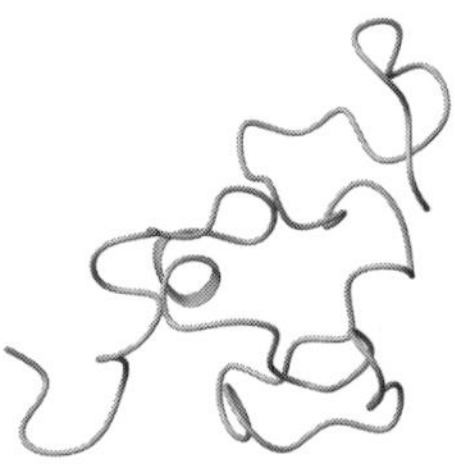

Figure 14. 1UTG Predict Structure.

heuristic in the prediction process, in contrast with Ab initio approaches like Rosetta[6] and the algorithms presented in[9], that have present relevant results to the PSP, using several heuristics and information from the protein structure databases, characterizing a semi Ab initio algorithm.

Another innovation of the proposed MOEA for PSP is the modelling using three objectives for the first time in the literature. The employment of a pure Ab initio strategy and the use of three objectives make inadequate comparisons of the results obtained by the proposed approach to results produced by other approaches.

It is important to highlight that, although a pure Ab initio approach has shown inferior results, the proposed approach can explore any region of

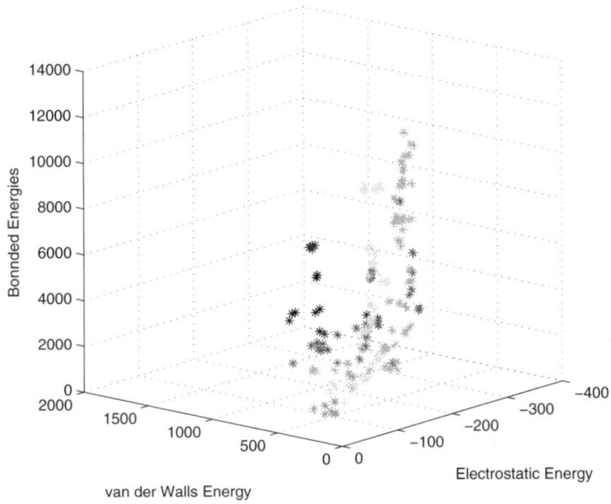

Figure 15. Pareto front of 1CRN protein.

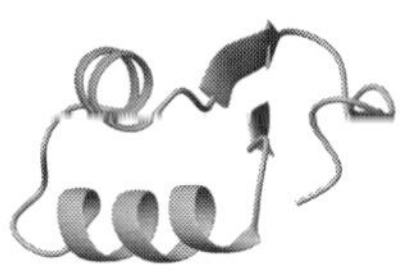

Figure 16. 1CRN Native Structure.

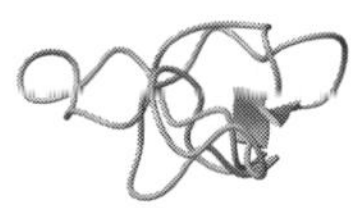

Figure 17. 1CRN Predict Structure.

the search space in order to find an adequate structure conformation even though there is no similar protein structure preciously known.

6. Conclusions

This paper has presented a multi-objective evolutionary approach for PSP. Proposed algorithm evaluates solutions obtained using three objectives: bonded energies (bending, angle, torsion, Urey-Bradley and improper), van der Waals energy and electrostatic energy. The energies arranged for each objective were grouped according to the classes of interactions among the atoms.

Results found by the proposed are not the best it can reach, since pure Ab initio model requires a lot of parameters, requiring a large period of time researching the refinement of the model. A example of modelling of the amino acid proline whose side-chain torsion angles have to be minimized for each protein since this amino acid has no side chain predefined conformation angles. It is important to highlight that inadequate conformations for proline can break an α-helix generating wrong predictions.

The NSGA-II algorithm is not the best GA to work with more than two objectives. Recently, a new GA approach was proposed for lead with more then 2 objectives[13]. The use of such approach for PSP should also improve the performance of the proposed approach.

Acknowledgment

TL has been supported by CNPq. PG, AD, and IS have been supported by FAPESP.

References

1. Bairoch, A. and Apweiler, R., The SWISS-PROT protein sequence database and its supplement TrEMBL Nucleic Acids Res,2000.
2. Baker, D. and Sali., A., Protein Structure Prediction and Structural Genomics, Science, 294:93, 2001.
3. Banner, D.W., Kokkinidis, M. and Tsernoglou, D., Structure of the ColE1 rop protein at 1.7 Å resolution. 1987 J. Mol. Biol. 196, 657.
4. Baxevanis, A. and Oullette, B., Bioinformatics - A pratical guide to the analysis of genes and proteins, 2001.
5. Bonneau, R. and Backer, D., Ab Initio Protein Structure Prediction: Progress and Prospects, Rev. Biophys Biomol Struct 2001, 2001.
6. Bonneau R, Tsai J, Ruczinski I, Chivian D, Rohl C, Strauss CEM, Baker D., Rosetta in casp4: progress in ab initio protein structure prediction. Proteins 2001; Suppl 5:119 –126.
7. Branden, C. and Tooze, J., Introduction to Protein Structure, Second Edition, 1999
8. Cornell, W. D. et al. 1995 A second generation force field for the simulation of proteins, nucleic acids, and organic molecules. J. Am. Chem. Soc. 117, 51795197.
9. Cutello, V. and Narzisi, G. and G. Nicosia, A Multi-objective Evolutionary Approach to the Proteins Structure Prediction Problem, Journal of The Royal Society Interface, 2005
10. Dandekar, T. and Konig, R., Computational methods for the prediction of protein folds, 1997.

11. Day, R. and Zydallis, J. and Lamont, G., "Solving the Protein Structure Prediction Problem Through a Multiobjective Genetic ALgorithm," Technical Proceedings of the 2002 International Conference on Computational Nanoscience and Nanotechnology, 2002
12. Day, R. O. and Lamont, G. B. and Pachter, R., Protein Structure Prediction by Applying an Evolutionary Algorithm, ipdps, p. 155a, International Parallel and Distributed Processing Symposium (IPDPS'03), 2003.
13. Deb, K. and Sundar, J. 2006 Reference point based multi-objective optimization using evolutionary algorithms. GECCO '06: Proceedings of the 8th annual conference on Genetic and evolutionary computation. 635–642.
14. Deb, K., Multi-objective Optimization Using Evolutionary Algorithms, John Wiley and Sons, Baffins Lane, Chichester, England, 2001.
15. Duan, Y. and Kollman, P. A., "Pathways to a Protein Folding Intermediate observed in a 1-microsecond simulation in aqueous solution," Science, 282:740, 1998.
16. Ezziane, Z. Applications of artificial intelligence in bioinformatics: A review, Elsevier, 1-10, 2006.
17. Fogel, D., An introduction to simulated evolutionary computation, 1994.
18. Goldberg, D. E., Genetic Algorithms in Search, Optimization and Machine Learning Reading, MA: Addison Wesley, 1989.
19. Inbar, Y., Benyamini, H., Nussinov, R. and Wolfson, H.J., Protein structure prediction via combinatorial assembly of sub-structural units, Bioinformatics, 158-168, 2003.
20. Karplus, M. and Shakhnovich, E. , Protein Folding, chapter protein Folding: Theorical Studies of Thermodynamics and Dynamics, 1992.
21. Momany, F. A., McGuire, R. F., Burgess, A. W. and Scheraga, H. A. 1975 Energy parameters in polypeptides VII, geometric parameters, partial charges, non-bonded interactions, hydrogen bond interactions and intrinsic torsional potentials for naturally occurring amino acids.
22. Mohan,K.S., Sheik,S.S., Ramesh,J., Balamurugan,B., Jeyasimhan,M., Mayilarasi,C. and Sekar,K., Conformational Angle DataBase - CADB 2.0, 2005, http://cluster.physics.iisc.ernet.in/cadb/.
23. Morize, I., Surcouf, E., Vaney, M. C., Epelboin, Y., Buehner, M., Fridlansky, F., Milgrom, E. and Mornon, J. P. Refinement of the C222(1) crystal form of oxidized uteroglobin at 1.34 Å resolution. 1987 J. Mol. Biol. 194, 725.
24. Schaffer, J. D., Multiple Objective Optimization with Vector Evaluated Genetic Algorithms, in Genetic Algorithms and their Applications: Proceedings of the First International Conference on Genetic Algorithms, pp. 93-100, 1985.
25. Srinivas, N. and Deb, K., Multiobjective optimization using non-dominated sorting in genetic algorithms, Tech. Rep., Dept. Mechanical Engineering, Kanput, India, 1993.
26. Starovasnik, M. A., Braisted, A. C. and Wells, J. A. Structural mimicry of a native protein by a minimized binding domain. 1997 Proc. Natl Acad. Sci. USA 94, 10 080.
27. Ticona, W. G. C., Aplicacao de Algoritmos Geneticos Multi-Objetivo para Alinahmento de Sequencias Biologicas. USP.pp 1 - 129, 2003.

28. Tuffery,P., Etchebest,C., Hazout,S. and Lavery, R., A New Approach to the Rapid Determination of Protein Side Chain Conformations, 1991, Journal of Biomolecular Structure e Dynamics, p.1267-1289 V.8.
29. Vullo, A., On the Role of Machine Learning in Protein Structure Determination, Department of Systems and Computer Science, University of Florence, 2002
30. Williams, R. W. and Teeter, M. M. Raman spectroscopy of homologous plant toxins: crambin and alpha 1- and beta-purothionin secondary structures, disulfide conformation, and tyrosine environment. 1984 Biochemistry 23, 6796.

THE NUMBER OF GENERATIONS BETWEEN BRANCHING EVENTS IN A GALTON–WATSON TREE AND ITS APPLICATION TO HUMAN MITOCHONDRIAL DNA EVOLUTION

ARMANDO G. M. NEVES, CARLOS H. C. MOREIRA

UFMG, Depto. de Matematica,
Caixa Postal 702, 30123-970 - B. Horizonte, MG, Brazil
E-mails: aneves@mat.ufmg.br, cmoreira@mat.ufmg.br

We have shown that the problem of existence of a mitochondrial Eve can be understood as an application of the Galton–Watson branching process and presents interesting analogies with critical phenomena in Statistical Mechanics. We shall review some of these results here. In order to consider mutations in the Galton–Watson framework, we shall derive a general formula for the number of generations between successive branching events in a pruned genealogic tree. We show that in the supercritical regime of the Galton–Watson model, this number of generations is a random variable with geometric distribution. In the critical regime, population in the Galton–Watson model is of constant size in the average. Serva worked on genealogic distances in a model of haploid constant population and discovered that genealogic distances between individuals fluctuated wildly both in time and in the realization of the model. Also, such fluctuations seem not to disappear when the population size tends to infinity. This phenomenon was termed lack of self-averaging in the genealogic distances. Although in our model, population is not strictly constant, we argue that the lack of self-averaging in the genealogic distance between individuals may be viewed as a consequence of being exactly at or near a critical point.

1. Introduction

The Galton–Watson process[1] is a well-known stochastic branching process, originally introduced for studying the survival probability of family names. Due to the fact that mitochondrial DNA (mtDNA), as family name, is inherited from only one of each individual's parents, its mother, the Galton–Watson process may be used for studying survival of mtDNA lineages and the phenomenon of existence of a mitochondrial Eve[2].

According to whether female population on average respectively decreases, is constant, or increases, we shall call the Galton–Watson process subcritical, critical or supercritical. Despite previous works denying this possibility[3], we showed[4,5] that a mitochondrial Eve may exist in an exponentially growing population, i.e. in the supercritical regime. Of course, a mitochondrial Eve is also allowed to exist in the critical regime for the Galton–Watson model and also in constant populations, as generally accepted[3]. As a consequence of the possible existence of a mitochondrial Eve in an exponentially growing population, we argue[4,5] that existence of an African mitochondrial Eve cannot be used as a validation argument for the Out of Africa model for human evolution.

Since the original work by Cann et al.[2], arguments based on the number of differences in the mtDNA in samples of individuals have also been used in order to support the Out of Africa model. In fact, the number of differences in the mtDNA of any pair of living humans is fairly low, being consistent with all of us being descendants of a single woman which lived 100-200 thousand years ago. This number of years is also roughly confirmed by fossils and archeological evidence[6].

Another recent achievement which seems again to corroborate the Out of Africa model is sequencing of the mtDNA of three neanderthal fossils by Krings et al.[7,8]. Whereas in a large sample of living humans the number of differences between pairs of individuals in a certain 360 base pair region in the HVR1 part of the control region is 8.0 ± 3.1, one of the fossils differed from modern humans in 27.2 ± 2.2 positions, a much higher count. The differences between the other neanderthal specimens sequenced and modern humans are similarly high. Krings et al.[7,8] thus concluded that modern humans and neanderthals, despite having coexisted in Europe and Asia for tens of thousands of years, are different species. According to general opinion, if there had been some genetic mixing between modern humans and neanderthals, the number of differences in their mtDNAs could not be so large.

In order to properly address the question of differences in the mtDNA of individuals descendant form a single individual, the mitochondrial Eve, we must include mutations in our Galton–Watson formalism. In this paper we shall consider again the same Galton–Watson model and explain its predictions on the genealogic distances between pairs of individuals.

By genealogic distance between two individuals we shall mean the time, measured in generations, elapsed since these two individuals had their most recent common maternal ancestor. The genealogic distance between two individuals is not amenable to direct experimental measure, but it should be roughly proportional to the genetic distance measured as the number of differences in some part of the mitochondrial genomes of the individuals. In order for this proportionality to be true, we must assume that mutations occur randomly at a certain known rate and mutations are neutral. Although there is some dispute on the correct value to be used for the mutation rate[9], it is generally agreed that it is fairly constant. In some regions in mtDNA it is also agreed that mutations are not deleterious nor advantageous.

Serva[10,11] criticized the conclusion, based on the number of differences in mtDNA, that modern humans and neanderthals are different species. He studied the genealogic distances between pairs of individuals in a model of a constant population with asexual reproduction. In such a model, he found out that the distribution of genealogic distances between pairs of individuals fluctuates wildly among different realizations of the process and has abrupt fluctuations in time within a single realization. He also showed that such fluctuations persist even in the limit of an infinite population. According to him, this *lack of self-averaging* is relevant in analyzing data on mtDNA extracted from neanderthals and comparing these data with modern human ones.

Although the model considered by Serva is not exactly the critical Galton–Watson model, these models have the common feature of population being constant in one case and constant on the average in the other. We agree with Serva's conclusions and in this work we shall analyze the distribution of genealogic distances between pairs of individuals in a Galton–Watson framework. We shall show that the lack of self-averaging in the genealogic distances may be viewed as a consequence of being exactly at or near the critical point in the Galton–Watson model.

Our main tool will be understanding the number of generations between branching events in a genealogic tree containing only the individuals in past generations that have left any descendants among individuals living nowadays. We shall obtain a general formula for this number of generations in the Galton–Watson model and show that in the supercritical regime, it is a random variable with geometric distribution. In the critical regime, the number of generations between branching events is approximately uniformly distributed in the part of the tree closer to the root.

This paper is organized as follows. In the next section we shall review some basic facts in the Galton–Watson model and derive our formula for the number of generations between branching events. In the following section we shall explore the consequences of this formula for the critical and slightly supercritical regimes in the Galton–Watson model. A conclusion section closes the paper.

2. The Galton–Watson model and the time between branching events in a genealogic tree

2.1. *Galton–Watson essentials*

In the Galton–Watson process, generations are treated as non-overlapping. As males do not contribute to passing mtDNA to further generations, we simply ignore them in our genealogic trees. So, when mentioning children of an individual in this paper, the individual will always be a woman and her children will always be the female ones.

The main feature of the model is that the number of (female) children of any woman is a random variable with a fixed probability distribution, independent of the number of children of other women. We shall denote q_r as the probability that a woman has r children, $r = 0, 1, 2, \ldots$ and we shall refer to the q_r's as the progeny probability distribution.

Galton–Watson genealogic trees may end at some generation if no woman at that generation has any daughter. The mtDNA lineage of a woman in the founding population will survive if and only if the Galton–Watson tree rooted at her will not end before present generation.

In order to calculate the survival probability for Galton–Watson trees, we introduce the generating function $S(x)$ for the female progeny distribution defined as

$$S(x) = \sum_{r=0}^{\infty} q_r x^r \,. \tag{1}$$

We shall make the convention that the individual at the root of a tree is at generation 0, her daughters, if any, are at generation 1 and so on. Let then $\overline{\theta}_n$ be the probability that a Galton–Watson tree ends at generation n or before. Accordingly, $\theta_n = 1 - \overline{\theta}_n$ is the probability that a tree will survive at least up to generation $n+1$. It can be seen[1,4] that

$$\overline{\theta}_n = S(\overline{\theta}_{n-1}) \,, \tag{2}$$

where the initial condition is

$$\overline{\theta}_0 = q_0 \,. \tag{3}$$

Let

$$m = \sum_{r=1}^{\infty} r\, q_r = S'(1) \tag{4}$$

be the mean number of female children per woman. In any case, by iteration of (2) with initial condition (3) there will always exist the *thermodynamic limit* $\theta = \lim_{n\to\infty} \theta_n$, with θ equal to 0 if $m \le 1$ and positive if $m > 1$. We thus see that the survival probability θ for a genealogic tree with infinite generations exhibits a second-order phase transition as in Statistical Mechanics, where parameter m acts as the inverse temperature. According to whether $m > 1$, $m = 1$ or $m < 1$, we say we are in the supercritical, critical or subcritical regime.

It can be shown[4] that in the subcritical and supercritical regimes, the convergence of θ_n to θ is exponential,

$$\theta_n - \theta \sim \mathrm{e}^{-n/\xi} \,, \tag{5}$$

with the *correlation time* being

$$\xi = -1/\ln[S'(1-\theta)] \,. \tag{6}$$

In the critical regime, the correlation time ξ diverges to infinity. In this case, instead of the exponential decay in (5), convergence of θ_n to $\theta = 0$ is polynomial if $S'''(1) < \infty$:

$$\theta_n \sim \frac{2}{S''(1)\, n} \,. \tag{7}$$

By using an estimate[12] of $W = 5,000$ women living at the same time of the mitochondrial Eve, we have shown[4,5] that a mitochondrial Eve may exist with reasonable probability if $1 \le m \le 1.0009$, regardless of the progeny distribution. We shall call the above range *mitochondrial Eve range*. For m close enough to 1, we may approximate all quantities with small errors by retaining only the smallest terms in powers of $m-1$. This approximation was termed[4,5] *small survival probability approximation*. We shall refer to the range of values for m in the supercritical regime in which we may use the above approximation as the *slightly supercritical* range. The slightly supercritical range contains the mitochondrial Eve range.

It was also shown[4,5] that in the small survival probability approximation we have

$$\xi \approx \frac{1}{|m-1|} \,. \tag{8}$$

In the mitochondrial Eve range, correlation times may thus have values ranging from some thousands of generations up to infinity.

2.2. *Pruned trees and the number of generations between successive branching events*

In the Galton–Watson model, the progeny distribution is the same for any woman. As a consequence, the probability θ_n that a genealogic tree survives up to some generation larger than n is the same as the probability that any subtree inside a tree survives to more than n generations after its root. In the critical and slightly supercritical regimes, values for θ_n are very small unless n is close to 1. This means that most subtrees in a critical or slightly supercritical Galton–Watson tree end after a small number of generations. Individuals in these extinct subtrees of course do not contribute to mtDNAs at the present generation.

We shall intend as a pruned tree, the tree we obtain when we delete from a genealogic tree all individuals which have left no descendants among the population at the present generation. In the critical or slightly supercritical regimes, due to the large probability of extinction for subtrees, pruned trees possess much less branching vertices than the genealogic trees from which they derive. At Figure 1, we show an example with a small number of generations.

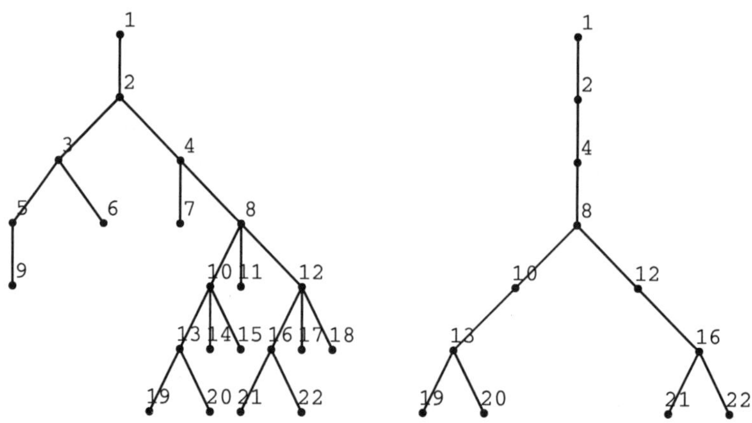

Figure 1. In the left hand side, we show a Galton–Watson tree, and at right hand side we show the pruned tree derived from it.

Let T_N be a tree obtained by realizing the Galton–Watson process for a period of N generations. Let l be the length of the stem of the pruned tree obtained from T_N, i.e. the number of generations between the root and the first branching point in the pruned tree. We shall now derive a formula for the *conditional* probability $P_N^{\rm surv}(t)$ that the random variable l has value equal to t generations, given that T_N survived up to generation N. The same symbol without the superscript will denote the (unconditional) probability for the same event. As the probability that T_N does not end before generation N is θ_{N-1}, then $P_N^{\rm surv}(t) = P_N(t)/\theta_{N-1}$.

As a first step, we shall compute the probability $\tau_N \equiv P_N(0)$ that the pruned tree obtained from T_N branches already at the root. This event will happen provided the woman at the root has at least two children and that at least two among the trees originating at the children of the woman at the root survive up to generation N. As the children of the woman at the root are already at generation 1, the trees originating at them and not ending before generation N must have at least $N-1$ generations after their roots. Thus

$$\tau_N = \sum_{l=2}^{\infty}\sum_{k=l}^{\infty} q_k \binom{k}{l} \theta_{N-2}^{l} (1-\theta_{N-2})^{k-l} , \tag{9}$$

where the summation index l labels the number of trees originating at generation 1 which survive up to generation N and the index k labels the number of children of the woman at the root. By interchanging the order of summation, using the binomial formula and expanding, we get

$$\begin{aligned}
\tau_N &= \sum_{k=2}^{\infty} q_k (1-\theta_{N-2})^k \sum_{l=2}^{k} \binom{k}{l} \left(\frac{\theta_{N-2}}{1-\theta_{N-2}}\right)^l \\
&= \sum_{k=2}^{\infty} q_k (1-\theta_{N-2})^k \left[\left(1+\frac{\theta_{N-2}}{1-\theta_{N-2}}\right)^k - 1 - k\,\frac{\theta_{N-2}}{1-\theta_{N-2}}\right] \\
&= \sum_{k=2}^{\infty} q_k - \sum_{k=2}^{\infty} q_k (1-\theta_{N-2})^k - \theta_{N-2} \sum_{k=2}^{\infty} k\, q_k\, (1-\theta_{N-2})^{k-1} \\
&= 1 - q_0 - q_1 - [S(1-\theta_{N-2}) - q_0 - q_1(1-\theta_{N-2})] \\
&\quad - \theta_{N-2} \left[S'(1-\theta_{N-2}) - q_1\right] .
\end{aligned}$$

Cancelling some terms, we finally have,

$$\tau_N = 1 - S(1-\theta_{N-2}) - \theta_{N-2}\, S'(1-\theta_{N-2}) . \tag{10}$$

In order to compute $P_N(t)$, notice that the event $l = t$ is the same as the woman at the root having at least one child and all subtrees originating at generation 1 ending before generation N, except for one, which must have a stem $t-1$ generations long. Thus

$$P_N(t) = \sum_{k=1}^{\infty} k\, q_k (1-\theta_{N-2})^{k-1} P_{N-1}(t-1) = S'(1-\theta_{N-2})\, P_{N-1}(t-1)\,.$$

Iterating the formula above we get

$$P_N(t) = \tau_{N-t} \prod_{j=1}^{t} S'(1-\theta_{N-1-j})$$

and finally

$$P_N^{\text{surv}}(t) = \frac{\tau_{N-t}}{\theta_{N-1}} \prod_{j=1}^{t} S'(1-\theta_{N-1-j})\,. \tag{11}$$

As (11) holds also for stem sizes of subtrees, then it may also be used to calculate the distribution of the number of generations between two successive branching events in a pruned tree.

3. Applications

In the supercritical regime Galton–Watson trees may be infinite generations long with positive probability. We may take limits when $N \to \infty$ in all quantities appearing in (11) and obtain that the probability that the stem length between two successive branching events in an infinitely long pruned tree is equal to t is given by

$$P^{\text{surv}}(t) = \frac{\tau}{\theta}\, S'(1-\theta)^t\,, \tag{12}$$

where of course $\theta = \lim_{N\to\infty} \theta_N$ and $\tau = \lim_{N\to\infty} \tau_N = 1 - S(1-\theta) - \theta\, S'(1-\theta)$. In other words, the number of generations l between branching events is a random variable with geometric distribution. The expected value of l is then easily seen to be

$$\mathbb{E}(l) = \frac{1}{1-S'(1-\theta)} = \frac{1}{1-e^{-1/\xi}}\,, \tag{13}$$

where we have used (6) in the last equality.

As $0 < S'(1-\theta) < 1$, we see that $P^{\text{surv}}(t)$ decays exponentially with t, so that the time between successive branching events in pruned supercritical trees is never too large. As a consequence, in a supercritical pruned tree with a large number of generations, there will be a large number of branching points. In a population at supercritical growth it is then expected that, given that the number of generations is sufficiently large, genealogic distances between individuals will increase steadily with time.

In the slightly supercritical situation, ξ is very large, so that

$$\mathbb{E}(l) \approx \xi \,. \tag{14}$$

In this case, $S'(1-\theta)$ is very close to 1 and it turns out that $P^{\text{surv}}(t)$ is virtually independent of t if $t \ll \xi$.

Consider now the mitochondrial Eve range, contained in the slightly supercritical regime. As the mitochondrial Eve lived about 10,000 generations ago, in that range ξ is of the order of magnitude of the number of generations of the part of the tree after the mitochondrial Eve, or even larger. If human population is really represented by a slightly supercritical Galton–Watson tree, then we may expect times between successive branching events in the tree more or less uniformly distributed, some of these times being close to the time elapsed since the mitochondrial Eve. In other words, along with short branches, there may also exist some very deep branches in the human genealogic tree, such branches linking individuals living nowadays almost directly to the mitochondrial Eve. Such a pruned genealogic tree representing all mankind can be seen e.g. as Fig. 3 in the original paper by Cann et al.[2] and it has exactly these features.

Another important feature of slightly supercritical trees is that the number of individuals in these deep branches is small, because the number of branching events along them is small. Branches with a small number of individuals at their tips are more likely to become extinct as time passes, even if they are very deep. When an event of extinction of a deep subpopulation occurs, an abrupt decrease in the mean genealogic distance between individuals results.

The phenomenon of abrupt negative variations in the mean genealogic distance between individuals was noticed by Serva[11], see e.g. Fig. 4 in the cited paper. It could explain why it is possible that modern humans and neanderthals may have been part of the same population, i.e. two subpopulations of the same species. In fact, neanderthals may be genetically distant from modern humans[7,8], but it is possible that at the time they lived, the mean genetic distance between pairs of humans, neanderthals included, was larger than it is nowadays.

In the critical regime, results should be similar to the ones obtained by Serva in the constant population model. As $\theta = 0$ in this regime, all subtrees, no matter how large, will eventually be extinguished if we wait long enough. So, we should still expect the same kind of abrupt negative fluctuations in the mean genealogic distance as seen in the slightly supercritical case.

Also, if N is large, then θ_N is almost independent of N, because, according to (7), it is tending slowly to 0. As a consequence, for large N, we see from (11) and (10) that $P_N^{\text{surv}}(t)$ is almost independent of t; times between successive branching events are uniformly distributed. This causes wild fluctuations in the mean genealogic distances in different realizations of the process. This is the phenomenon of *lack of self-averaging* observed by Serva.

4. Conclusion

We have given a qualitative description of some features in the human genealogic tree as seen for example in Fig. 3 at the paper by Cann et al.[2]. Although a more complete quantitative treatment would be desirable, this description supports our belief that the Galton–Watson model can be used to understand important aspects of human evolution – existence of a mitochondrial Eve and the distribution of genetic distances between pairs of individuals in mtDNA.

Although our description remains consistent with the Out of Africa model, it is consistent also with the Multiregional Evolution hypothesis[13] in which humans living nowadays may have had genetic influences from neanderthals and other extinct hominids without showing these influences in their mtDNA.

As noticed first by Serva[10,11], we agree that the large genetic distance between mtDNA of neanderthals and living humans cannot be used as an argument for concluding that neanderthals and modern humans did not belong to the same interbreeding population. To his analysis we add the explanation that the large fluctuations in the mean genealogic distances between individuals also happen in the Galton–Watson model and are a consequence of being exactly at or near a critical point.

References

1. T. E. Harris, The Theory of Branching Processes, Dover, New York (1989).
2. R. L. Cann, M. Stoneking, A. C. Wilson, *Nature* **325**, 31 (1987).
3. J. C. Avise, J. E. Neigel, J. Arnold, *J. Molec. Evol.* **20**, 99 (1984).
4. A. G. M. Neves, C. H. C. Moreira, *Physica A* **368**, 132 (2006).
5. A. G. M. Neves , C. H. C. Moreira, in BIOMAT 2005, R. P. Mondaini and R. Dilão (eds.), World Scientific (2005).
6. C. Stringer, *Nature* **423**, 692 (2003).

7. M. Krings, A. Stone, R. W. Schmitz, H. Krainitzki, M. Stoneking, S. Paabo, *Cell* **90** (1), 19 (1997).
8. M. Krings, C. Capelli, F. Tschentscher, H. Geisert, S. Meyer, A. von Haeseler, K. Grossschmidt, G. Possnert, M. Pauvonic, S. Paabo, *Nature Genetics* **26** (2), 144 (2000).
9. T. J. Parsons, D. S. Muniec, K. Sullivan et al, *Nature Genetics* **15**, 363 (1997).
10. M. Serva, *Physica A* **332**, 387 (2004).
11. M. Serva, *J. Stat. Mech. Theory Exp.* P07011 (2005).
12. N. Takahata, *Mol. Biol. Evol.* **10** (1), 2 (1993).
13. A. G. Thorne, M. H. Wolpoff, *Sci. Am.* **266** (4), 76 (1992).

EXPLORATIONS IN INSECT SOCIALITY: TOWARDS A UNIFYING APPROACH

PAULO SÁVIO DA SILVA COSTA

Lecturer of Computer Science, Departamento de Ciências Exatas e Tecnológicas, Universidade Estadual de Santa Cruz, Km 16 Rodovia Ilhéus-Itabuna, Ilhéus, BA, 45662-000, Brazil

Email: psscosta@uesc.br

An important trend in the study of insect societies has taken place in the last 30 years. Traditional theories of social organization such as adaptive demography and worker castes have been revised in the light of new evidence of the dynamic nature of regulatory processes underlying social behaviour. However, this new trend is still markedly reductionist in its methods. This paper advocates a unifying approach to the study of insect sociality, illustrated by an agent-based computational model allowing detailed investigations of the dynamics of insect social behaviour. The model incorporates key components of social organization and their underlying mechanisms, both at the individual and colony levels. The model was applied to field studies of behavioural plasticity in red harvester ants, in order to evaluate its performance when applied to a concrete problem in insect sociobiology. Simulation experiments reproduced several aspects of harvester ant social organization and produced insights into the dynamics of collective responses to changing ecological conditions. The results suggest that temporal patterns of colony resource allocation may be more complex than currently believed. We found a non-linear relationship between ecological stress and the colony's response strategy, revealing a significant event in the temporal dynamics of the system behaviour: the collapse of the relative priorities of communal tasks. We present a testable hypothesis about the hierarchy of task priorities in harvester ants, as well as suggestions for an experimental procedure for testing the hypothesis. The paper concludes with a discussion of the prospects and pitfalls of computational approaches to the study of insect societies.

1. Introduction

Insect societies embody the key to a fundamental problem in biology: that of establishing the mechanisms of social coordination, shaped by natural selection in the course of a species' evolutionary history, that enabled the transition from independent organisms to integrated societies. Although much is known about individual insects, the same cannot be said about their societies. Only recently have we started to unveil important genetic and ontogenetic determinants of insect sociality. Traditional theories of social organization have been abandoned or revised in the light of new evidence of the interconnected nature of various regulatory mechanisms underlying social behaviour. Much effort has been made to account for this new evidence, but such efforts are still markedly reductionist in their methods. Colony behaviour cannot be described purely in terms of individual traits and behaviours; it must take into account the interplay between such diverse factors as ecological stress, intracolony genotypic diversity, hierarchies of tasks, and distributed acquisition and dispersal of information. This paper advocates a unifying approach to the study of insect societies, illustrated by a computational model that successfully reproduced several results of field studies of behavioural plasticity in harvester ants, producing new insights into the colony's ability to cope with ecological stress.

2. Computational model of insect societies

2.1. *Definitions*

The activities collectively carried out by workers are referred to as *tasks*. Each task is associated with a global *stimulus level*. A colony is modelled as a *network* whose *units* correspond to individual workers. Each unit is associated with a set of *stimulus response thresholds*, one for each task. A unit can be either *idle* or engaged in a given task. Simulations are executed in discrete time steps, or *cycles*. At each cycle a unit is randomly chosen, and its response thresholds compared to the corresponding stimulus levels. If all thresholds are higher than the associated stimulus levels, the unit becomes or remains inactive, otherwise it performs the task whose stimulus level departs the most from its response threshold. The survival requirements of a colony are expressed in terms of stimulus *containment* parameters and differential *effects* of task execution. Changing environmental conditions are obtained by exposing networks to different sets of *disturbances*. A simulation consists of creating a network with a given response threshold composition, exposing it to a series of disturbances, and analysing its behaviour. A colony is then evaluated in terms of general performance indices as well as fine-grained indicators of division of labour.

2.2. *Stimulus perception*

Much of the information flow in an insect society is based on cues: a worker perceives the effects of its nestmates' actions on the shared nest environment and acts accordingly[1]. Information flow based on signals can greatly improve a colony's efficiency and overall fitness, but it is an expensive commodity, requiring special-purpose behaviours in the part of the sender and sophisticated reception mechanisms in the part of the receiver. For the most part, information flow in an insect colony travels from the group to the individual[2]. For this reason, the present model provides facilities for information flow via cues only. Since individuals do not directly exchange information about their sensory perception or internal states, they must be able to perceive colony needs by other means. This is done by associating each communal task with a global stimulus level that is continuously updated to reflect the activity of workers.

2.3. *Response thresholds*

A network unit is characterized by its current activity state and response threshold set, representing an insect's genetic tendency to perform each communal task. The network threshold composition summarizes the colony-level distribution of the genetic influences on individual patterns of activity. Each task has its own threshold composition, specified as follows:

Range: Lower and upper limits for each threshold. The wider the threshold composition range, the greater the colony's genotypic diversity.

Size: Number of thresholds, equal to the number of network units.

Distribution: Type of random distribution to use: normal and uniform. The standard deviation of normal distributions is equal to 1·0.

Redundancy: Number of copies of each random value in the distribution. The larger the redundancy, the smaller the colony's genotypic diversity.

2.4. *Disturbances*

Changing environmental conditions are simulated by means of disturbances. During a simulation, the network is exposed to zero or more disturbances, applied at regular intervals, and collectively referred to as a disturbance set. The specification of a disturbance set consists of four parameters:

Type: A *task stimulus disturbance* affects global stimulus levels; when applied to a particular task, it causes a variation in the global stimulus level associated with that task. A *network unit disturbance* affects network units; when applied to a particular task, it causes the removal of that task from the behavioural repertoire of some of the units engaged in it.

Scope: Depending on the disturbance type, the scope refers either to the number of tasks whose associated stimulus levels are altered, or the number of tasks removed from the repertoire of network units.

Value: For a *task stimulus* disturbance, this is the amount by which the stimulus level of the disturbed tasks is altered. For a *network unit* disturbance, this is the number of units whose behavioural repertoire is restricted.

Period: Number of time steps between two consecutive disturbances. Disturbances are applied periodically; a simulation is divided into phases, and a *simulation phase* is the interval between two consecutive disturbances.

2.5. *Effects of unit activity*

Differential effects of unit activity on stimulus levels are encapsulated in the *effects matrix*, which can be seen as an abstract representation of the ergonomics of social life in the colony. Suppose there are two tasks in a simulation: foraging (Fg) and nest maintenance (Nm). Each worker is either engaged in one of the tasks, or idle (Id). At this level of abstraction, foraging is synonymous with production of energy, and nest maintenance with the removal of refuse. On average, idle workers consume little energy and produce little refuse, but they do not produce energy or remove refuse. Nest maintenance workers consume more energy and produce more refuse than idle workers; they also remove refuse at a higher rate than they produce it. Foragers consume more energy and produce more refuse than nest maintenance workers; they also produce energy at a higher rate than they consume it. These 'metabolic rates' are summarized in Table 1.

Table 1. Differential effects of unit activity on task stimuli. Tasks are foraging (Fg) and nest maintenance (Nm). Workers can be either active or idle (Id).

Task	Energy		Effect on
	Consumption	Production	**Fg** stimulus
Id	1	0	$1 - 0 = 1$
Fg	3	8	$3 - 8 = -5$
Nm	2	0	$2 - 0 = 2$

Task	Refuse		Effect on
	Production	Removal	**Nm** stimulus
Id	1	0	$1 - 0 = 1$
Fg	3	0	$3 - 0 = 3$
Nm	2	8	$2 - 8 = -6$

Effect		on	
		Fg	**Nm**
of	**Id**	1	1
	Fg	−5	3
	Nm	2	−6

These figures do not reflect metabolic rates of individual workers, which may vary due to such factors as physical caste, developmental stage and behavioural profile. Rather, they should be seen as colony averages, and need not be constant. They could change as a result of qualitative variations in the environment, such as the nutritional value of food brought back by foragers. Many perennial colonies experience periodic changes in colony size, age distribution and ratio between reproductives and workers, which in turn have an impact on the relative costs, benefits and priorities of tasks.

The general form of the effects matrix is given by

$$\mathcal{E} = \begin{bmatrix} \mathcal{E}_{0,1} & \mathcal{E}_{0,2} & & \mathcal{E}_{0,n} \\ \mathcal{E}_{1,1} & \mathcal{E}_{1,2} & \cdots & \mathcal{E}_{1,n} \\ & \vdots & \ddots & \\ \mathcal{E}_{n,1} & \mathcal{E}_{n,2} & & \mathcal{E}_{n,n} \end{bmatrix}$$

where $\mathcal{E}_{i,j}$ is the effect of a unit performing task i on the stimulus level of task j, $i \in \{0, \cdots, n\}$, $j \in \{1, \cdots, n\}$ and n is the number of tasks.

In order to specify the particular values in this matrix, several constraints have to be taken into account. The execution of a given task should produce two effects: a decrease in the stimulus level of that task, and perhaps an increase in the stimulus levels of all other tasks:

$$\mathcal{E}_{j,j} < 0 \ , \ \mathcal{E}_{i,j} \geq 0 \ (i \neq j) \tag{1}$$

$\mathcal{E}_{0,j}$ represents the impact of inactive units on the stimulus level of task j. However small the amount of resources consumed by idle workers in an insect colony, they always contribute to an increase in the need for some tasks (eg. foraging). Therefore $\mathcal{E}_{0,j}$ must not be negative, otherwise the stimulus level of task j will be continuously reduced by inactive units:

$$\mathcal{E}_{0,j} \geq 0 \tag{2}$$

When all units are active, the total increase or decrease in stimulus levels due to units' activity must satisfy the following condition:

$$-\sum_{j=1}^{n} \mathcal{E}_{j,j} \geq \sum_{\substack{i,j=1 \\ i \neq j}}^{n} \mathcal{E}_{i,j} \tag{3}$$

In other words, at each time step the overall reduction in stimulus levels must at least compensate for the simultaneous increase, otherwise stimulus levels of all tasks will rise continuously. In the presence of inactive units, stimulus levels may still stabilize, depending on the average number of inactive units and the particular values in the effects matrix.

2.6. *Task priorities*

Whenever a unit is selected for updating, if all stimulus levels are below the unit's response thresholds then it becomes or remains inactive. If only one stimulus level is above its response threshold, the unit becomes or remains engaged in that task. However, if two or more stimulus levels are greater than the unit's thresholds, the task with the highest stimulus level is chosen.

When the threshold distributions for the various tasks have different means, this results in a hierarchy of priorities: the lower the threshold distribution mean, the lower the task priority. The hierarchy of priorities is flexible to the extent that the corresponding threshold distributions overlap.

2.7. *Stimulus containment*

In order to survive, an insect colony must monitor its environment and keep certain needs at acceptable levels. For example, brood rearing in honey bees depends on thermoregulatory mechanisms with rigid requirements[4–6]. Honey bees cannot allow large temperature fluctuations in the brood area of the nest for any appreciable period of time, at the risk of causing developmental abnormalities and death. They are very sensitive to both the intensity and the duration of temperature deviations from the optimal range.

In this model this is equivalent to the colony's ability to keep stimulus levels within limits. Critical activities may be regarded as more sensitive to deviations of the corresponding stimulus from a given range. Such sensitivity has two independent components: the intensity and the duration of variations in stimulus levels. These are modeled by *stimulus containment parameters*, or triplets of the form $\{target, range, resistance\}$. *Target* is the optimal stimulus level for the corresponding task. *Range* is the amount of variation in the stimulus level that the colony can comfortably tolerate. The tighter the containment range, the more sensitive the colony to stimulus level fluctuations around its optimal value. *Resistance* is a constant representing the colony's tolerance to long term variations in the stimulus level. The smaller the resistance, the more sensitive the colony to sustained stimulus levels outside the optimal range. The stimulus containment index for a given task is given by

$$I_{ctm} = \exp(-\frac{\kappa}{\tau}\sum_{t=1}^{\tau}\Delta S_t) \; ; \; \Delta S_t = \begin{cases} S_t - S_{max} & \text{if } S_{max} < S_t \\ 0 & \text{if } S_{min} \leq S_t \leq S_{max} \\ S_{min} - S_t & \text{if } S_t < S_{min} \end{cases}$$

where κ is the resistance constant, τ the length of the evaluation period in execution cycles, and S_t the task's stimulus level at time t.

2.8. *Stimulus convergence*

Stimulus containment indices indicate the colony's ability to keep stimulus levels within an acceptable range, but they say nothing about its ability to maintain stable stimulus levels. This stability is translated into stimulus convergence indices, allowing the assessment of the colony's ability to cope with changing environments. The stimulus convergence index of a given task is calculated over a number of execution cycles, or evaluation period. It is defined in terms of the variation of the stimulus level around the mean:

$$I_{cvg} = \frac{\overline{S} - kS_{std}}{\overline{S}}$$

where $\overline{S}$ is the stimulus mean, S_{std} the standard deviation and k an arbitrary constant.

2.9. *Division of labour*

An execution cycle consists of choosing a unit at random, changing its activity state if necessary, and updating stimulus levels of all tasks accordingly. A unit may switch tasks if the stimulus levels of some or all tasks have changed since it was last chosen. The temporal patterns of task switching are used to quantify three indicators of division of labour: *team sizes*, *task affinity*, and *task fidelity*.

2.10. *Team size convergence*

The term *team* refers to a group of workers engaged in the same task. No distinction is made between individual workers; the only distinguishing features of worker teams are their sizes and their rate of change. Team sizes are indicators of the amount of resources the colony has allocated to each task; their rate of change after a disturbance is an estimate of the strength of the link between environmental conditions and division of labour.

The team size convergence index of a given task is defined as

$$I_{team} = \frac{\overline{T} - kT_{std}}{\overline{T}}$$

where $\overline{T}$ is the mean value of the team size over the evaluation period, T_{std} the standard deviation and k an arbitrary constant.

2.11. *Task affinity*

Division of labour is difficult to quantify due to its intrinsically dynamic nature and to variations between conspecific colonies. Team size convergence rates reflect the global trend in the colony's response to disturbances, but frequent changes in worker activity states will go unnoticed if the numbers engaged in each task remain constant. A more discriminant indicator is needed to assess the stability of the colony's resource allocation strategy.

The model keeps record of the *affinity* of units for tasks and their *fidelity* to them. Task affinity is defined as the relative amount of time a unit spends on a given task. It is a measure of specialization: the more a unit performs a task, the stronger its affinity for the task and the more specialized in it. Suppose the activity state of unit 2 evolved as shown below:

$$0 \Rightarrow 1 \Rightarrow 0 \Rightarrow 1 \Rightarrow 0 \Rightarrow 2 \Rightarrow 2 \Rightarrow 2 \Rightarrow 2 \Rightarrow 1 \Rightarrow 0 \Rightarrow 1 \Rightarrow 0 \Rightarrow 1 \Rightarrow 1 \Rightarrow 0 \Rightarrow 3$$

This sequence lists the unit's activity state at each time step during a simulation, thus reflecting the unit's temporal pattern of activity. The affinity indices of this unit are shown in Table 2.

Table 2. Quantifying task affinity of network units. There are 3 tasks in this example; task 0 corresponds to idleness.

Unit 2	Task			
	0	1	2	3
1				
2	5	6	4	1
3				
⋮				
n				

Task	Time doing this task	Affinity index
0	5	5/16 = 0.3125
1	6	6/16 = 0.3750
2	4	4/16 = 0.2500
3	1	1/16 = 0.0625
Total	16	1.0000

2.12. *Task fidelity*

Task fidelity is a measure of the stability of a unit's activity state. The more the unit switches tasks, the lower its task fidelity. As another example, suppose unit 18 switched tasks according to the sequence below:

$$0 \Rightarrow 1 \Rightarrow 0 \Rightarrow 1 \Rightarrow 0 \Rightarrow 2 \Rightarrow 2 \Rightarrow 2 \Rightarrow 2 \Rightarrow 1 \Rightarrow 0 \Rightarrow 1 \Rightarrow 0 \Rightarrow 1 \Rightarrow 1 \Rightarrow 0 \Rightarrow 3$$

This sequence does not reflect the unit's temporal pattern of activity in the same way as it did in the previous example. Here, the sequence signals *switching events*: the unit was initially inactive (task 0); when it was first selected it switched to task 1; then it was selected again and went back to performing task 0; etc. This unit's fidelity indices are shown in Table 3.

2.13. *Behavioural categories*

Task affinity and fidelity indices of individual units are combined into network indices: the network affinity index reflects the collective trend in unit

Table 3. Quantifying task fidelity of network units. Numbers in shaded areas are switches to the same task. *Undef* indicates that the unit never had a chance to switch from task 3, therefore nothing can be said about its fidelity to that task.

unit 18		to task 0	to task 1	to task 2	to task 3
from task	0	0	4	1	1
from task	1	5	1	0	0
from task	2	0	1	3	0
from task	3	0	0	0	-

Unit 18			
Task	Switches to		Fidelity
	same task	all tasks	index
0	0	$0+4+1+1=6$	$0/6=0$
1	1	$5+1+0+0=6$	$1/6=0.1667$
2	3	$0+1+3+0=4$	$3/4=0.75$
3	0	$0+0+0+0=0$	$0/0=$ *Undef*

specialization, while the network fidelity index reflects the stability of the units' patterns of activity. Affinity and fidelity indices of units are also used to quantify division of labour, which results from the interaction between the colony's genotypic composition and the ecological context. Individual workers change their behaviour as colony needs change, but their capacity to change is differential rather than uniform[7]. Variations in individual response thresholds may give rise to colony-level phenomena such as the emergence of elite and reserve worker forces. Network units are classified according to their activity level and temporal pattern of task switching.

2.13.1. *Activity level*

The first classification system is based on the activity levels of units, ie. their task affinity indices. There are six categories: *inactive*, *average*, *active*, *reserve*, *specialist* and *elite*. These categories are not mutually exclusive; each unit is assigned to at least one category, but possibly to others as well.

Classification with regard to a single task

The first three categories apply to all combinations of units and tasks. A unit is *inactive* with respect to a given task if it spends little or no time on that task. Likewise, the unit is *active* with respect to that task if it spends a large proportion of time on the task. If neither condition applies, the unit is said to be *average* with respect to the task. If $\mathcal{A}_{u,t}$ is the affinity of unit u for task t then unit u is classified as *inactive* if $0 \leq \mathcal{A}_{u,t} \ll 1$, *average* if $0 < \mathcal{A}_{u,t} < 1$, or *active* if $0 \ll \mathcal{A}_{u,t} \leq 1$. These inequalities are re-written in terms of two arbitrary constants, $\mathcal{A}_0$ and $\mathcal{A}_1$: unit u is classified as *inactive* if $0 \leq \mathcal{A}_{u,t} \leq \mathcal{A}_0$, *average* if $\mathcal{A}_0 < \mathcal{A}_{u,t} < \mathcal{A}_1$, or *active* if $\mathcal{A}_1 \leq \mathcal{A}_{u,t} \leq 1$.

Classification with regard to all tasks

A *reserve* unit hardly ever performs any task for any considerable length of time: if $\mathbb{T}$ is the set of all tasks, then u is a *reserve* unit if $0 \leq \mathcal{A}_{u,t} \leq \mathcal{A}_0$ for all $t \in \mathbb{T}$. Similarly, a unit is said to be a *specialist* in task t if it is *active* with respect to task t and *inactive* with respect to all others, ie. if $\mathcal{A}_1 \leq \mathcal{A}_{u,t} \leq 1$ and $0 \leq \mathcal{A}_{u,t'} \leq \mathcal{A}_0$ for all $t' \in \mathbb{T} - \{t\}$.

Classification with regard to other units

An *elite* unit is a specialist in a task performed only by specialists: if $\mathbb{T}$ is the set of all tasks and $\mathbb{U}$ the set of all units, then u is an *elite* unit if $\mathcal{A}_1 \leq \mathcal{A}_{u,t} \leq 1$ and $0 \leq \mathcal{A}_{u,t'} \leq \mathcal{A}_0$ for all $t' \in \mathbb{T} - \{t\}$, and $0 \leq \mathcal{A}_{u',t} \leq \mathcal{A}_0$ for all $u' \in \mathbb{U} - \{u\}$.

2.13.2. *Task switching*

The second classification system is based on the patterns of task switching of network units, ie. their task fidelity indices. There are seven categories: *unstable*, *regular*, *stable*, *evasive*, *reliable*, *committed* and *dedicated.*

Classification with regard to a single task

Figure 3 (p. 307) shows a unit with an undefined fidelity index for task 3, so it cannot be classified according to its task fidelity as that information does not exist (a rare occurrence). If the fidelity index is available, the unit is said to be *unstable* with respect to a given task if it switches out of it frequently. Alternatively, the unit is *stable* with respect to that task if it tends to continue performing the task whenever it switches to it. If neither condition holds true, the unit is said to be *regular* with respect to the task.

If the fidelity of unit u to task t is given by $\mathcal{F}_{u,t}$ then it is classified as *unstable* if $0 \leq \mathcal{F}_{u,t} \ll 1$, *regular* if $0 < \mathcal{F}_{u,t} < 1$, or *stable* if $0 \ll \mathcal{F}_{u,t} \leq 1$. These inequalities are re-written in terms of two arbitrary constants, $\mathcal{F}_0$ and $\mathcal{F}_1$: unit u is classified as *unstable* if $0 \leq \mathcal{F}_{u,t} \leq \mathcal{F}_0$, *regular* if $\mathcal{F}_0 < \mathcal{F}_{u,t} < \mathcal{F}_1$, or *stable* if $\mathcal{F}_1 \leq \mathcal{F}_{u,t} \leq 1$.

Classification with regard to all tasks

An *evasive* unit is one that hardly ever settles on any task: u is an *evasive* unit if $0 \leq \mathcal{F}_{u,t} \leq \mathcal{F}_0$ for all $t \in \mathbb{T}$ for which $\mathcal{F}_{u,t}$ is defined. A unit is said to be *reliable* if it displays high fidelity to all tasks that it has ever performed: $\mathcal{F}_1 \leq \mathcal{F}_{u,t} \leq 1$ for all $t \in \mathbb{T}$ for which $\mathcal{F}_{u,t}$ is defined. Finally, a unit is said to be *committed* to a given task if it has a high fidelity to that task and avoids all others: $\mathcal{F}_1 \leq \mathcal{F}_{u,t} \leq 1$ and $0 \leq \mathcal{F}_{u,t'} \leq \mathcal{F}_0$ for all $t' \in \mathbb{T} - \{t\}$.

Classification with regard to other units

A *dedicated* unit is one that is committed to a task which is performed entirely by committed units: u is a *dedicated* unit if $\mathcal{F}_1 \leq \mathcal{F}_{u,t} \leq 1$ and $0 \leq \mathcal{F}_{u,t'} \leq \mathcal{F}_0$ for all $t' \in \mathbb{T} - \{t\}$ and $0 \leq \mathcal{F}_{u',t} \leq \mathcal{F}_0$ for all $u' \in \mathbb{U} - \{u\}$.

3. Case study: Task switching in red harvester ants

This section describes the application of the computational model to Gordon's field studies of behavioural plasticity in harvester ants, which focused on the relationship between ecological events[8–11], colony priorities and task switching by exterior workers. Those studies produced several results, a subset of which are summarized below:

(1) The numbers of ants engaged in the various tasks follow a clear temporal pattern, or 'daily round'.
(2) Perturbations to one task are always followed by changes in the temporal patterns of execution of other tasks.
(3) The colony-level response to perturbations has two components: reallocation of active workers engaged in other tasks, and recruitment of inactive workers. Their relative contributions are not understood, and may vary according to the circumstances[10].
(4) Nest maintenance appears to be the most labile of external tasks.
(5) When either foraging or nest maintenance is perturbed, these tasks have reciprocal priorities — an increase in foraging is accompanied by a decrease in nest maintenance, and vice versa.
(6) When foraging and nest maintenance are perturbed simultaneously, foraging takes precedence over nest maintenance.
(7) Simultaneous perturbations to two tasks have synergistic effects.

The purpose of the simulations was to determine how closely the model can reproduce a biological system behaviour, particularly the interplay between environmental variability and colony resource allocation. The objectives were: (1) to characterize network responses to disturbances as a function of recruitment (allocation of inactive units) and task switching (reallocation of active units), and (2) to determine how the temporal patterns of resource allocation are affected by the relative priorities of tasks, the scope and intensity of disturbances, and the network recovery time.

3.1. *Experimental design*

create network
for D_s **in** $\{1,2\}$
    for V **in** $\{250, 500, 1000, 2000, 4000\}$
        $D_v = V \ / \ D_s$
        $D_p = 1500$
        $E_c = 4 \times D_p$
        run network with $\{D_s, D_v, D_p, E_c\}$
        extract data from output:
            team sizes
            stimulus levels
            task switching statistics
            affinity and fidelity indices of units
            indicators of division of labour
    endfor
endfor

where V is the total increase in all stimulus levels
D_s is the disturbance scope
D_v is the disturbance value
D_p is the disturbance period
E_c is the length of the simulation in execution cycles

Figure 1. Algorithmic representation of the experimental design.

3.1.1. *Number of tasks*

The number of tasks performed in insect colonies varies greatly from one species to another. The identification of behavioural acts themselves (and therefore of a behavioural repertoire) can be highly subjective, and comparisons among species of such repertoires are often problematic[12,13]. This reveals a certain arbitrariness in the stipulation of distinct tasks into which individual workers are to be classified.

Three tasks were used in the simulations, in line with the field studies of harvester ants which focused primarily on foraging, patrolling, and nest maintenance. Colony members in some other species of social insects are also organized into three roles[14], suggesting this is not an unrealistic choice.

3.1.2. *Network size*

The sizes of insect colonies also vary enormously. Colonies of social wasps can have as few as 20 (*Belonogaster junceus*) and 30 (*Mischocyttarus drewseni*) adult members. Ant colonies, on the other hand, reach up to 22 million workers (*Anomma wilverthi*). Bee colonies, intermediate in size, also vary considerably: colonies of bumblebees vary between 100 and 400 workers, and honey bee colonies between 10,000 to 80,000 workers[15].

Three sets of simulations were performed, with network sizes of 200, 1000 and 2000. These simulations produced qualitatively similar results; this paper reports the results obtained with the smaller network size.

3.1.3. *Response threshold distributions*

The units' response thresholds for each task were drawn from a separate normal distribution of random numbers with the following parameters:

Task	Distribution parameters		
	mean	range	std dev
1	800	50	1.0
2	500	50	1.0
3	200	50	1.0

Each threshold distribution was obtained by drawing 200 random numbers from a normal distribution with mean 0.0 and standard deviation 1.0 and then scaling all numbers in the distribution in order to match the range of stimulus thresholds for the corresponding task. Distribution means and ranges were arbitrarily chosen; a different set of values could equally have been used, so long as the resulting distributions did not overlap. The purpose was to establish a clear hierarchy of task priorities (see Sec. 2.6), with task 1 having the highest priority and task 3 the lowest.

3.1.4. *Effects matrix*

The effects matrix used is shown below. These values were chosen because (1) they comply with the constraints listed in Sec. 2.5, and (2) they result in various proportions of idle units, depending on the disturbance parameters of a given simulation. Thus it is possible to determine if and how the global redistribution of resources is affected by the existence of uncommitted units.

Effect		on task		
		1	2	3
	0	1	1	1
of	1	−3	1	1
task	2	1	−3	1
	3	1	1	−3

3.1.5. *Disturbances*

Changing environmental conditions were simulated by applying different sets of disturbances to task stimulus levels. Two disturbance parameters were varied: intensity, or value (D_v), and number of tasks affected, or scope (D_s). For double disturbances ($D_s = 2$) the intensity was halved, so the overall stress put on the network by single and double disturbances varied within the same range ($250 \leq D_v \leq 4000$). Ten sets of disturbance parameters were obtained by combining 5 values of D_v and 2 values of D_s. The period of disturbances (D_p) was kept constant. The network was run 10 times, once with each set of disturbance parameters. The length of all runs was equal to 4 times the disturbance period, resulting in 4 phases per run. Disturbances were made at the start of each new phase, so each task was disturbed exactly once (for $D_s = 1$) or twice ($D_s = 2$) per run.

The values of D_s and D_p match the disturbances described in Gordon's studies. There, disturbance intensity was constant and assumed to be of similar magnitude as naturally occurring events. Here, the values of D_p and D_v cover a range of possibilities, from mild to severe disturbances. The are two reasons for this: (1) We can observe the effects of disturbances in different contexts, depending on how much time the network had to recover from a previous disturbance. (2) We may be able to generalize the effects of disturbances on network behaviour under more extreme conditions.

3.2. *Results*

In the following discussion, the term 'team t' refers to all network units engaged in task t, and the term 'idle team' refers to all inactive units.

3.2.1. *Resource allocation following single disturbances*

Figures 3 and 4 (pp. 323–324) show the temporal pattern of global resource allocation following each single disturbance. They reveal that, for mild to moderate disturbances ($D_v \leq 1000$), the patterns of resource allocation change gradually as disturbances become stronger, but disturbances to different tasks produce qualitatively different results.

Figures 3a to 3c show the effects of mild disturbances to each task separately ($D_v = 250$, $D_s = 1$). In Fig. 3a, when the stimulus for task 1 is increased, more units are allocated to that task. Newly allocated units come from all other teams — idle units as well units previously engaged in the other two tasks. The largest contributor of new recruits is the idle team, which steadily decreases in size for twice as long as team 3, the second largest contributor. Task 2 is hardly affected. The relative contributions of teams performing affected tasks, ie. teams 2 and 3, mirror the relative task priorities: the lower the task priority, the more units are diverted from it and reallocated to the disturbed one. In Fig. 3b, when task 2 is disturbed a similar situation is observed: the network responds by allocating increasingly more units to that task. This time, however, new recruits come from two sources only — team 3 and the idle team. Again, the largest contributor is the idle team, steadily decreasing in size for twice as long as the second largest contributor of new recruits, but now task 1 is not affected. In Fig. 3c, when task 3 is disturbed the pattern becomes apparent: new recruits come entirely from the idle team; tasks 1 and 2 are unaffected.

The observed trend is consistent with variations in stimulus levels of all tasks and can be summarized as follows. For small disturbances,

1a. the network responds by allocating new units to the disturbed task;
2a. new units come from one or more sources (idle units, the largest contribution, and units previously engaged in tasks with lower priority than the disturbed one); and
3a. tasks with higher priority than the disturbed one are not affected.

Figures 3d to 3f show the effects of moderate disturbances ($D_v = 500$, $D_s = 1$). The results are similar to those obtained with small disturbances, but now all tasks are affected, regardless of their relative priorities. The corresponding variations in stimulus levels confirm this. The interpretation of this result is slightly different from before:

1b. the network responds by allocating new units to the disturbed task;
2b. new units come from all teams, regardless of the relative priorities of the undisturbed tasks; and
3b. the lower the tasks priority, the more units are drawn away from it.

These results can be rephrased as follows:

Perturbations to one task are always followed by changes in the temporal patterns of execution of other tasks.

This matches Gordon's observation listed as item 2 on p. 309. Secondly, the results obtained for $D_v = 250$ and $D_v = 500$ can be restated as follows:

The network response to disturbances has two components: reallocation of active units engaged in other tasks, and recruitment of inactive units. The relative contributions of these two components vary according to the intensity of disturbances.

This closely matches Gordon's observation listed as item 3 on p. 309. Finally, item 3b above can be rewritten as

Task 3 is the most labile of all tasks.

If tasks 1 to 3 are taken to represent foraging, patrolling and nest maintenance respectively, the previous sentence becomes nearly identical with Gordon's observation listed as item 4 on p. 309.

3.2.2. *Resource allocation following combined disturbances*

Figures 5 and 6 (pp. 325–326) show the effects of double disturbances on the temporal patterns of unit allocation. It should be noted that disturbances to different pairs of tasks produce qualitatively different results. Figures 5d to 5f show that, for mild disturbances ($D_v = 250$), the allocation of units follows a wave pattern. When tasks with the highest priorities are disturbed (Fig. 5d), units are redirected to all tasks one at a time; the higher the task priority, the sooner its needs are met. In the first wave, units are drawn away from all groups and reallocated to task 1. When the corresponding stimulus returns to normal levels, the second wave begins. Units are now redirected from task 1 to task 2, which continues to absorb units from the other two teams. By the time the corresponding stimulus returns to normal, team 3 has been seriously depleted; the third wave then starts. Units start to be reallocated from task 2 to task 3, but now all units are engaged in some task. When the stimulus associated with task 3 returns to normal, the surplus units are no longer needed and become inactive. Figures 5e and 5f show similar wave patterns, but now the task with highest priority, task 1, is hardly affected. This is true whether this task is disturbed (phase 3) or not (phase 4). Figures 5b, 5e and 5h show that when tasks 1 and 3 are disturbed simultaneously, the numbers engaged in task 1 increase at the expense of all other teams. In other words:

When tasks 1 and 3 are disturbed simultaneously, task 1 takes precedence over task 3.

If these tasks are replaced by foraging and nest maintenance respectively, this result matches exactly Gordon's observation listed as item 6 on p. 309.

3.2.3. *Synergistic effects of combined disturbances*

Figure 2 (p. 315) shows the non-additive effects of combined disturbances. Bar graphs indicate the total number of task switching events in two separate simulations with equal value of combined disturbances; the disturbance parameters were $\mathcal{S}_1 : \{D_v = 500,\ D_s = 1\}$ and $\mathcal{S}_2 : \{D_v = 250,\ D_s = 2\}$. Switching events were recorded separately for each pair of tasks, and data for all units were pooled together. Each solid triangle in Fig. 2 groups four pairs of tasks of the form (*from*,*to*). The number below a triangle represents the *to* task, and the 4 numbers above it, the *from* tasks. Each of the 3 graphs refers to a different pair of disturbed tasks. The following discussion applies to graph a only; the other two were obtained in a similar manner.

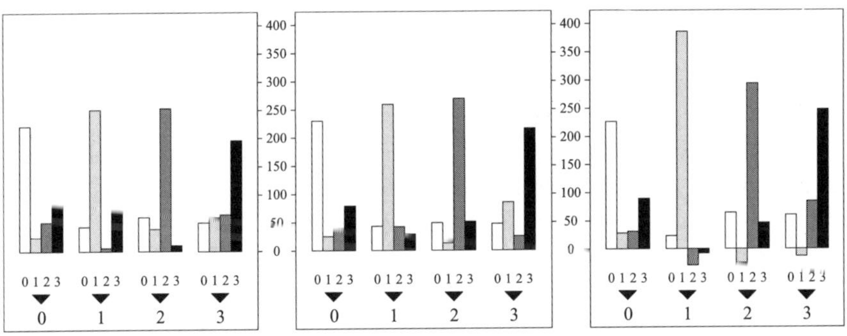

Figure 2. Synergistic effects of combined disturbances on task switching events.

For each pair of tasks, the number of switching events in phases 2 and 3 of the first simulation ($\mathcal{S}_1$) were added together, yielding the total effect of separate disturbances to tasks 1 and 2. This was then subtracted from the number of switching events in phase 2 of $\mathcal{S}_2$, when tasks 1 and 2 were disturbed simultaneously. This procedure was repeated for each pair of tasks, and the results plotted as a bar graph. If the effects of the combined disturbances had been of similar magnitude as the combined effects of separate disturbances, all bars would be of similar height and close to 0. Instead, the graphs show that combined disturbances to each pair of tasks had non-additive effects. This is consistent with results obtained for other values of D_v and confirms Gordon's observation listed as item 7 on p. 309.

3.2.4. *Stability of the task priority hierarchy*

Figure 5 (p. 325) shows that increasingly severe disturbances cause gradual changes in the temporal patterns of unit allocation, but only up to a point. An important transition occurs for some value of D_v between 1000 and 2000. From then on, unit allocation patterns become qualitatively identical in all phases and for all disturbance values (Fig. 6, p. 326). The same is observed when disturbances are made in combination; the transition takes place for some value of D_v between 500 and 1000. In both cases, the qualitatively distinct patterns are characterized by a common circumstance: the first disturbance causes a recruitment for more units than actually exist; all units become engaged in some task, after which no unit ever becomes inactive again. Regardless of the disturbance scope, the distinct patterns reflect the network's inability to restore the task eliciting stimuli quickly enough to their normal levels. The network becomes overloaded, ie. still in the process of bringing stimulus levels back to normal when struck by a new disturbance. Further periodic disturbances cause all stimuli to rise increasingly higher above their normal levels.

Network overload marks a significant event in the system behaviour. In Fig. 5i (p. 325) the relative priorities of all tasks are reflected in the diversion of units from task 3 to task 2 in spite of both having been simultaneously disturbed, while task 1 was mostly unaffected. In Fig. 6c, however, the temporal patterns of unit allocation are significantly different. Moreover, the relative priorities of the disturbed tasks (2 and 3) are no longer apparent. Numbers engaged in these two tasks follow identical temporal trajectories, in opposite direction to that followed by numbers performing task 1. The hierarchy of task priorities has collapsed.

This collapse was not simply due to a particularly severe disturbance. In Fig. 6a the variations in numbers engaged in tasks 1 and 2 show that, when the first disturbance is made, their relative priorities are still in effect. Two disturbances later, however, the hierarchy disappears (Fig. 6c).

3.3. *Discussion*

3.3.1. *Reproducing biological system behaviour*

The simulations focused on correlating changes in external conditions with global patterns of unit allocation to tasks. The results were interpreted in the context of Gordon's field studies of behavioural plasticity in harvester ants. Several of her observations about colony behaviour and social organization were successfully reproduced. However, the simulated system behaviour evolved in a way that can only be validated against field studies with more extensive monitoring procedures than those used in Gordon's study. The implications for empirical studies of insect social organization are discussed below.

3.3.2. *Suggesting a conceptual approach for empirical work*

The simulation results suggest that it may be useful to refine the concept of task priorities. In Gordon's studies the hierarchy of task priorities represents a fixed colony trait; the relative priorities of tasks are not seen as contingent on short term fluctuations in ecological conditions. As the simulation results indicate, this may not be the case. Under severe stress, the hierarchy of task priorities may collapse, and descriptions of system behaviour that make reference to such hierarchies become inadequate.

This difficulty may be avoided by breaking the concept of task priorities into two components: (1) At the individual level, the relative priorities of tasks reflects that individual's pre-disposition to perform some tasks but not others. Depending on the circumstances, it may be appropriate to regard the individual-level hierarchy of task priorities as a fixed trait. The present model takes this approach; each unit has a fixed set of response thresholds, and their relative values determine that unit's priorities. This is equivalent to regarding an individual's response thresholds as being determined entirely at the genetic level. However, this approach may be inadequate in a study concerned with such issues as learning or hormonal regulation of behavioural plasticity. (2) At the colony level, the relative priorities of tasks reflects the collection of hierarchies operating 'inside' each colony member. Consequently, they might be modulated by the same factors affecting an individual's behaviour, such as variations in task eliciting stimulus levels.

The distinction between individual and colony level priorities may also help the analysis of group effects such as the collapse of priority hierarchies.

3.3.3. *Suggesting a methodological approach for empirical work*

Measuring disturbance effects

Gordon[8] states that "one crucial question left unresolved concerns the duration of response to perturbations (...) When the recovery time of a colony is known, it will be possible to test the effects of perturbations of increasing magnitude." In other words: how lasting and far reaching are the colony's responses to different kinds and intensities of perturbations? Our results indicate that answers to this question must be qualified in terms of the relative priorities of tasks, and whether the colony's reserve force is large enough to fully absorb the impact of recurrent perturbations.

Consider the network recovery times. Let team i be the set of units performing task t_i, and $\mathcal{T}_i$ the time it takes for team i to return to its normal size following a disturbance. Table 4 (p. 318) shows that, in any phase, $\mathcal{T}_i$ increases with disturbance intensity, as expected. If the tasks are numbered $t_1, \cdots t_n$ in decreasing order of priority and n is the number of tasks, then the data shows that a disturbance to task t_i will always affect teams i to n. However, teams 1 to $i-1$ will also be affected if the network becomes overloaded, even if only temporarily.

The implication for empirical studies is that an accurate quantification of the effects of disturbances to an insect colony requires more than 'before and after treatment' measurements. It requires a procedure for monitoring system behaviour with much higher resolution in order to detect non-linearities such as the overloading phenomenon.

Table 4. Recovery times for teams 1, 2 and 3, in execution cycles.

D_v	Phase	Single disturbances				Double disturbances			
		Task	$\mathcal{T}_1$	$\mathcal{T}_2$	$\mathcal{T}_3$	Tasks	$\mathcal{T}_1$	$\mathcal{T}_2$	$\mathcal{T}_3$
250	2	1	300	450	**500**	1 2	330	400	**520**
	3	2	0	**400**	600	1 3	400	**110**	640
	4	3	**0**	0	500	2 3	**0**	320	510
500	2	1	470	720	**720**	1 2	340	600	**700**
	3	2	140	**590**	820	1 3	400	**200**	800
	4	3	**0**	440	700	2 3	**0**	500	710
1000	2	1	960	960	**1300**	1 2	700	900	**1300**
	3	2	600	**990**	1390	1 3	550	**1000**	1220
	4	1	**500**	910	1300	2 3	**120**	900	1300
2000	2	1	1320	–	–	1 2	1320	–	–
	3	2	–	–	–	1 3	–	–	–
	4	3	–	–	–	2 3	–	–	–

Note: Missing entries indicate that the corresponding team never returned to its normal size due to the depletion of the idle team following a severe disturbance.

Determining the relative priorities of tasks

Our simulations indicate how field experiments with ant colonies could reveal the relative priorities of tasks. In Fig. 5d the numbers engaged in all tasks follow a distinctive wave pattern (see Sec. 3.2.2). All team sizes suffered a temporary increase, but team sizes reached their peaks at different times: the higher the task priority, the sooner the group performing that task reached its peak. The hierarchy of task priorities is also revealed in Table 4 (see the numbers in bold typeface). For any set of double disturbances, the recovery times of teams engaged in undisturbed tasks in each phase are inversely proportional to the task priorities. If $\mathcal{T}_{i,p}$ is the recovery time of team i in phase p, then $\mathcal{T}_{1,4} < \mathcal{T}_{2,3} < \mathcal{T}_{3,2}$, ie. t_1 has the highest priority and t_3 the lowest.

3.3.4. *Making a testable hypothesis*

The phenomenon of network overload suggests a non-linear relationship between environmental stress and the colony's response mechanism. The non-linearity becomes manifest when the entire colony's reserve force becomes engaged, at which point the colony's hierarchy of task priorities collapses. This observation leads to the following testable hypothesis:

The relative priorities of communal tasks in a harvester ant colony are kept as long as there are idle workers ready to engage in needed tasks. If, under ecological stress, the colony's entire worker force becomes engaged, further disturbances will cause the hierarchy of task priorities to collapse.

4. Some lessons

The model reproduced qualitative aspects of Gordon's field studies of behavioural plasticity in harvester ants. However, it is remarkably difficult to replicate the ant colonies' exact behaviour, for the reasons discussed below.

4.1. *Cycles of activity*

The numbers of workers performing various tasks in harvester ant colonies follow distinctive temporal patterns. These so-called daily rounds vary not only from one species to another, but also between conspecific colonies[16–18]. Such activity cycles may show different levels of responsiveness to changes in environmental conditions, but they are rooted in the *constancy* or *regularity* of crucial ecological events. All animal life is affected by many periodically recurring phenomena of survival relevance. In harvester ants, cyclic patterns in the execution of certain tasks but not others are intimately related to regular, short term variations in the colony's habitat (ambient light, temperature, feeding habits of predators, etc.) Thus, patrolling is mostly done at the start of the morning activity period, while brood care is done round-the-clock. This has important implications for the very process of modeling insect social organization. The existence of activity cycles sets limits to the range of natural phenomena that can be accurately represented in a computational model. It would not be appropriate to simply implement a mechanism of cyclic activity in the model and then observe the various effects of disturbances. Without an adequate understanding of the relevant ecological context, there would be no basis for a meaningful correlation between disturbances and the natural processes that gave rise to the cyclic patterns of activity in the first place.

4.2. *Population sampling and bias*

Gordon's field studies[8,9,11] focused on a small subset of the colonies' worker force. She detected changes in activity states of marked individuals, as well as variations in temporal patterns of task execution. However, only exterior tasks were observed. By limiting the investigation to selected tasks and ignoring the remaining ones, important aspects of the colony's strategy of resource allocation may have gone undetected. Gordon's methodological approach was appropriate for the questions addressed in those studies, but it is not entirely compatible with the experimental framework provided — and assumed — by the present computational model.

Matching the simulations' experimental design to that of Gordon's studies is further complicated by the process of data acquisition. In field experiments, only a small subset of the colony population is marked and observed. There are obvious constraints to the number of ants that can be tracked by an observer, as well as practical difficulties in accessing the interior of the nest without disrupting the colony's functioning[23]. An ant colony can contain in excess of 1,000,000 workers[15]; in *Pogonomyrmex* colonies, external workers comprise only about 10 to 20% of the total colony population[24,25]. Perhaps the problem of sampling a significant proportion of the colony population could be minimized by using larger samples, but those could easily be biased as particular regions of the nest and categories of workers are easier to access than others. Gordon[10] states that "I assume it to be unlikely that, in many colonies, the distributions of large samples of marked ants would show similar trends, consistent across experiments, which masks different, hidden trends in the distribution of unmarked ants." Perhaps this assumption should be revised; we now know that workers of the army ant *Leptothorax unifasciatus* are faithful to specific positions in their nest, tending to move within zones of limited area[26–29].

In a model where observations and measurements can be applied to all system components with equal frequency and accuracy, attempts to reproduce empirical studies of biological systems are faced with a great difficulty. In order to select a subset of system components for observation so as to match experimental procedures of an empirical work, one needs to identify the ecological factors underlying the inevitable bias introduced by population sampling. This may require a level of understanding of the biological system that is not available at the time the study is carried out.

4.3. *Resolution of experimental designs*

The experimental designs of empirical investigations have a much lower resolution than that tipically afforded by a computational model. Empirical works are often based on 'before and after treatment' measurements, while a computer simulation is able to detect changes in the system behaviour on a moment-by-moment basis. Thus Gordon compares the numbers of workers engaged in various tasks before and after a given perturbation, while the present simulations keep track of the effects of disturbances on team sizes as a function of time. In order to reproduce the snapshot-like nature of Gordon's measurements, it would be necessary to determine an appropriate synchronization point, so that team sizes before and after that point could be pooled together and then compared. However, without an understanding of the cyclic activity patterns underlying the pre- and post-disturbance colony behaviour, it is not possible to meaningfully choose a specific moment in time for taking such pooled measurements.

5. Conclusion

Computer models can be very useful in the study of living systems. Given adequate resources, one can build computational models of complex ecosystems, simulate their evolution, test theories, help resolve conflicts between theory and empirical data, and clarify discrepancies between sets of observations. However, sound interdisciplinary studies of insect societies jointly conducted by biologists and computer scientists are relatively rare. Such studies often fail to provide plausible and consistent links between the computational model and the biological system, ultimately limiting the depth of insight they can produce. A truly useful model must do more than reproduce aspects of an insect colony; it must at the very least generate insights into how new knowledge may be gained from further experiments.

Acknowledgments

This work was funded by the Brazilian Research Council (CNPq), grant 202191/89-3.

References

1. J. E. Lloyd, *Ann. Rev. Entomol.* **28**, 131–160 (1983).
2. T. D. Seeley, *The Wisdom of the Hive: The Social Physiology of Honey Bee Colonies*, Harvard Univ. Press (1995).
3. R. E. Page, S. D. Mitchell, *Phil Sci. Assoc.* **2**, 289–298 (1991).
4. M. V. Brian, *Social Insects: Ecology and Behavioural Biology*, Chapman and Hall (1983).
5. M. L. Winston, *The Biology of the Honey Bee*, Harvard Univ. Press (1987).
6. B. Heinrich, *E. Behav. Ecol. Sociobiol.* (Eds. B. Hölldobler and M. Lindauer), 393–406 (1985).
7. R. E. Page and G. E. Robinson, *Adv. Insect Physiol.* **23**, 117–169 (1992).
8. D. M. Gordon, *Anim. Behav.* **34** (5), 1402–1419 (1986).
9. D. M. Gordon, *Anim. Behav.* **35** (3), 833–834 (1987).
10. D. M. Gordon, *Anim. Behav.* **38** (8), 194–204 (1989).
11. D. M. Gordon, *Oxford Surveys in Evolutionary Biology* **6**, 55–72 (1989).
12. P. Jaisson, D. Fresneau and J. P. Lachaud, *Interindividual Behavioral Variability in Social Insects* (Ed. R. L. Jeanne), 1–51. Westview Press (1988).
13. M. H. Villet, *S. Afr. J. Zool.* **26** (4), 182–187 (1991).
14. M. H. Villet, *S. Afr. J. Zool.* **25** (4), 254–259 (1990).
15. E. O. Wilson, *The Insect Societies*, The Belknap Press (1971).
16. D. M. Gordon, *Psyche* **90** (4), 413–423 (1983).
17. D. M. Gordon, *J. Kans. Entomol. Soc.* **56** (3), 277–285 (1983).
18. D. M. Gordon, *Insectes Sociaux* **31** (1), 74–86 (1984).
19. C. R. Gallistel, *The Organization of Learning*, The MIT Press (1990).
20. J. Aschoff, *Biological Rhythyms*, Plenum (1981).
21. D. S. Farner, *Ann. Rev. Physiol.* **47**, 65–82 (1985).
22. T. D. Seeley, *Honeybee Ecology: A Study of Adaptation in Social Life*, Princeton Univ. Press (1985).
23. E. O. Wilson, *Behav. Ecol. Sociobiol.* **7**, 157–165 (1980).
24. W. P. MacKay, *Psyche* **88**, 24–74 (1981).
25. W. P. MacKay, *J. Kans. Entomol. Soc.* **56**, 538–542 (1983).
26. A. Sendova-Franks and N. Franks, *Bull. Math. Biol.* **55** (1), 75–96 (1993).
27. A. Sendova-Franks and N. Franks, *Proc. R. Soc. Lond. B* **256** (1347), 305–309 (1994).
28. A. Sendova-Franks and N. Franks, *Anim. Behav.* **50** (7), 121–136 (1995).
29. A. Sendova-Franks and N. Franks, *Behav. Ecol. Sociobiol.* **36** (4), 269–282 (1995).

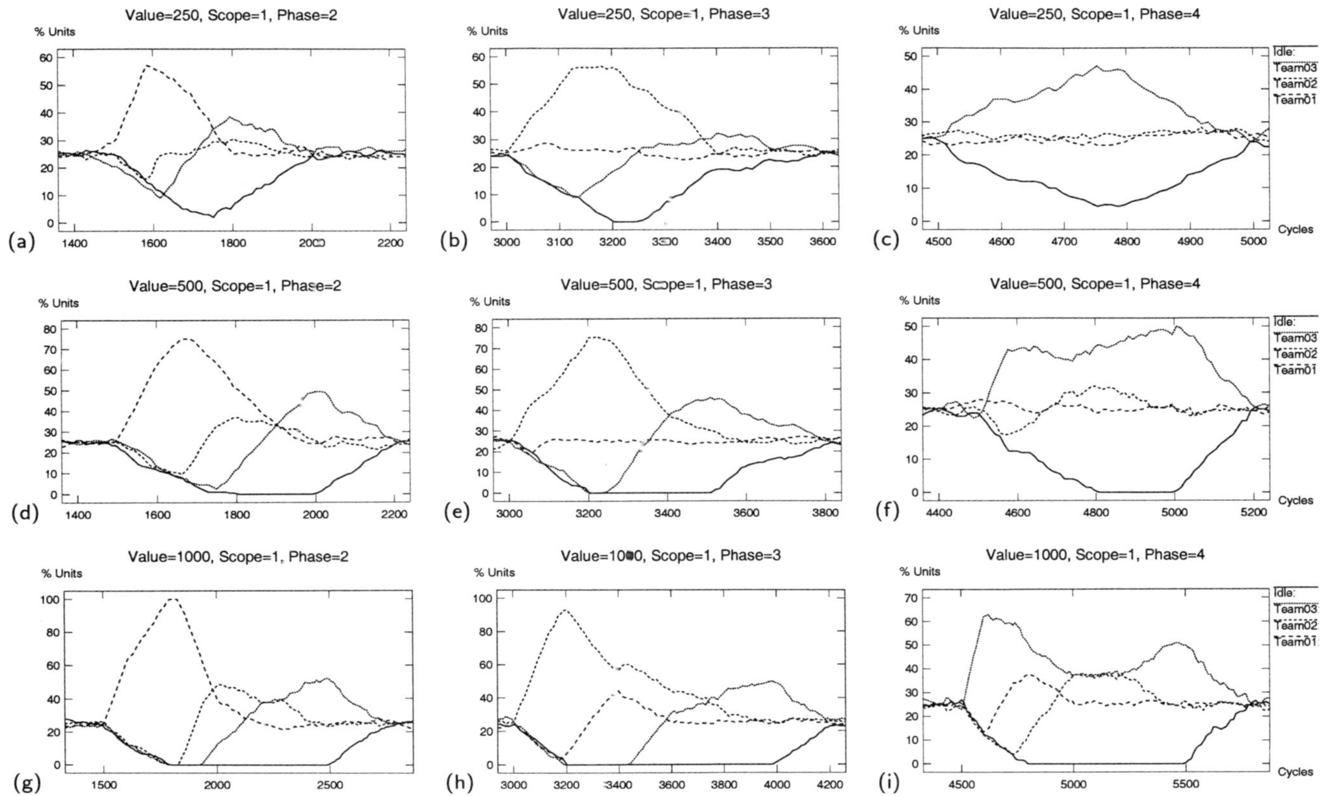

Figure 3. Temporal patterns of resource allocation. Disturbed tasks: 1 in phase 2, 2 in phase 3, and 3 in phase 4. See Sec. 3.2.1 (p. 312) for discussion.

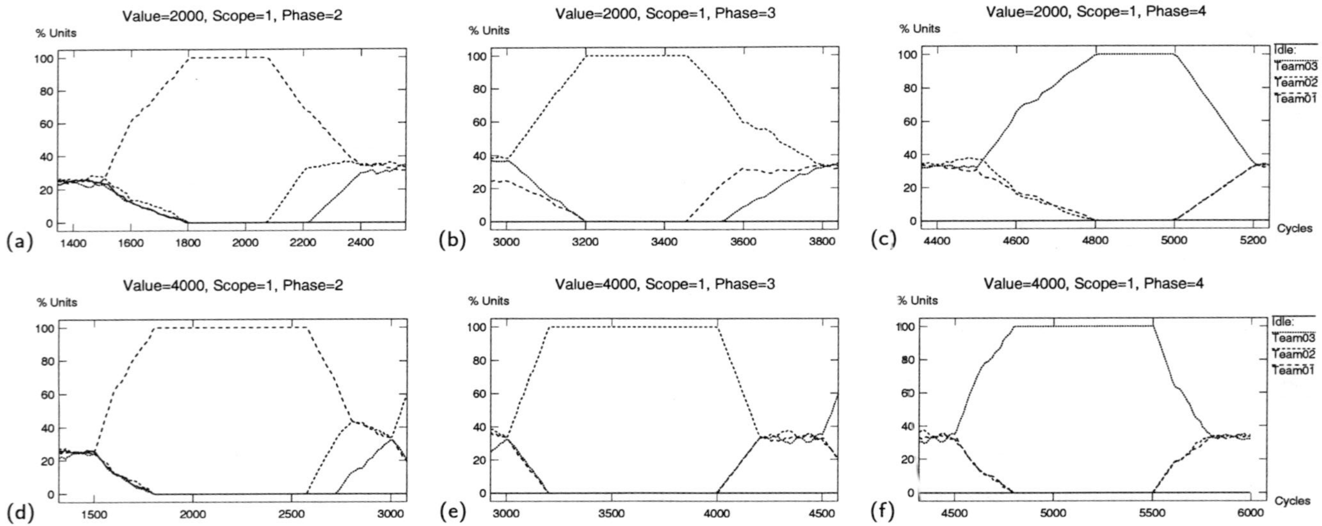

Figure 4. Temporal patterns of resource allocation. Disturbed tasks: 1 in phase 2, 2 in phase 3, and 3 in phase 4. See Sec. 3.2.1 (p. 312) for discussion.

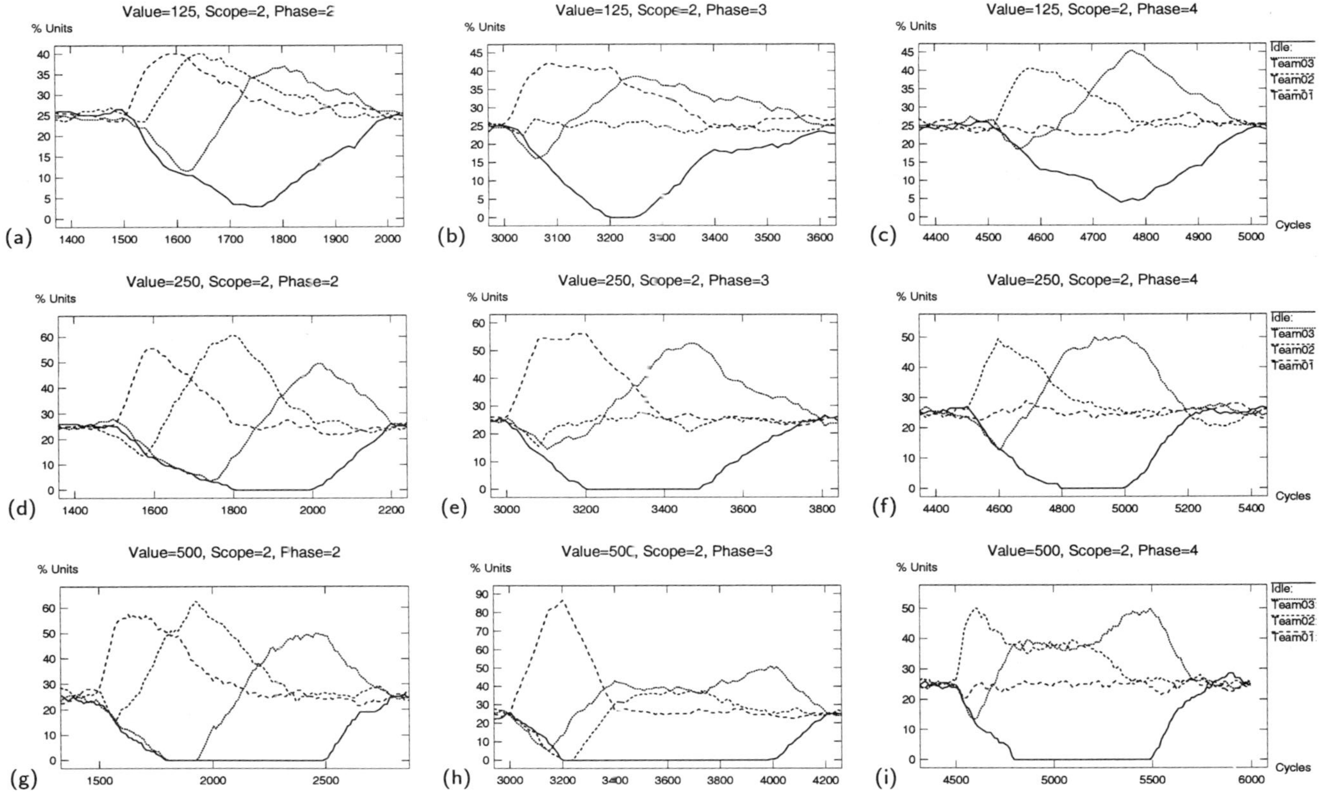

Figure 5. Temporal patterns of resource allocation. Disturbed tasks: 1 and 2 in phase 2, 1 and 3 in phase 3, and 2 and 3 in phase 4. See Sec. 3.2.1 (p. 312) for discussion.

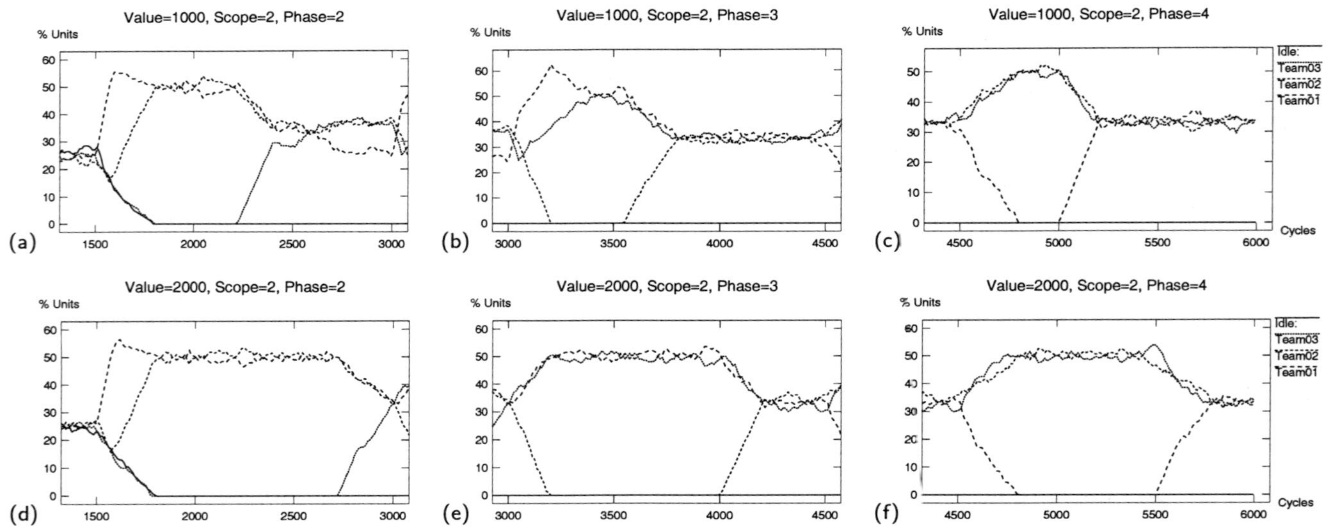

Figure 6. Temporal patterns of resource allocation. Disturbed tasks: 1 and 2 in phase 2, 1 and 3 in phase 3, and 2 and 3 in phase 4. See Sec. 3.2.1 (p. 312) for discussion.

A STAGE-STRUCTURED FINITE ELEMENT MODEL FOR THE POPULATION DYNAMICS OF TWO INTERTIDAL BARNACLES WITH INTERSPECIFIC COMPETITION

ANA PAULA RIO DOCE, REGINA ALMEIDA, MICHEL COSTA

Laboratório Nacional de Computação Científica - LNCC/MCT
Av. Getúlio Vargas, 333.
Petrópolis - RJ - Brasil - CEP: 25651-075
riodoce@lncc.br, rcca@lncc.br and michel@lncc.br

Marine life cycles commonly include a planktonic larvae phase that is transported by ocean currents. In this work, to assess the relationship between the physical processes that disperse larvae and the intensity of species interaction in the benthic habitat, we use a two dimensional stage-structured finite element model that introduces an interspecific competition for space among adults of the barnacles *Balanus glandula* (dominant species) and *Chthamalus* (*Chthamalus dalli* and *Chthamalus fissus* - subordinate species) which inhabit the rocky intertidal zone of North American Pacific coast. The main objective of this work is to characterize the effect of some idealized current patterns combined with hierarchical competition for space among adults on their relative abundances. A four stage-structured representation of the species life cycle is used. Numerical simulations showed that coastal flows may affect the adult distribution and the population interaction strength of the dominant on the subordinate species, that decreases as the velocity speed of the flow field increases. This behavior yields different population distribution patterns at the coast depending on the current pattern.

1. Introduction

In coastal zones, most marine benthic invertebrates have a complex life cycle in which the adult phase is preceded by a planktonic larval phase. The passive nature of larval dispersal of these species results in a larval transport governed by hydrodynamic processes like advection and eddy diffusion. Such transport, that may carry larvae away from their release site before they return to the coast in order to recruit, is advantageous if it ensures the maintenance of population among spatially isolated habitats and a rapid recolonization of habitats after an eventual local extinction. Moreover, the recruitment of other substrates reduces the competition with dense populations of adults [2, 1]. Spatially and temporally variable recruitment of these species has emerged as an important cause either of the variability in dynamics or intertidal community structure [3]. The term recruitment used in this article includes both the larval settlement processes and the early survival of settlers [4]. The recruitment depends on a variety of physical and biological factors, which include spawning, larval dispersal and survival, larval settlement, metamorphosis, post-settlement events, interspecific and intraspecific competitions [1]. The ecological value of recognizing the correlation between these factors and coastal recruitment motivates the development of models for the intertidal population dynamics. Several of them have focused on the importance of larval dispersal phase and are significant to determine the role of physical oceanographic processes on fluctuations of intertidal populations. They combined a planktonic larval phase and a benthic adult phase to analyze the barnacle population dynamics in both space and time [6]. For example, in 1988, Roughgarden et al. [7] presented a stage model that included mesoscale processes related to the transport of larvae. The other processes included in the model were settlement onto bare substrate, mortality and reproduction of adult, and mortality and dispersion of larvae. In this version, larvae are produced by the adults, transported in the water column by advection (only alongshore flow) and eddy-diffusion, and may settle where they strike the coastline. On settling, it is assumed that larvae metamorphose into adults immediately. The model shows that the adult population initially distributed on a small area can spread to adjacent places, of suitable habitat, through the larval dispersion. Possingham and Roughgarden [8], in 1990, improved the previous work, showing how the strength of the alongshore flow field, the amount of available suitable habitat and initial conditions interact to influence species persistence. Later, in 1996, Alexander and Roughgarden [9] presented the hypothesis that recruitment pulses result from the approach and eventual collision of upwelling fronts with the intertidal zone. The results showed how the interaction of offshore and alongshore advection, eddy-diffusion, and the location of the upwelling front determine the dynamics of coastal barnacle populations. Connoly and Roughgarden [10], in 1998, introduced in [9] a competition for space among two barnacles species and larval transport to explain the effect of recruitment variation on the intensity and importance of species interactions during coastal upwelling events in the North American Pacific coast. In 2000, Gaylord and Gaines [5] extended the model [7] so as to better explore the ocean circulation pattern in the recruitment processes of the marine species that have a planktonic life-history phase. They studied four simple representations of current patterns found in association with biogeographic boundaries worldwide (alongshore flow, converging flow, diverging flow and eddy-circulation). The fundamental issue introduced in the latter model was a more realistic representation of the life cycle of the barnacle species in which the larvae are produced in yearly pulses and, in addition to the larval and competent adults stages, it also included the larval pre-competency and the non-reproducing juvenile stages.

The main objective of this work is to characterize the effect of some current patterns combined with competitive interactions for space on the relative abundances of the barnacles *Balanus glandula* (dominant species) and *Chthamalus* (subordinate species) along the coast. To this end, we use the stage-structured representation of the barnacle life cycle introduced in [5] and a hierarchical competition for space, as developed in [10]. Among acorn barnacles, competitive dominant species may settle on, overgrow, or undercut subordinates. This means that the recruitment of *Balanus glandula* barnacle would have a negative effect on *Chthamalus* abundance by increasing mortality from competition and decreasing the amount of available substrate [10]. We are considering that the *Chthamalus* species includes the acorn barnacles *Chthamalus dalli* and *Chthamalus fissus* (as in [10])

All of the previous models mentioned here were numerically solved using the Finite Difference Method [7, 8, 9, 10, 5] . In this work, we follow another approach using the Finite Element Method [11, 12]. Besides having a solid theoretical foundation, the finite element method can handle complicated geometries, general boundary conditions and variable or non-linear properties in a relatively easy way as compared with finite difference method [12]. Thus, the proposed methodology allows flexibility in dealing with complex domains and is easily combined with an adaptive mesh procedure. The adaptive finite element method can be used to simulate the change in the domain when upwelling and relaxation events are present [13, 14] and to save computational efforts by adapting an unstructured finite element mesh to improve accuracy. We consider a two-dimensional model that includes a larval transport defined by an idealized velocity field $v \equiv (v_x, v_y)$). The discrete problem of the larval transport is built based on the Galerkin method using piecewise linear finite element interpolation functions in space and the implicit Euler method in time. The dynamics of the adult barnacles at the coast are discretized using the implicit Euler method. The discrete larvae transport equations and the adults dynamics form a set of nonlinear algebraic equations. We use a predictor-corrector procedure to solve this problem based on the one developed in [13, 14]. In the following sections we present the mathematical model, the numerical methodology, simulation of some scenarios and, finally, we discuss the numerical results.

2. Mathematical Model

2.1. *Population Dynamics Model*

In this section, a mathematical model is briefly presented to describe the population dynamics of two intertidal barnacles species in a two-dimensional space $\left(\mathbf{x} \equiv (x, y) \in \Re^2\right)$ and in time $\left(t \in (0, T]\right)$. These organisms have two main life phases: a sessile adult phase $(B \equiv B(y,t))$ restricted to the coastline, and a larval phase $(L \equiv L(x,y,t))$ influenced by the pattern of ocean circulation [9].

The model is built based on a single population stage-structured model developed in [5] and on the competition model formulated in [10]. The main assumptions utilized here are described as follow. The coast is considered a straight line and coincides with axis y. Only the central part of the coast is formed by a suitable substratum for the settlement of larvae [5, 8], and the population of adults is located along this habitat, as shown in the figure 1. The larvae are produced by the reproducing adults population of both dominant and subordinate species; they are released in the water column and passively transported by eddy-diffusion k and advection (characterized by the velocity field $v \equiv (v_x, v_y)$) while they develop. These larvae may settle where they strike the suitable coast. This model approximates the interactions previously described by allowing the dominant to settle on, and replace, the subordinate. The subordinate can settle only on free space [3]. Hereafter, we use the subscript $i = 1, 2$ to denote the dominant and the subordinate species, respectively.

The model assumes that the larvae remain homogeneously distributed in a water column of constant depth where they die at the density-independent larvae mortality rates λ_i. Along the coast adults mortality rates μ_i are also independent of density. Larvae settle on the coast at a rate that is proportional to the product of the amount of free space available at time t, defined as $F(y,t)$, and the larval concentration in the water column immediately adjacent to the shoreline at time t, $L_i(0,y,t)$. The constant of proportionality that quantifies the amount of larvae to settle onto the shore is c_i. The amount of free space available in the suitable substratum per unit length of coastline is represented by

$$F(y,t) = A(y) - a_1 B_1(y,t) - a_2 B_2(y,t), \tag{1}$$

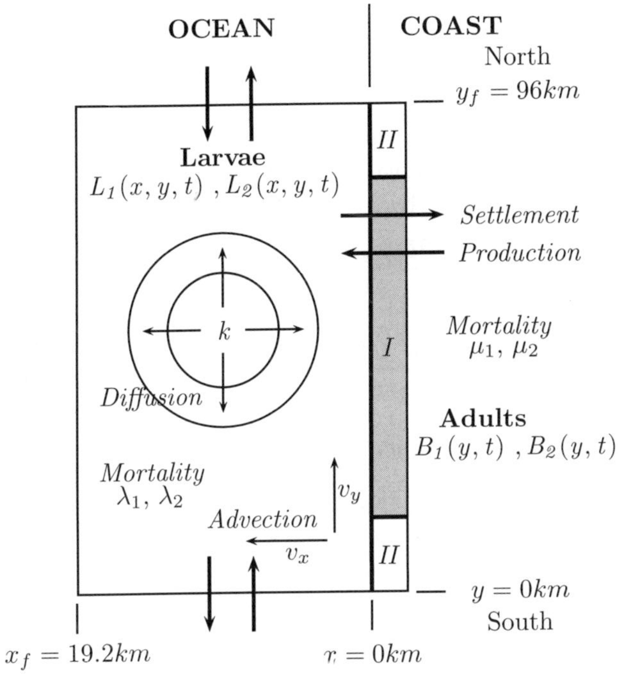

Figure 1. Schematic representation of the model (I) suitable habitat; (II) unsuitable habitat.

where $A(y)$ is the total available free space for recruitment at position y per unit length of coastline; $a_1B_1(y,t)$ and $a_2B_2(y,t)$ quantify the amount of occupied space by dominant and subordinate adults, respectively, at position y and time t per unit length of coastline. In this expression, a_i is the average basal area of an adult and $B_i(y,t)$ is the adult population size of species i per unit length of coastline at time t.

With these assumptions, the rates of change of the number of dominant and subordinate adult barnacles on the coast at position y with respect to time are, respectively, expressed by

$$\frac{dB_1(y,t)}{dt} = c_1L_1(0,y,t)\left[F(y,t) + a_2B_2(y,t)\right] - \mu_1B_1(y,t), \qquad (2)$$

$$\frac{dB_2(y,t)}{dt} = c_2L_2(0,y,t)F(y,t) - c_1L_1(0,y,t)a_2B_2(y,t) - \mu_2B_2(y,t). \quad (3)$$

As initial conditions, it is assumed that the adult populations on the coast are known at the initial time ($B_i(y,t=0) \equiv B_{i;0}(y)$).

The transport of larvae is governed by the advection due to currents in the water column and eddy-diffusion, which is assumed constant. As the larvae of all species depend on the same process scale, $k_1 = k_2 = k$. With these assumptions, the rate of change of the larval populations, $L_i(x,y,t)$, with respect to time is described by:

$$\frac{\partial L_i(x,y,t)}{\partial t} + v_x(x,y,t)\frac{\partial L_i(x,y,t)}{\partial x} + v_y(x,y,t)\frac{\partial L_i(x,y,t)}{\partial y}$$

$$= k\left(\frac{\partial^2 L_i(x,y,t)}{\partial x^2} + \frac{\partial^2 L_i(x,y,t)}{\partial y^2}\right) - \lambda_i L_i(x,y,t)\,, \tag{4}$$

$$x \in (0, x_f), \quad y \in (0, y_f), \quad t > 0,$$

where x_f is the frontal boundary that acts as a convergence zone where barnacle larvae accumulate [9]. We assumed that the initial distributions of larvae are $L_i(x,y,t=0) = L_{i;0}(x,y)$.

Some hypotheses are assumed for the establishment of the boundary conditions. Along the coast, larvae settle and are produced by the reproducing adult populations. The larval flux at the coast (in an outwardly normal direction) is given by the difference between the rate at which larvae enter the larval population (larval production $m_i B_i(y,t)$) and the rate at which they leave the larval population (settlement). Again, the subordinate can settle only on free space (at rate $c_2 L_2(0,y,t)F(y,t)$), whereas the dominant can also settle on space occupied by the subordinate ($c_1 L_1(x=0,y,t)\,[F(y,t) + a_2 B_2(y,t)]$). Therefore, the dominant and subordinate boundary conditions are, respectively,

$$\left.\left(v_x(x,y,t)L_1(x,y,t) - k\frac{\partial L_1(x,y,t)}{\partial x}\right)\right|_{x=0}$$

$$= m_1\, B_1(y,t) - c_1 L_1(x=0,y,t)\,[F(y,t) + a_2 B_2(y,t)]\,. \tag{5}$$

$$\left.\left(v_x(x,y,t)L_2(x,y,t) - k\frac{\partial L_2(x,y,t)}{\partial x}\right)\right|_{x=0}$$

$$= m_2\, B_2(y,t) - c_2 L_2(x=0,y,t)F(y,t). \tag{6}$$

The front $x = x_f$ is a reflecting boundary for both species such that

$$\left(v_x(x,y,t)L_i(x,y,t) - k\frac{\partial L_i(x,y,t)}{\partial x} \right)\Bigg|_{x=x_f} = 0. \tag{7}$$

The diffusive larval flux at the south ($y = 0$) and north ($y = y_f$) boundaries are null and are expressed by

$$-k\frac{\partial L_i(x,y,t)}{\partial y}\Bigg|_{y=0} = 0 \quad \text{and} \quad -k\frac{\partial L_i(x,y,t)}{\partial y}\Bigg|_{y=y_f} = 0. \tag{8}$$

It should be remarked that the equations (2,3) and (4) are coupled through the coastal boundary conditions (5,6).

The four-stages model proposed in [5] is used in this work. In the first stage, larvae are produced in yearly pulses at the beginning of the reproductive season. This feature more accurately represents patterns observed in nature for many species, where offspring are produced seasonally rather than continuously, as considered in [10]. The released larvae initially enter a pre-competency stage that lasts about 21 days. During this period larvae are transported by currents and dispersed by turbulent diffusion but cannot settle even if they encounter a suitable habitat. After that, larvae transition to a competency stage, during which those larvae that reach suitable shoreline settle, exiting the water column and joining the benthic populations. This stage is finite and lasts as long as the pre-competency stage. Afterwards, all larvae that remain in the water column at the end of the competence period die. The settled larvae pass through a several-month juvenile stage that lasts until the following reproductive season, when they spawn for the first time.

The physical and biological parameters used in this work are defined in Table 1 [5, 3]. The biological parameters for the dominant competitor (barnacle *Balanus glandula*) and subordinate (barnacles *Chthamalus dalli* and *Chthamalus fissus*) are based on the field studies of the rocky intertidal zone of California, USA. The stage-structured model implies that the larval production rate (adult fecundity) (m_i) and larval settlement coefficient (c_i) are functions of time. The larval production rate is a pulse at the beginning of each reproductive season; otherwise m_i vanishes. The larval settlement coefficient is zero except during the competency period. For the subordinate species, the larval production rate, m_2, was calculated based on the

Balanus glandula larval production rate reported in [5]. The eddy-diffusion coefficient (k) yields the dispersion of the larvae in the open ocean due to turbulence and depends on the phenomenon scale. Experiments following dye diffusing in the open ocean on the appropriate spatial and temporal scales for barnacle larvae estimate $k = 10\,cm^2/s$ [16, 17].

Table 1: Physical and biological parameters.

Parameters	Nomenclature	Value
Larvae per $100m^2$	$L_i\,(x, y, t)$	
Adult per $100m$ length of coastline	$B_i\,(y, t)$	
Free space available for settlement per $100m$ length of coastline	$F\,(y, t)$	
Adult basal area	a_1	$1 \times 10^{-4} m^2$
	a_2	$5 \times 10^{-5} m^2$
Mortality rate of larvae	λ	$5.6 \times 10^{-7} s^{-1}$
Mortality rate of adults	μ	$2.2 \times 10^{-8} s^{-1}$
Larval production rate	$m_1\,(t)$	0 or $3.2 \times 10^{-3} s^{-1}$
	$m_2\,(t)$	0 or $2 \times 10^{-2} s^{-1}$
Larval settlement coefficient	$c_i\,(t)$	0 or $5 \times 10^{-5} s^{-1}$
Larval pre-competency duration	d_1	$1.8 \times 10^6 s$
Larval competency duration	d_2	$1.8 \times 10^6 s$

3. Numerical Method

We use the finite element method to solve numerically the dominant and subordinate problems defined by equations (4, 5, 7, 8) and (4, 6-8). This methodology is showed in details in [14], regarding accuracy and convergence when dealing with only one species. For completeness, it is briefly presented here. The discrete problems are built based on the Galerkin method using piecewise linear finite element interpolation functions in space and the finite difference method (implicit Euler method) to approximate the time derivative. Thus, defining $S = H^1(\Omega)$, the usual Sobolev space of order 1, and $S^h \subset S$, the discrete variational formulations associated to (4, 5, 7, 8) and (4, 6-8) can be writen, respectively, as: find $L^h_{i;n} \in S^h$ ($i = 1, 2$, where the subscripts 1 and 2 denote the dominant and the subordinate species), for a given time t_n, $n = 1, ..., N$, such that

$$\int_\Omega L^h_{1;n}\,\phi^h\,d\Omega - \Delta t \int_\Omega v_x L^h_{1;n} \frac{\partial \phi^h}{\partial x} d\Omega + \Delta t \int_\Omega v_y L^h_{1;n} \frac{\partial \phi^h}{\partial y}\,d\Omega$$

$$+k\ \Delta t\left(\int_{\Omega}\frac{\partial L_{1;n}^h}{\partial x}\frac{\partial \phi^h}{\partial x}d\Omega+\int_{\Omega}\frac{\partial L_{1;n}^h}{\partial y}\frac{\partial \phi^h}{\partial y}d\Omega\right)+\lambda_1\Delta t\int_{\Omega}L_{1;n}^h\phi^h\ d\Omega$$

$$-\Delta t\int_0^{y_f}\left[\left(m_1B_{1;n}-c_1L_{1;n}^h(A-a_1B_{1;n})\right)\phi^h\right]_{|x=0}dy$$

$$=\int_{\Omega}L_{1;n-1}^h\ \phi^h\ d\Omega \tag{9}$$

$$\int_{\Omega}L_{2;n}^h\ \phi^h\ d\Omega-\Delta t\int_{\Omega}v_xL_{2;n}^h\frac{\partial \phi^h}{\partial x}d\Omega+\Delta t\int_{\Omega}v_yL_{2;n}^h\frac{\partial \phi^h}{\partial y}\ d\Omega$$

$$+k\ \Delta t\left(\int_{\Omega}\frac{\partial L_{2;n}^h}{\partial x}\frac{\partial \phi^h}{\partial x}d\Omega+\int_{\Omega}\frac{\partial L_{2;n}^h}{\partial y}\frac{\partial \phi^h}{\partial y}d\Omega\right)+\lambda_2\ \Delta t\int_{\Omega}L_{2;n}^h\phi^h\ d\Omega$$

$$-\Delta t\int_0^{y_f}\left[\left(m_2B_{2;n}-c_2L_{2;n}^h(A-a_2B_{2;n}-a_1B_{1;n})\right)\phi^h\right]_{|x=0}dy$$

$$=\int_{\Omega}L_{2;n-1}^h\ \phi^h\ d\Omega \tag{10}$$

The dominant and subordinate adult distributions at the coast are discretized using the implicit Euler method in equations 2 and 3, yielding

$$B_{1;n}(y)=\frac{B_{1;n-1}(y)+\Delta t\,c_1\,A\,L_{1;n}^h(0,y)}{\left[\Delta t\,c_1\,a_1\,L_{1;n}(0,y)+\Delta t\,\mu_1+1\right]}, \tag{11}$$

$$B_{2;n}(y)=\frac{B_{2;n-1}(y)+\Delta t\,c_2\ L_{2;n}^h(0,y)\left[A-a_1B_{1;n}(y)\right]}{\left[\Delta t\,c_2\,a_2\,L_{2;n}^h(0,y)+\Delta t\,c_1\,a_1\ L_{1;n}^h(0,y)+\Delta t\,\mu_2+1\right]}, \tag{12}$$

The solutions of (11) and (12) depend on knowing the density of larvae at the coast, $L^h_{i;n}(0,y), i = 1,2$. To solve the non-linear systems of equations (9, 11) and (10, 12) we apply the following predictor-corrector algorithm. This algorithm involves two phases to be performed in each time-step n, $n = 1, ..., nt$, where nt is the total number of time steps. For the two problems, this iterative procedure starts with an initial guess for the values $B_{i;n}(y)$ and $L_{i;n}(x,y)$ at every grid point (x,y), denoted by $B^0_{i;n}$ and $L^0_{i;n}$, where the superscript stands for the number of the iterative step. These guessed values are plugged into equation (11) or (12). Afterwards, the updated values $B^j_{i;n}(y)$ are plugged into equation (9) and (10), and the values $B^j_{i;n}(y)$ and $L^j_{i;n}(x,y)$ are used as new guesses for the next iterative step $(j+1)$. The iterative procedure is repeated n_c times, where n_c stands for the number of iterations in the corrector loop. For transient problems, the convergence at each time is reached when the difference between two subsequent solutions is given by $\left|B^j_{i;n}(y) - B^{j-1}_{i;n}(y)\right| < tol$, where tol is the prescribed tolerance [13, 14, 15].

```
1. B_{i;0}; L_{i;0}; t_0; T_final and nt                       # initial data
   Δt = T/nt; t = t_0
2. n = 1, nt                                                   # time looping
      t = t + Δt
      B^0_{i;n} = B_{i;n-1}; L^0_{i;n}(0,y) = L_{i;n-1}(0,y)   # predictor phase
      3. j = 1, n_c                                            # corrector looping
                                                                 Dominant Species
            determine B^j_{1;n} using B_{1;n-1} and L^{j-1}_{1;n}      by  eq. (11)
            determine L^j_{1;n} using L_{1;n-1} and B^j_{1;n}          by  eq. (9)
            IF |B^j_{1;n}(y) - B^{j-1}_{1;n}(y)| < tol GOTO 4
      4. B_{1;n} = B^j_{1;n} ; L_{1;n} = L^j_{1;n}

      5. j = 1, n_c                                            # corrector looping
                                                                 Subordinate Species
            determine B^j_{2;n} using B_{2;n-1} , L^{j-1}_{2;n} ,
                           B_{1;n} and L_{1;n}                 by  eq. (12)
            determine L^j_{2;n} using B^j_{2;n} , B_{1;n},
                           L^{j-1}_{2;n} and L_{1;n}           by  eq.(10)
             IF |B^j_{2;n}(y) - B^{j-1}_{2;n}(y)| < tol GOTO 6
      6. B_{2;n} = B^j_{2;n} ; L_{2;n} = L^j_{2;n}
```

Algorithm 1. Predictor-corrector.

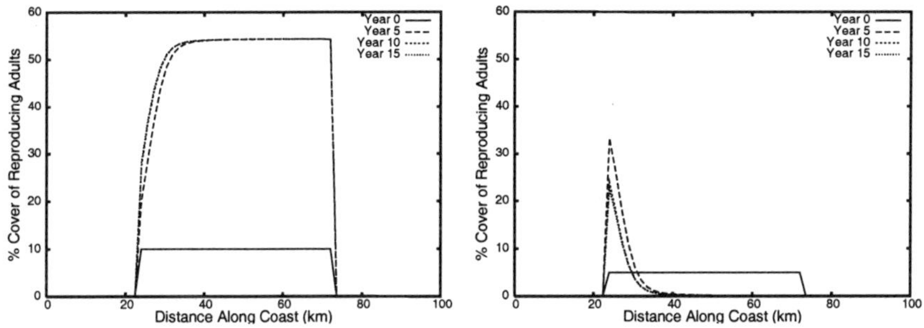

(a) dominant species

(b) subordinate species

Figure 2. Alongshore Flow ($0.4cm/s$): adult shoreline distribution in the suitable habitat.

4. Numerical Results

In this section, we present the numerical results obtained by applying the previous methodology. In all experiments we use $tol = 10^{-3}$ as the prescribed tolerance, $\Delta t = 8h$ as the time step size and a quadrilateral finite element mesh with 30x60 elements. Moreover, we consider that no larvae are present in the ocean initially and 100 000 reproducing adults/$100m$ of coastline of each species are homogeneously distributed along the suitable habitat. To determine the influence of ocean circulation on the competitive interaction in the benthic habitat, we consider two different idealized flow field representations. The first flow field represents a northward unidirectional current parallel to the coastline (alongshore flow). The second flow describes a unidirectional current perpendicular to the shoreline (offshore flow). The role of the flow speed is also investigated by considering three velocity speed levels ($0.25cm/s$, $0.4cm/s$ and $1cm/s$) for each flow field. In the following simulations, the adult population distributions refer to the end of the juvenile stage for each considered year.

4.1. *Alongshore Flow*

Figures 2(a-b), 3(a-b) and 4(a-b) illustrate the dominant and subordinate adult shoreline distributions resulting from the uniform northward alongshore flow. At the flow speed of $0.4cm/s$, figure 2(a) shows that the dominant adult population grows very fast from the initial condition

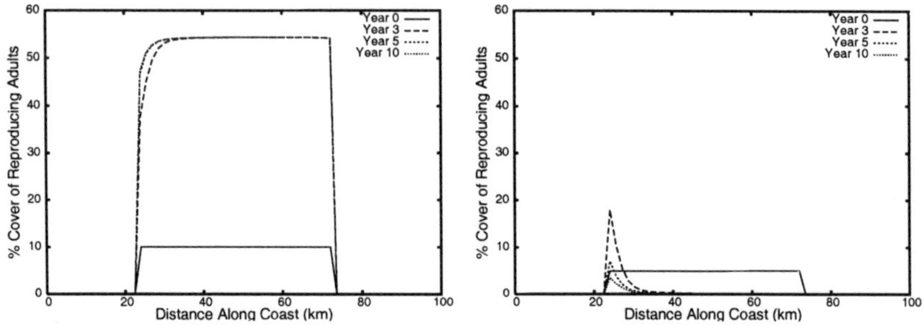

(a) dominant species (b) subordinate species

Figure 3. Alongshore Flow ($0.25cm/s$): adult shoreline distribution in the suitable habitat.

(10% cover) to an equilibrium level (approximately 55% cover), reached at about 10 years. For the subordinate species, we can observe a very different profile in figure 2(b). The subordinate adult population declines from the initial condition (5% cover) towards extinction except in the southern part of the habitat. Due to the slightly lower level of the dominant species on that part of the habitat, the subordinate density grows and declines to an equilibrium level after 15 years. On the rest of the suitable substrate, the dominant species inhibits the settlement of subordinate larvae.

A decrease in the velocity speed has similar consequences in the dominant and subordinate adult shoreline distributions because the slower flows favor the *Balanus glandula* settlement, as shown in figures 3(a-b) for $0.25cm/s$ flow. However, the equilibrium abundance of the subordinate species at the southern part of the suitable substrate is lower than the previous case. This happens because the population interaction strength of the dominant on the subordinate species increases with the decrease in the velocity speed. When velocity speed increases to $1.0cm/s$ the larvae are swept rapidly downstream such that the dominant population slides towards this direction while it declines to extinction (figure 4(a)). The subordinate population benefits from the dominant decay and increases while its distribution also slides downstream, as shown in figure 4(b).

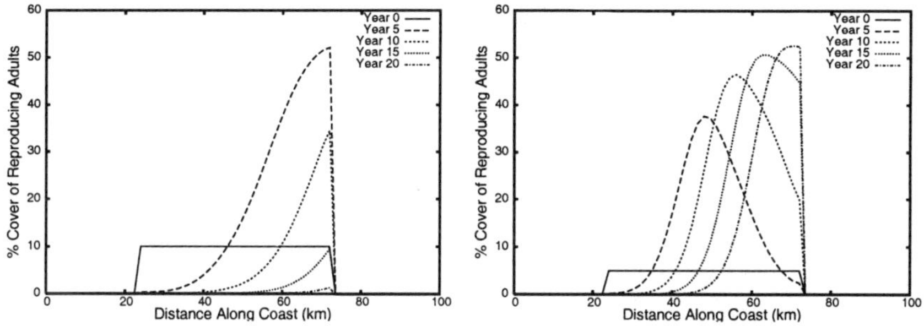

(a) dominant species

(b) subordinate species

Figure 4. Alongshore Flow ($1.0cm/s$): adult shoreline distribution in the suitable habitat.

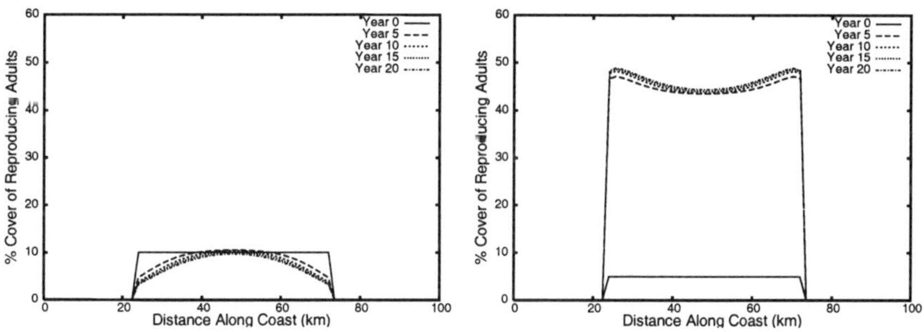

(a) dominant species

(b) subordinate species

Figure 5. Offshore Flow ($0.4cm/s$): adult shoreline distribution in the suitable habitat.

4.2. *Offshore Flow*

Figures 5(a-b), 6(a-b) and 7(a-b) show the dominant and subordinate adult shoreline distributions resulting from the offshore flow. For a velocity speed of $0.4cm/s$, the dominant population decreases as a result of larval transport (figure 5(a)). While this occur, the subordinate population increases, as shown in figure 5(b), because the population interaction strength of the dominant on the subordinate species decreases.

A small decrease in the velocity speed significantly affects the adult

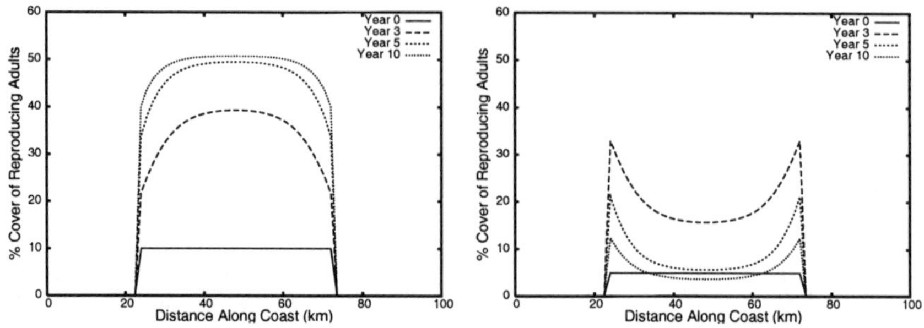

(a) dominant species

(b) subordinate species

Figure 6. Offshore Flow ($0.25cm/s$): adult shoreline distribution in the suitable habitat.

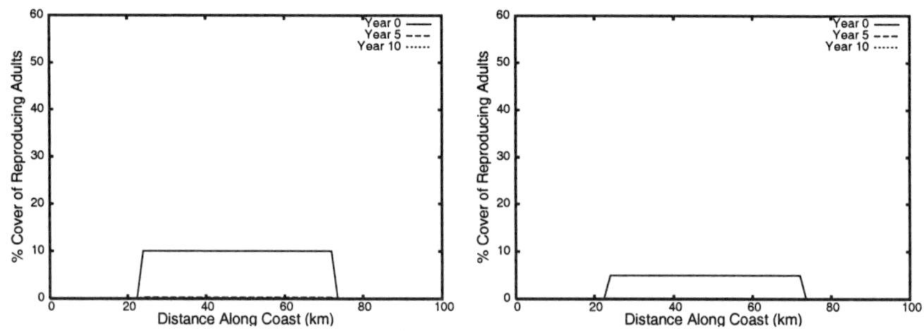

(a) dominant species

(b) subordinate species

Figure 7. Offshore Flow ($1.0cm/s$): adult shoreline distribution in the suitable habitat.

population distributions (figures 6(a-b)). This occurs because the larvae are pushed slower offshore ($0.25cm/s$) than in the previous case. Under this condition, the larval concentration near the coast (competent stage) results on higher recruitment of the dominant species, which rapidly increases (6(a)). This species concentrates primarily in the central part of the suitable habitat, allowing highest abundance of the subordinate species at the upstream and downstream limits of the suitable habitat(figure 6(b)). Because of the flow pattern, larvae are carried faster away from the coast when the velocity speed is $1.0cm/s$ so as the adult population declines rapidly to extinction, for both the dominant and subordinate species (figures 7(a-b)).

5. Discussion

In this work we present a finite element methodology to solve the population dynamics of two intertidal barnacles with complex life cycle that introduces an interspecific competition for space among adults. These species have a planktonic larval stage that disperse in the water column by eddy-diffusion and advection. These physical processes strongly influence the adult distribution and abundance at the coast. We present several scenarios that allow the understanding of the qualitative effects of idealized flow patterns, velocity speed magnitude and adult competition on the population dynamics at the coast. Our results show that coastal flows may affect the adult distribution and the population interaction strength of the dominant on the subordinate species, that decreases as the velocity speed of the flow field increases. These results qualitatively agree with [10], which show that a latitudinal gradient in upwelling intensity in the northeast Pacific produces a gradient in the intensity of species interaction in rocky intertidal communities. These gradients match field data for upper intertidal barnacles *Balanus glandula* and *Chthamalus* (*Chthamalus dalli* and *Chthamalus fissus*) between central California and northern Oregon. The model presented in this article is a step towards more complex models that can assimilate real velocity fields, complex geometries as well as more complex organism interactions. Thus, the main value of the proposed numerical methodology is to provide a reliable tool to study the distribution and abundance of sessile organisms with a planktonic larval stage.

Acknowledgments

The first author would like to thank the fellowship provided by project GEOMA – "Rede Temática de Pesquisa em Modelagem da Amazônia". The second author thanks the Brazilian Government, through the Agency CNPq, for the financial support provided. This work was also supported by the project PRONEX/FAPERJ E-26/171.199/2003.

References

1. C. Ellien, E. Thiébaut, F. Dumas, J. Salomon and P. Nival., *Journal of Plankton Research* **26(2)**, pp. 117-132 (2004).

2. A. S. Barnay, C. Ellien, F. Gentil and E. Thiébaut, *Helgol Mar. Res.* **56**, pp. 229-237 (2003).
3. S. R. Connolly and J. Roughgarden, *Ecological Monographs* **69(3)**, pp. 277-296 (1999).
4. M. L. Carrol *Journal of Experimental Marine Biology and Ecology* **199**, pp. 285-302 (1996).
5. B. Gaylord and S. D. Gaines, *The American Naturalist* **155(6)**, pp. 769-789 (2000).
6. A. S. Pfeiffer-Hoyt and M. A. Mcmanus, *Journal of Plankton Research* **27(12)**, pp. 1211-1228 (2005).
7. J. Roughgarden, S. D. Gaines and H. P. Possingham, *Science* **241**, pp. 1460-1466 (1988).
8. H. P. Possingham and J. Roughgarden, *Ecology* **71(3)**, pp. 973-985 (1990).
9. S. E. Alexander and J. Roughgarden, *Ecological Monographs* **66(3)**, pp. 259-275 (1996).
10. S. R. Connolly and J. Roughgarden, *The American Naturalist* **151(4)**, pp. 311-326 (1998).
11. C. Hirsch, *Numerical Computational of Internal and External Flows* **vol(1)**, John Wiley and Sons (1988).
12. T. J. R. Hughes. *The Finite Element Method - Linear Static and Dynamic Finite Element Analysis.* Dover Publications, Inc.(2000).
13. A. P. C. Rio Doce, R. C. Almeida and M. I. S. Costa. *Proceedings of the Fourth Brazilian Symposium on Mathematical and Computational Biology/First International Symposium on Mathematical and Computational Biology*, **vol(1)**, pp.56-76 (2005).
14. A. P. C. Rio Doce, R. C. Almeida and M. I. S. Costa. *Proceedings of the 2005 International Symposium on Mathematical and Computational Biology/BIOMAT 2005*, pp.1-27 (2006).
15. A. P. C. Rio Doce, R. C. Almeida and M. I. S. Costa. *Anais do XXVI CILAMCE/ Iberian Latin American Congress on Computational Methods in Engineering*, CD-ROM (2005).
16. A. Okubo, *Deep-Sea Reasearch* **18**, pp. 789-802 (1971).
17. A. Okubo and S. A. Levin, *Diffusion and Ecological Problems - Modern Perspectives*, Springer, second edition (2001).

ADVANCES IN A THEORY OF IMPULSIVE DIFFERENTIAL EQUATIONS AT IMPULSE-DEPENDENT TIMES, WITH APPLICATIONS TO BIO-ECONOMICS*

FERNANDO CÓRDOVA-LEPE†

Instituto de Ciencias Básicas, Universidad Católica del Maule,
3605 San Miguel Av., Talca, Chile
E-mail: fcordova@ucm.cl

In this article we study bio-economics of renewable resources which define processes with impulsive dynamical behavior. The biomass is modelled by an ordinary differential equation, but in a sequence of punctual times the biomass jumps by harvest. We consider that the catches are dependent of an effort parameter, the size of the biomass and the time. We suppose that the waiting time for the next harvest instant is a function of the amount captured. We establish the introductory elements for a theory of this new type of impulsive differential equations. It is shown in an example that under logistic growth and impulsive harvest it is always possible to fix a function of the length of closed seasons that determines a sustainable regulation.

1. Introduction

1.1. *Impulsive General-Production Model*

Many evolutionary processes are characterized by the fact that at certain moments they present an abrupt change of state. These processes are conditioned to short-term perturbations whose duration is not significant in comparison with the duration of all the process. So that, it is very natural to assume that these perturbations are instantaneous, this is, they are in the form of impulses. The impulsive differential equations (IDE), this is, ordinary differential equations (ODE) having impulsive effects, appear as a natural description of the evolution of real phenomena.

In this work, we present advances in the construction of a theory of IDE that allows to model some natural evolutionary systems with impulsive

*This work is supported by grant 8.1.6.04 of Universidad Católica del Maule.
†Work partially supported by FIBAS 2306 of Universidad Metropolitana de Ciencias de la Educación.

dynamical behavior. These processes appear in management of marine fisheries, in fact, if a fish population grows continuously and it is submitted to instantaneous harvest, with the lengths of the season closings depending on the size of the capture, we are in presence of impulsive dynamics of our interest. We study the long-term effect on the biomass of the population, of a harvest policy based on continuous prohibition with only a discrete set (without accumulation points) of open access instants; though with the time between harvest moments a function of the captured amount.

The excellent introduction to the management of renewable resources given by Clark[4] is well-known. There, with respect to the metered stock-recruitment models he presents, without say it, some examples of impulsive models. In fact, in these models the parent stock P_k of the k-th generation impulsively gives rise to an initial number of larvae, which decreases continuously (by an ODE modelling a relative mortality) toward the ultimate number of young fish survivors, or the k-th recruitment R_k. An impulsive harvest H_k diminishes the population to the next parent stock P_{k+1}. Thus the Ricker[13] (1954) and the Beverton-Holt[2] (1957) stock-recruitment classical models, can be considered the first, in the bio-economic literature, wherein we can find implicity IDE.

A graphical discussion that assumes the harvest centered in punctual instants is possible to find it in the book of Johnson[10].

From the beginning of the new century many modellers started to introduce some bio-economic examples that consider the harvest as an impulse. Impulsive harvest policy at periodic fixed times is considered in the articles of Córdova and Pinto[6]; and Zhang, Shuai and Wang[20]. Dong, Chen and Sun[7] construct a model with a predator that harvests impulsively. Zhang, Chen and Sun[19] obtain the maximum sustainable yield for seasonal harvesting. Optimal impulsive harvesting for fish polulations is treated in the papers of Zhang, Xiu and Chen[22]; and Zhao, Zhang and Yang[23].

Let us consider a stock whose size at time $t \geq t_0$, certain starting time $t_0 \in \Re$, is denoted by $x = x(t)$. We have in mind a fish population, with $x(t) \in]0, \infty[^n$ representing the total structured biomass in n age groups. The population varies continuously while there is not authorization to capture, depending on its own biology and habitat, according to an ordinary differential law of growth $x'(t) = F(x)$. Moreover, in a sequence of times $\{t_k\}_{k \geq 0}$, as a result of instantaneous harvesting, we have $x(t_k^+) = x(t_k) - H(E, x)$. Here $x(s^+)$ is the limit by the right as $t \to s$. The parameter E is a measure of the exerted effort, that can be considered n-dimensional, so its components E_i, $i = 1, \cdots, n$, are the efforts applied in the i-th age group. If the impulse times are predetermined, we are in the world of the classical IDE. In this work we consider that the impulses occur at variable times. More precisely, for each impulse time t_k, $k \geq 0$, we suppose the next one t_{k+1} determined by a function $G : (0, \infty) \to (0, \infty)$ of the harvested amount, $\|H(E, x(t_k))\|$. So-that, this general-production model will be denoted

$$\begin{cases} x'(t) &= F(x), & t \neq t_k, \\ x(t^+) &= x(t) - H(E, x(t)), & t = t_k, \\ t_{k+1} &= t_k + G(\|H(E, x(t_k))\|). \end{cases} \tag{1}$$

We will call this new kind of IDE an *Impulsive Differential Equation at Impulse-Dependent Times* (IDE-IDT).

Among the authors of studies that incorporate in different ways the impulsive phenomenon to a continuous dynamics are: Bajo[1]; Berezansky and Braverman[3]; Gao and Chen[8]; Qi and Fu[12]; Tarafdar and Watson[17]; and Wang and Wen[18]. Our idea comes to introduce another natural option.

The purpose is to prove the existence of effort levels that permit a sustainable biomass evolution and an optimization of the yield under this regulation.

In Sec. 2, we will prove, for two of these impulsive regulations, but in one dimension, the existence of levels that permit a sustainable biomass evolution. Later, for the first example, the obtained sustainable rate of production is optimized as a function of the effort rate.

Given a level of effort E and a first harvest time t_0, there are two problems about the equilibrium. How do we know if a solution of (1) exists? How do we find explicitly a periodic stable solution of (1)? Answering these questions it persists the problem of the static optimization, this is, how do we determine the expression of the periodic trajectory and the value of the effort and the harvest at the Maximum Sustainable Yield (MSY)? We prove for Malthusian growth and Logistic growth of the biomass, under the assumption of catch per unit time proportional to the biomass, and for a wide domain of the effort parameter, that there exists a *curve of length of closure seasons* which determines a stable periodic solution, this is, a sustainable yield.

In Sec. 3, we will establish some elements for an introductory theory of IDE-IDT. What are we going to understand formally about this type of equations and their solutions? This is a first thing we answer. Advances of results of existence, uniqueness and continuation of solutions are showed. Some directions to develop notions of stability of the impulse points and stability of solutions are given, always related to future applications in bio-economics.

2. Bio-economic Results

2.1. *Malthusian Case*

We consider this first case only with illustrative intentions. We want to show how the impulsive dynamics acts on the biomass, $N(t)$, of a population that is a renewable resource. Supposing the population in a closed homogeneous habitat, with the biomass increasing to a constant per capita rate r, $r > 0$, we have in absence of harvest the most basic model of growth, the Malthusian Model.

Let us take a harvest policy that limits the access to a sequence of instants t_k, $t_k < t_{k+1}$, $k = 0, 1, \cdots$. The harvested amount at each time t_k, this is $N(t_k) - N(t_k^+)$, is a proportion h, $0 < h < 1$, of $N(t_k)$. In principle, h is a function of the displayed effort in this instant.

Our main objective is to suppose that the length of the k-th closure season, $]t_k, t_{k+1}[$, $k = 0, 1, \cdots$, is a function of the catches. We will assume that this interval of time is inversely proportional to the impulsive capture at moment t_k.

The combination (of these three dynamic rules) determines piecewise-continuous trajectories of the population that follows an evolution law that summarizes them in one. We obtain the following first simple example of an IDE-IDT,

$$\begin{cases} N'(t) = rN(t), & \text{if } t \neq t_k \\ N(t^+) = (1-h)N(t), & \text{if } t = t_k \\ t_{k+1} = t_k + \frac{\alpha}{hN(t_k)}, \end{cases} \tag{2}$$

where $(t, N) \in [t_0, \infty[\times]0, \infty[$, some $t_0 \geq 0$, and α is a positive constant that establishes the waiting time per each harvest unit.

This is a new equation in the sense of considering the difference between impulse instants $t_{k+1} - t_k$ a function of the state variable $N(t_k)$. This type of IDE does not appear yet in mathematical literature and will be formally presented in Sec. 3. Hence, now we are going to concentrate the analysis only in the specific example (2) which still does not need a formal theory.

Observe that the assumed regulatory tool establishes that more capture implies less waiting time for the next moment of access, and of inverse manner to a small capture follows a long closed season. This is a conservative rule because corresponds to the implicit supposition that catch per unit of effort (CPUE) is proportional to the biomass stock. In fisheries biology, the basic measure of abundance is the CPUE, it is a direct index of stock abundance. We consider $h = qE$, i.e., the fishing intensity level exerted on the stock is a constant E, $0 < E < 1$, in every instant of impulsive harvest. So, the parameter q is the maximum fraction of the biomass that can be captured and, as its analogous continuous model that was developed by Schaefer[16], it can be called the coefficient of catchability. We note here, in contrast with the continuous model, that E is not a rate per a unit time.

The IDE-IDT (2) is associated to a difference equation that determines the behavior of the piecewise continuous trajectories that satisfy (2). If we consider the question for the functional relation between $N(t_{k+1})$, the biomass just before the $(k+1)$-th capture, and $N(t_k^+)$, the biomass after the k-th capture, it is obtained a recurrence relation. Note that $N(t_k^+)$ is equal to $(1-qE)N(t_k)$, and that this biomass increases at a per capita rate r during $t_{k+1} - t_k = \frac{\alpha}{qEN(t_k)}$ units of time, so that

$$N(t_{k+1}) = (1-qE)N(t_k)\exp\left(\frac{r\alpha}{qEN(t_k)}\right), \quad k = 0, 1, \cdots. \tag{3}$$

If we denote $N(t_k)$ by N_k, $k = 0, 1, \cdots$, then (3) defines the discrete and recursive dynamical system

$$N_{k+1} = F(N_k), \quad k = 0, 1, \cdots, \tag{4}$$

where $F(N) = (1-qE)N\exp(r\alpha/qEN)$, $N > 0$.

There is a biunivocal relation between solutions of (2) and solutions of (4), particularly constant solutions of (4) are periodic solutions of (2) and vice versa.

2.1.1. *Analysis of Steady States*

In the system (4) the factor $(1-qE)$ contracts N, but factor $\exp(\frac{R\alpha}{qEN})$ expands it, so that dynamics has possibilities of not being monotone.

Theorem 2.1. *For any effort E, $0 < E < 1$, there exists a unique periodic biomass trajectory $N_p : [t_0, \infty[\to]0, \infty[$ that satisfies (2) with period $-\frac{1}{r}\log(1-qE)$, the constant length of the closed seasons.*

Proof. It is easy to prove that the function F in (4) has a unique fixed point at

$$N^* = \left(\frac{r\alpha}{qE}\right)\frac{1}{\log\left(\frac{1}{1-qE}\right)}. \tag{5}$$

This steady state N^* of the discrete system (4), determines a initial size of the population so that if the dynamics begins here, at time t_0, the evolution law (2) gives us the following periodic future trajectory

$$N_p(t) = \frac{(1 - qE)N^* \exp(r(t - t_0))}{\exp\left(\frac{k\alpha}{qEN^*}\right)}, \quad t \in]s_k, s_{k+1}], \tag{6}$$

where $s_k = t_0 + \frac{k\alpha}{qEN^*}$, $k = 0, 1, \ldots$. □

Now we study the stability of this periodic solution.

Theorem 2.2. *If $E < \frac{1}{q}(1 - e^{-2})$ the periodic solution N_p of (6) is globally asymptotically stable.*

Proof. Since $F'(N) = \frac{F(N)}{N}\left[1 - \frac{r\alpha}{qEN}\right]$ and $F''(N) - \frac{F(N)}{N}\left(\frac{r\alpha}{qEN}\right)^2 > 0$, we infer that F' is an increasing function and that there exists a global minimum of F at $\frac{r\alpha}{qE}$. For other side, $F(0^+) = \infty$ and $\frac{F(N)}{N} \to 1 - qE$ as $N \to \infty$, then $F'(0^+) = -\infty$ and $F'(\infty) = 1 - qE < 1$.

There are two cases.

a) If $F\left(\frac{r\alpha}{qE}\right) \geq \frac{r\alpha}{qE}$, this is, if $E \leq \frac{1}{q}(1 - e^{-1})$, then $N^* \geq \frac{r\alpha}{qE}$. Since F' is increasing we have $0 = F'\left(\frac{r\alpha}{qE}\right) \leq F'(N^*) \leq 1 - qE$. Therefore, $|F'(N^*)| < 1$, so that N^* is an attractor (globally) of the dynamics of (4).

b) If $F\left(\frac{r\alpha}{qE}\right) < \frac{r\alpha}{qE}$, this is, if $E > E_1 = \frac{1}{q}(1 - e^{-1})$, then $N^* < \frac{r\alpha}{qE}$, i.e., $-\infty < F'(N^*) < 0$. Note that by (5)

$$F'(N^*) = 1 - \left(\frac{R\alpha}{qE}\right)\frac{1}{N^*} = 1 - \log\left(\frac{1}{1 - qE}\right), \tag{7}$$

therefore, N^* is an attractor (globally) of (4), if $-1 < F'(N^*) < 0$, this is, if $E_1 < E < E_2 = \frac{1}{q}(1 - e^{-2})$.

When $E \in]0, E_2[$, we have $N_k \to N^*$ as $k \to \infty$, for all $N_0 > 0$. So that, from any initial condition t_0, $N_0 \geq 0$, the respective solution $N(t)$ of (2) converges to the periodic solution $N_p(t)$, in the sense that for all $\varepsilon, \eta > 0$, there exists $\delta > 0$ such that if $|N_0 - N^*| < \delta$, then $|N(t) - N_p(t)| < \varepsilon$, for all $t > t_0$ such that $|t - s_k| > \eta$. This is an adapted definition of convergency to avoid compare the solutions near the impulsive times of N_p.

When $E \geq E_2$, the population never is under $F\left(\frac{r\alpha}{qE}\right)$, this is, $e(1 - qE)\left(\frac{r\alpha}{qE}\right) > 0$ is an inferior bound. □

2.1.2. *Static Optimization*

If $E < \frac{1}{q}(1-e^{-2})$, we have a sustainable yield qEN^* every $\frac{\alpha}{qEN^*}$ units of time. So the sustainable yield per unit time is

$$Y(E) = \frac{1}{\alpha}(qEN^*)^2. \tag{8}$$

There is an effort E each $\frac{\alpha}{qEN^*}$ units of time, therefore, the effort per unit of time $\tilde{E}$ is

$$\tilde{E} = \frac{1}{\alpha}qE^2N^*. \tag{9}$$

In some manner this value is associated with the variable cost per unit time, this is, the cost of to do an effort E each $\frac{\alpha}{qEN^*}$ units of time. Then the sustainable yield-effort curve that interest optimize to get MSY is

$$Y(\tilde{E}) = q\tilde{E}N^*. \tag{10}$$

Theorem 2.3. *The sustainable yield per unit time, $Y(\tilde{E})$, increases if the effort per unit time, $\tilde{E}$, increases to its maximum value $\frac{r}{\alpha}$.*

Proof. From (5) and (9) we have $\tilde{E} = \frac{rE}{\log\left(\frac{1}{1-qE}\right)}$, thus

$$\tilde{E}'(E) = -\frac{r}{(1-qE)\log^2(1-qE)}[(1-qE)\log(1-qE)-qE]. \tag{11}$$

Putting $v = \frac{1}{1-qE}$ in the well known inequality $\log(v) < v-1$ for $v > 1$, we get $\tilde{E}'(E) < 0$. Then $E'(\tilde{E}) < 0$, moreover if $E \to 0^+$ then $\tilde{E} \to \left(\frac{R}{q}\right)^-$.

Since $Y'(\tilde{E}) = Y'(E)E'(\tilde{E})$, but

$$Y'(E) = \frac{2qr^2\alpha}{(1-qE)\log^3(1-qE)} < 0. \tag{12}$$

we have proved that $Y'(\tilde{E}) > 0$. Therefore, we get greater sustainable yield per unit time if $\tilde{E} \to \left(\frac{r}{q}\right)^-$. □

We conclude that stable yields per unit time more optimal are gotten if the effort $E \to 0^+$. Considering that the time between harvest instants of the stable periodic solution $\Delta s_k = -\frac{1}{r}\log(1-qE) \to 0$, when $E \to 0^+$, we observe that the continuous harvesting policy is superior to this impulsive harvesting policy. These fact, but for an impulsive model at fixed times, was point out in Zhang, Shuai and Wang[20] and also in the similar works of Córdova and Pinto[6].

2.2. *Logistic Case*

Now the production model to consider is

$$\begin{cases} N'(t) &= r(1 - N/K)N, \quad & t \neq t_k, \\ N(t^+) &= N(t) - h \cdot N(t), \quad & t = t_k, \\ t_{k+1} &= t_k + G(h \cdot N(t_k)), \end{cases} \tag{13}$$

where r, $r > 0$, is the intrinsic growth rate, and K, $K > 0$, is the environmental carrying capacity, or saturation level. This is, we take as the natural growth law the Logistic continuous model.

In (2) we gave the function of length of closure seasons, now we are asking us by the appropriated timing policy of harvest. Our problem is to find $G : (0, \infty) \to (0, \infty)$ that permits, for a constant impulsive effort given, the existence of a sustainable yield.

2.2.1. *Existence of Steady States*

Theorem 2.4. *For any constant proportion of harvest h, $0 < h < 1$, there exists a function G, i.e., a length of closure season that determines a unique periodic trajectory of the biomass, with period the length of the closed seasons.*

Proof. For a solution $N : (0, \infty) \to (0, K)$ of (13) we observe that $N(t_{k+1})$ is the value of the solution of the ordinary equation $N'(t) = r(1 - N/K)N$, with initial condition $N(t_k^+)$ at the initial time $t = t_k$. So that

$$N(t_{k+1}) = \frac{N(t_k^+)K}{N(t_k^+) + (K - N(t_k^+)) \exp\left(r(t_{k+1} - t_k)\right)}. \tag{14}$$

Since $N(t_k^+) = (1 - h)N(t_k)$ and $t_{k+1} - t_k = G(h \cdot N(t_k))$, and considering the change of variables $u_k = \frac{N(t_k)(1-h)}{K}$,we get a recurrence relation that takes the form

$$u_{k+1} = \frac{u_k p}{u_k + (1 - u_k) \exp[-rG(\frac{K(1-p)}{p} u_k)]}, \tag{15}$$

where $p = 1 - h$.

The domain of the variable N that is interesting for us is $[0, K]$, then for the new variable u it is $[0, p]$.

For the curve of length of closure seasons we will consider the decreasing function

$$G(H) = -\alpha \ln\left(\frac{H}{K(1 - p)}\right) + \beta, \quad H \in [0; K(1 - p)], \tag{16}$$

where α and β are some positive constants. Note that β determines a minimum length of the closure seasons and α is a parameter to control the decreasing rate of the function.

Considering this function G, the recurrence (15) takes the form

$$u_{k+1} = \frac{u_k p}{u_k + (1-u_k)\left(\frac{u_k}{p}\right)^{r\alpha} e^{-r\beta}}. \tag{17}$$

Now, we are interested in the non null steady state of (17), so we look for $u \in (0,p]$ such that $u + (1-u)\left(\frac{u}{p}\right)^{r\alpha} e^{-r\beta} = p$, expression that can be putted in the form

$$\left(\frac{u}{p}\right)^{r\alpha} e^{-r\beta} = \frac{p-u}{1-u}. \tag{18}$$

The left said of (18) is a decreasing function that in 0 is p and in p is 0. The right side of (18) is 0 in 0 and increase to $e^{-r\beta}$ in p. So, it always there exists a unique fixed point $\hat{u} \in (0,p)$.

We have to observe that $\hat{u}$ defines a unique positive periodic solution of the system (13). □

2.2.2. *Stability of the Steady States*

Theorem 2.5. *The unique periodic trajectory, with period the length of the closed seasons, is a stable solution of (13), i.e., it defines a sustainable yield, if* $h < 1 - \left(\frac{r\alpha-1}{r\alpha+1}\right) e^{-\frac{r\beta}{r\alpha+1}}$.

Proof. It is enough to prove that the steady state $\hat{u}$ of (17) is an attractor, for that we will analyze the derivative of the function $F(u) = up/[u + (1-u)\left(\frac{u}{p}\right)^{r\alpha} e^{-r\beta}]$, in $\hat{u}$.

Considering the identity

$$\frac{u^2 F'(u)}{r\alpha\eta p} = \frac{u^{r\alpha}(u-q)}{\left[1 + \frac{1-u}{u} u^{r\alpha}\eta\right]^2}, \tag{19}$$

where $\eta = e^{-r\beta}/p^{r\alpha}$ and $q = 1 - \frac{1}{r\alpha}$, we have three cases to analyze.

a) If $r\alpha < 1$, we have $q < 0$, then from (19) we have $F'(u) > 0$ for all $u \in]0,p[$. Since F is increasing, $F'(0^+) = \infty$ and $F(0^+) = 0$, moreover the fact of the uniqueness of the fixed point $\hat{u}$, it implies $0 < F'(\hat{u}) < 1$. Therefore $\hat{u}$ is an attractor.

b) If $r\alpha = 1$, then from (19) we have $F(u) = p/(1 + (1-u)e^{-r\beta}/p)$. So, we get $F'(u) = e^{-r\beta}/(1 + (1-u)e^{-r\beta}/p)^2 > 0$, and

$$F''(u) = \frac{2e^{-r\beta}/p}{(1 + (1-u)e^{-r\beta}/p)^3} > 0, \tag{20}$$

in $[0, p]$. From (20), we can conclude that function F' is increasing in $[0, p]$ and then

$$0 < F'(\hat{u}) < F'(p) = \frac{e^{-r\beta}}{\left(1 + \frac{1-p}{p}e^{-r\beta}\right)^2} < e^{-r\beta} < 1. \tag{21}$$

In other words $\hat{u}$ is an attractor.

c) If $r\alpha > 1$, then $q > 0$, so that (19) implies $F'(u) < 0$ if $0 < u < q$ and $F'(u) > 0$ if $u > q$. Then in q the function F has a minimum.

If $F(q) \geq q$, then $q \leq \hat{u} < 1$. Since $F(1) = p < 1$, the uniqueness of $\hat{u}$ as fixed point implies $0 \leq F'(u) < 1$. Then $\hat{u}$ is an attractor.

If $F(q) < q$, then $0 < \hat{u} < q$. We can note that

$$F'(u) = \frac{\eta r\alpha}{p} \frac{u^{r\alpha}(u-q)}{u^2}[F(u)]^2. \tag{22}$$

Then

$$F'(\hat{u}) = \frac{\eta r\alpha}{p}\hat{u}^{r\alpha}(\hat{u} - q) < 0. \tag{23}$$

If we define $g(v) = \frac{\eta r\alpha}{p}v^{r\alpha}(v-q)$, $0 \leq v \leq q$, then we observe that $g(0) = g(q) = 0$ and its minimum is in $v = \frac{q}{2-q}$. So $-1 < F'(\hat{u}) < 0$ if

$$0 > g\left(\frac{q}{2-q}\right) = -e^{-r\beta}\left(\frac{r\alpha - 1}{p(r\alpha + 1)}\right)^{r\alpha+1} > -1. \tag{24}$$

Therefore $\hat{u}$ is an attractor if

$$h < 1 - \left(\frac{r\alpha - 1}{r\alpha + 1}\right)e^{-\frac{r\beta}{r\alpha+1}}. \tag{25}$$

Note that this is only a sufficient condition.

It is important to note that the attractor fixed point of the discrete dynamical system (13), associated with (11), determines a periodic solution of the IDE-IDT that also is stable in a sense to clarify in Sec. 3.2. □

In Córdova and Montenegro[5] it was considered (13) but with $G(H(E, N)) = \alpha/H(E, N)$ and $H(E, N) = qEN$. In that work we get similar conclusions, but using numerical methods.

3. Mathematical Advances

In this section we formalize the IDE-IDT. We define the concept of solution. We give conditions for the local existence of solutions and for the continuation of solutions. Moreover, we give some adapted definitions of stability.

We consider the IDE (1) with $(t,x) \in [0,\infty) \times \Re^n$ and the following conditions:

i) $F : \Re^n \to \Re^n$ is continuous.
ii) $\frac{\partial F}{\partial x_i} : \Re^n \to \Re^n$ is continuous for each $i = 1, \cdots, n$.
iii) $H(E,\cdot) : \Re^n \to \Re^n$ is continuous for each $E \in [0,\infty)^n$.
iv) $G : (0,\infty) \to (0,\infty)$ is continuous.

We propose the following definition of an IDE-IDT solution.

Definition 3.1. Given $E \in [0,\infty)^n$, a function $x : [t_0, \beta) \to \Re^n$, $0 \le t_0 < \beta \le \infty$, with a set T_x of discontinuities without accumulation points in it (discrete set), $t_0 \in T_x$, is a solution of (1) if:

(i) For all $t \in [t_0,\beta) \setminus T_x$, we have $x'(t) = F(x(t))$.
(ii) For all $t \in T_x$, we have: $x(t^-) = x(t)$ and $x(t^+) = x(t) - H(E, x(t))$.
(iii) When t' and t'', $t' < t''$, are two consecutive elements of T_x, then $t'' - t' = G(x(t'))$.

3.1. *Existence and Continuation of Solutions*

Theorem 3.1. *Given a pair $(t_0, x_0) \in [0,\infty) \times \Re^n$, the initial value problem (IVP) formed by (1) and condition $x(t_0) = x_0$ has at most one solution defined in $[t_0, \alpha)$, some $\alpha > t_0$.*

Proof. For a fixed $E \in [0,\infty)^n$ we define $x_0^+ = x_0 - H(E, x_0)$. Note that the ordinary IVP

$$\begin{cases} y'(t) = F(y) \\ y(t_0) = x_0^+, \end{cases} \tag{26}$$

has a unique solution $y : (t_0 - \beta, t_0 + \beta) \to \Re^n$, for some $\beta > 0$. Let α be the smallest of the positive numbers $t_0 + \beta$ and $t_0 + G(\|H(E, x_0)\|)$, then the function $x : [t_0, \alpha) \to \Re^n$ defined as

$$x(t) = \begin{cases} x_0 \text{ if } t = t_0 \\ y(t) \text{ if } t_0 < t < \alpha, \end{cases} \tag{27}$$

is a local solution of the IVP (1) with $x(t_0) = x_0$. □

In this way it is not difficult to have existence and uniqueness of solutions, but we are principally interested (for sustainable yield) in solutions with unbounded domain. So when $\alpha < \infty$ the problem of continuation of solutions is important.

Theorem 3.2. *A solution $x : [t_0, \alpha) \to \Re^n$, $\alpha < \infty$, of the IVP (1) with $x(t_0) = x_0$, has a continuation beyond α if only if*

i) $x(\alpha^-)$ exists, and
ii) $T_x \cap [t_0, \alpha)$ is a finite set.

Proof. Let $x : [t_0, \alpha) \to \Re^n$ be a solution that satisfies i) and ii). Let t_k be the maximum of $T_x \cap [t_0, \alpha)$, then there are two possibilities $\alpha < t_k + G(\|H(E, x(t_k))\|)$ or $\alpha = t_k + G(\|H(E, x(t_k))\|)$. Note that $\alpha > t_k + G(\|H(E, x(t_k))\|)$ implies $t_k + G(\|H(E, x(t_k))\|) \in T_x \cap [t_0, \alpha)$, but this contradicts de definition of t_k.

If $\alpha < t_k + G(\|H(E, x(t_k))\|)$, then since $x(\alpha^-)$ exists we can consider the ordinary IVP formed by $y' = F(y)$ and $y(\alpha) = x(\alpha^-)$ to prolong before of $t_k + G(\|H(E, x(t_k))\|)$.

If $\alpha = t_k + G(\|H(E, x(t_k))\|)$, we define $x(\alpha) = x(\alpha^-)$ and consider the ordinary IVP formed by $y' = F(y)$ and $y(\alpha) = x(\alpha) - H(E, x(\alpha))$, to prolong x before of $\alpha + G(\|H(E, x(\alpha))\|)$ defining $x(t) = y(t)$, if $t > \alpha$.

Inversely if $\tilde{x} : [t_0, \tilde{\alpha}) \to \Re^n$, $t_0 < \alpha < \tilde{\alpha}$, is a continuation of $x : [t_0, \alpha) \to \Re^n$, then $x(t) = \tilde{x}(t)$ for all $t \in [t_0, \alpha)$. It is clear that $x(\alpha^-)$ exists because $\tilde{x}(\alpha) = x(\alpha^-)$.

The set $T_x \cap [t_0, \alpha)$ is bounded, if it is infinite, then it must have an accumulation point $s \in [t_0, \alpha]$. If $s < \alpha$, then necessarily $s \notin T_x$, in this case x is continuous in s and the impulsive dynamic finished, this is a contradiction with the fact that the solution exist until just before α. Then we have $s = \alpha$. But this can not be because if α is an accumulation point of T_x and $\alpha \in T_x$ then, it is also an accumulation point of $T_{\tilde{x}}$, $T_x \subset T_{\tilde{x}}$, but $T_{\tilde{x}}$ is a discrete set. In an other words $T_x \cap [t_0, \alpha)$ is a finite set. □

3.2. *Stability of Solutions*

Associated with the IDE-IDT (1) there is a discrete system. If x is a solution of (1) with t_k and t_{k+1} two of its consecutive impulse times, then we can express a recursive relation between $x(t_{k+1})$ and $x(t_k)$. This relation is given by the integral equivalent version of the IVP

$$\begin{cases} y' = F(y) \\ y(t_k) = x_k^+, \end{cases} \tag{28}$$

where $x_k^+ = x(t_k) - H(E, x(t_k))$.

We know that

$$x(t) = x_k^+ + \int_{t_k}^{t} F(\varphi(s; t_k, x_k^+))ds, \quad t \in (t_k, t_k + G(\|H(E, x(t_k))\|)], \tag{29}$$

where $\varphi(s; t_k, x_k^+)$ is the solution of the ordinary IVP (28).

If in (29) we do $t = t_{k+1}$ we get a recurrence of the form $x(t_{k+1}) = T(x(t_k))$. For the IDE-IDT (2) and (13) we prove that T has a unique fixed point. We will denote this fixed point $\hat{x}$, and the periodic solution, defined by $\hat{x}$ and with starting time τ, will be denoted x_τ.

If $\tau = t_k$ and t is such that $t - \tau$ is positive and less than the minimum between $\{G(\|H(E, x(\tau))\|)$ and $G(\|H(E, \tilde{x})\|)\}$, then

$$|x(t) - x_\tau(t)| \leq |x_k^+ - \hat{x}^+| + \int_{\tau}^{t} |F(\varphi(s; \tau, x_k^+)) - F(\varphi(s; \tau, \hat{x}^+))|ds, \tag{30}$$

where $\hat{x}^+ = \hat{x} - H(E, \hat{x})$.

For a solution of (2) this last inequality (30) takes the form

$$|x(t) - x_\tau(t)| \leq |x_k^+ - \hat{x}^+|e^{r(t-t_k)} \leq |x_k^+ - \hat{x}^+| \exp\left(rG(|H(E, x_k^+)|)\right). \tag{31}$$

So that, if $\hat{x}$ is a fixed point which is an attractor and $x(t_k)$ is attracted for some k big enough, the continuity of G and $H(E, \cdot)$ implies that for all $\varepsilon > 0$ there exists $j > k$ such that $|x(t) - x_\tau(t)| < \varepsilon$ for all $t \in (t_j, t_j + \min\{G(|H(E, x(t_j))|), G(|H(E, \hat{x})|)\}$. For a solution of (13), it is also possible to get similar comparisons with the corresponding periodic solution defined by the fixed point, when it exists.

For defining stability concepts we need to know how to measure nearness between solutions. For a pair of solutions $x^1, x^2 : [t_0, \infty) \to \Re^n$ of (1), that are piecewise continuous functions with discontinuities at different times, we have to define a kind of metric for controling the proximity between them. The main problem is the comparison in the neighborhoods of the impulse times. In the above two examples it is impossible to do punctual comparisons in the times among $t_{k_n} + G(|H(E, x(t_{k_n}))|)$ and $t_{k_n} + G(|H(E, \hat{x})|)$. A chance is to work, for example, with the notion of quasi-stability of IDE at Variable Times solutions introduced in Samoilenko and Perestyuk[15].

Definition 3.2. A solution x of (1) such that $x(t_0) = x_0$, is stable if for each $\varepsilon > 0$ and $\eta > 0$ there exists $\delta > 0$ such that $|x_0 - \tilde{x}_0| < \delta$ implies

$|x(t) - \tilde{x}(t)| < \varepsilon$, for all $t \geq t_0$, $|t - t_x^k| > \eta$ and $k \geq 0$, where $\tilde{x}$ is a solution of (1) with $\tilde{x}(t_0) = \tilde{x}_0$ and $\{t_x^k\}$ are the impulse times of x.

In order to control nearness between impulse points we can explore some notions of stability of impulse points by means of the distance $D_k(x^1, x^2) = |t_k^1 - t_k^2| + \|x^1(t_k^1) - x^2(t_k^2)\|$, where $\{t_k^1\}$ and $\{t_k^2\}$ are the sequences of impulse times of the functions x^1 and x^2 respectively. With this metric we introduce the following definition.

Definition 3.3. A solution $x : [t_0, \infty) \to \Re^n$, with $x(t_0) = x_0$, $(t_0, x_0) \in [0, \infty) \times \Re^n$, is of stable impulse points if $\forall \varepsilon > 0$, $\exists \delta > 0$, such that, for each solution $y : [t_0, \infty) \to \Re^n$, $y(t_0) = y_0$, $(t_0, y_0) \in [0, \infty) \times \Re^n$, the inequality $|x_0 - y_0| < \delta$ implies $D_k(x, y) < \varepsilon$, for certain k and its successors.

We affirm that quasi-stability and our stability of impulse points are independent concepts, but technically this last one is easier to be worked. We conjecture that with the control in the impulse points and some usual conditions of stability on F, with the continuity of H and G, it is possible to prove quasi-stability.

References

1. I. Bajo, *Pulse acumulation in impulsive differential equations with variable times.* J. of Math. Analysis and Applications. 216(1) 211-217, (1997).
2. R.J.H. Beverton and S.J. Holt, *On the dynamics of exploited fish populations. Ministry of Agriculture, Fisheries and Food (London),* Fish. Invest. Ser. 2(19), (1957).
3. L. Berezansky and E. Braverman, *On impulsive difference equations and applications.* J. of Dif. Equa. and Appl. 10(9), 851-868, (2004).
4. C.W. Clark, *Mathematical Bio-economics: the optimal management of renewable resources.* 2nd edition, Wiley, New York, (1990).
5. F. Córdova-Lepe and M. Montenegro, *Sustainable impulsive harvesting with next capture time inversely proportional to the present catch.* Actas Taller de Biomatemática and BIOMAT-V, Ufro Valdivia-Chile, Jan. (2005).
6. F. Córdova and M. Pinto, *Bioeconomía Matemática, explotación y preservación.* CUBO, U. de La Frontera. 4(1) 49-64, (2002).
7. L. Dong, L. Chen and L. Sun, *Extinction and permanence of the predator-prey system with stocking of prey and harvesting of predator impulsively.* Math. Methods Appl. Sci. 2a(4), 415-425, (2006).
8. SJ. Gao and LS. Chen, *The effect of seasonal harvesting on a single-species discrete population model with stage structure and birth pulses.* Chaos Solitons and Fractals 24(4), 1013-1023, (2005).
9. S. Guenettes and TJ. Pitcher, *An age-structure model showing the benefits*

of marine reserves in controlling overexploitation. Fisheries Research 39(3), 295-303, (1991).

10. F. Johnson, *Fisheries. Harvesting life from water.* Kendall, Hunt Publishing Company. (1989).
11. B. Lui and L. Chen, *The periodic Lotka-Volterra model with impulsive effect.* Math. Medicine and Biology - A journal of the IMA. 21(2), 129-145, (2004).
12. JG. Qi and XL. Fu, *Comparison principle for impulsive differential systems with variable times.* Indian J. Pure Ap. Mat. 32(9), 1395-1404, (2001).
13. W.E. Ricker, *Stock and recruitment.* Journal of the Fisheries Research Board of Canada. 11, 559-623, (1954).
14. R.J. Rowley, *Marine Reserves in Fisheries Management.* Aquatic Conservation-Marine and Freshwater Ecosystems. 4(3), 233-254, (1994).
15. A.M. Samoilenko and N.A. Perestyuk, *Impulsive differential equations.* World Scientific Series on Nonlinear Science, Series A, vol. 14. (1995).
16. Schaefer, *Some aspects of the dynamics of populations important to the management of commercial marine fisheries.* Bull. Inter-Amer. Trap Tuna Comm. 1, 25-26, (1954).
17. E. Tarafdar and PJ. Watson, *Periodic solutions of impulsive differential equations.* Dynam. of Continuous Discrete and Impulsive Systems. 6(2), 301-306, (1999).
18. YL. Wang and CB. Wen, *Existence of solutions nth-order impulsive differential equations with variable times.* Dynam. Cont. Dis. Ser. A. 13. 554-562. Part2 Suppl. S (2006).
19. Y. Zhang, L. Chen and L. Sun, *Maximum sustainable yield for seasonal harvesting in fishery management.* Liu, Xin zhi (ed) et al., IDS and applications. Proceedings of the 1st international conference, Shangai, China, Aug. 7–9, (2004). Waterloo: Watorn Press. 311-316 (2004).
20. XY. Zhang, ZS. Shuai and K. Wang, *Optimal impulsive harvesting policy for single population.* Nonlinear Analysis-Real World Appl. 4(4), 639-651, (2003).
21. Y. Zhang and JT. Sun, *Stability of impulsive delay differential equations with impulses at variable times.* Dynam. Syst. 20(3), 323-331, (2005).
22. Y. Zhang, Z. Xiu and L. Chen, *Optimal impulsive harvesting of a single-species with Gompertz law of grow.* J. Biol. Syst. 14(2), 303-314 (2006).
23. L. Zhao, Q. Zhang and Q. Yang, *Optimal impulsive harvesting for fish populations.* J. Syst. Sci. Complex. 16(4), 466-474, (2003).

MULTISTABILITY ON A LESLIE-GOWER TYPE PREDATOR-PREY MODEL WITH NONMONOTONIC FUNCTIONAL RESPONSE

BETSABÉ GONZÁLEZ-YAÑEZ, EDUARDO GONZÁLEZ-OLIVARES*, JAIME MENA-LORCA†

Grupo de Ecología Matemática, Instituto de Matemáticas
Pontificia Universidad Católica de Valparaíso
E-mails: betsabe.gonzalez@ucv.cl, ejgonzal@ucv.cl, jmena@ucv.cl

In this work, a bidimensional continuous-time differential equations system is analyzed which is derived from a Leslie type predator-prey model by considering a nonmonotonic functional response (or Holling type IV or Monod Haldane). This functional response is employed to explain a class of prey antipredator strategies and we study how it influences in bifurcation and stability behavior of model. System obtained can have one, two or three equilibrium point at interior of the first quadrant, but here we describe the dynamics of the particular cases when system has one or two equilibrium points. Making a time rescaling we obtain a polynomial differential equations system topologically equivalent to original one and we prove that for certain subset of parameters, the model exhibits biestability phenomenon, since there exists an stable limit cycle surrounding two singularities of vector field one of these stable. We prove that there are conditions on the parameter values for which the unique equilibrium point at the first quadrant is stable and surrounded by two limit cycles, the innermost unstable and the outhermost stable. Also we show the existence of separatrix curves on the phase plane that divide the behavior of the trajectories, which have different ω-limit sets, and we have that solutions are highly sensitives to initial conditions. However, the populations always coexist since the singularity $(0,0)$ is a nonhyperbolic saddle point.

1. Introduction

In this work we present the first part of analysis of a deterministic continuous predator-prey model considering two important aspects for describe the interaction: a logistic type growth function for predator and a functional response of predators of non-monotonic type. The first aspect characterize

*Partially supported by Fondecyt project 1040833 and DI. PUCV project 124794/2004
†Partially supported by DI. PUCV project 124704/2006

Leslie type predator-prey models[23] or logistic predator prey model[6,36] or Leslie-Gower model[4,21], in which the conventional environmental carrying capacity K_y is a function of the prey quantity x, that is, dependent of the available resources[27]. Here we consider that $K_y = nx$, proportional to prey abundance as in the May-Holling-Tanner model[3,32].

Although Leslie models can leads anomalies in their predictions [36], because it predicts that even at very low prey density, when the consumption rate by individual predator is essentially zero, predator population can nevertheless increase, if predator prey ratio is very small[36], this models are recently employed[18]. This form of modeling differs from a more common Gause type model[12,16,44] in which the predator equation is based in mass action principle, because the numerical response is dependent of functional response.

It is well known that *functional response of predators* or *consumption rate function* refers to the change in the density prey attached per unit time per predator as the prey density changes[12,31]. In most predator-prey models considered in the ecological literature, the predator response to prey density is assumed to be monotonic increasing, the inherent assumption being that the more prey in the environment, the better off the predator [13].

However, there is evidence that indicates that this need not always be the case, for instance when a type of antipredator behavior (APB) exists. *Group defence* is one of this, and the term is used to describe the phenomenon whereby predators decrease, or even prevented altogether, due to the increased ability of the prey to better defend or disguise themselves when their number are large enough[13,31,40,41] and in this case a non-monotonic functional response is better. For example, lone musk ox can be successfully attached by wolves, however large herds of them can be attached but with rare success.

An another manifestation of APB in which a non-monotonic functional response can be used, is the phenomenon of *aggregation*, a social behavior of prey, in which prey congregate on a fine scale relative to the predator, so that the predator's hunting is not spatially homogeneous[35], such as succeed with miles long schools of certain class of fishes. In this case, a primary advantage of schooling seems to be confusion of predator when it attacks. The more important benefits of aggregation than group defence is an increased in wariness. Moreover, aggregation can be both decrease in vulnerability to attack and increase the time group member can devote to activities other than surveillance[35].

Also related examples of non-monotone consumption occur at the microbial level where evidence indicates that when faced with overabundance of nutrient the effectiveness of the consumer can begin to decline. This is often seen when micro-organism are used for waste discomposing or for water purification, phenomenon that is called *inhibition*[13,31,41].

The functional response curves, in particular a non-monotonic curve that it shows in Figure 1, has an upper bound on the rate of predation per individual predator at some prey density, in contrast to the old Lotka-Volterra model which had assumed a linear relationship between prey density and the rate of predation over the entire range of prey densities[35].

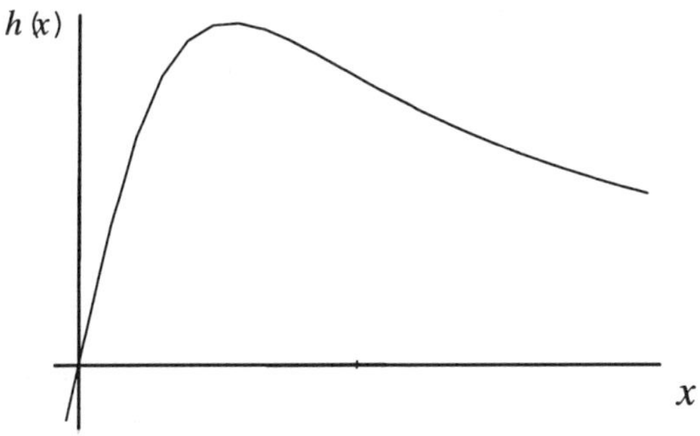

Figure 1. A non-monotonic functional response.

Here we use the function $h(x) = qx/(x^2 + a)$ also employed in[14,15,31,40,41,42] and correspond to Holling type IV functional response[35] which is generalized as $h(x) = qx/(x^2 + bx + a)$ in[7,43,44] for a Gause model.

This generalized expression is derived by Collings[10], which affirms that this type of functional response seems a reasonable possibility if it is assumed that prey and webbing densities are directly related. In[19] it is affirmed that the corresponding Gause model has an unique stable limit cycle, but in[44], it is proved the existence of two limit cycle. Moreover the system exhibit bifurcation of cusp-type with codimension two[3,17] or Bogdanović-Takens bifurcation[22,43,44]. The same model is modified in [25] considering delay and in[28] for a reaction diffusion system.

The model here studied is partially analyzed by Collings[10] and Li and Xiao[24]. Collings employs a computer program to determine the behavior of system. In[24], the same model is analyzed in $\mathbb{R}_2^+ = \{(x, y) : x > 0, y \geq 0\}$, since it is not well-define at $x = 0$. Using a topologically equivalent polynomial system[2,8,33] to original one, we show that the singularity $(0, 0)$, is a nonhyperbolic saddle point.

We determine conditions for existence of one, two and three equilibrium points or singularities at interior of the first quadrant and we demonstrate that when a unique positive singularity there exist, then a subset of parameters there is for which two (infinitesimal) limit cycles can surround these singularity (multiple Hopf bifurcation), the innermost unstable and the outhermost stable which it answers partially the question formulated in[9].

In[24], by simulation is shown the possibility of existence of two limit cycles, one of this surrounding both singularities and the other around only one of them. We prove that this last limit cycle is non-infinitesimal[5], and it can not obtained by Hopf bifurcation. Then, it can coexist an repellor node with an unstable limit cycle that appear because an homoclinic of the other singularity is broken and a stable limit cycle surrounding both singularities.

2. The Model

In this work we denote by $x = x(t)$ and $y = y(t)$ the prey and predator population size of respectively (measured in biomass, density or number), assuming that they vary continuously with time, uniform distribution over space, neither age or sex structure and as variables as parameters are of the deterministic nature. We consider that the functional response is non-monotonic and then the Leslie type model (or Leslie-Gower model)[4,6,21,23] is expressed by the following autonomous bidimensional differential equations system:

$$X_\mu : \begin{cases} \frac{dx}{dt} = \left(r\left(1 - \frac{x}{K}\right) - \frac{q\,y}{a + x^2} \right) x \\ \frac{dy}{dt} = s\left(1 - \frac{y}{n\,x}\right) y \end{cases} \tag{1}$$

System (1) is of Kolmogorov type[12,26] and all the parameters are positives, that is, $\mu = (r, K, a, q, s, n) \in \mathbb{R}^6_+$ with $a < K$ and have the following meanings:

- r and s are the intrinsic growth rates or biotic potential of the prey and predators respectively.
- K is the prey environment carrying capacity.
- q is the predator maximum consumption rate per capita, i.e., the maximum number of prey that necessary can be eaten by a predator in each time unit.

- $\sqrt{a}$ is the amount of prey for which the predation effect is maximum.
- n is a measure of the quality of the food that provides the prey and it is converted into predators birth.

System (1) is defined at the set

$$\Omega = \{(x, y) \in \mathbb{R}^2 |\, x > 0, y \geq 0\} = \mathbb{R}^+ \times \mathbb{R}_0^+$$

The equilibrium point of system (1) or singularities of vector field X_μ are $P_K = (K, 0)$ and the intersection points between the isoclines

$$r\left(1 - \frac{x}{K}\right) - \frac{q\, y}{a\, +\, x^2} = 0 \;\text{ and }\; y = nx.$$

We note that system (1) is not defined at $O = (0, 0)$, but it can possible to make a continuous extension to this point. For simplify the calculus, following the methodology used in[1,14,15,16,32,37] we make the change of variables and time rescaling given the function $\varphi : \hat{\Omega} \times \mathbb{R} \longrightarrow \Omega \times \mathbb{R}$ such that,

$$\varphi(u, v, \tau) = \left(Ku, Knv, \frac{u}{r}\left(\frac{a}{K^2} + u^2\right)\tau\right) = (x, y, t)$$

where $\hat{\Omega} = \{(u, v) \in \mathbb{R}^2 |\, u \geq 0 \text{ and } v \geq 0\}$ and it has that

$$\det D\varphi(u, v, \tau) \;=\; \frac{\nu(a + K^2 u^2)}{r} > 0,$$

that is, φ is diffeomorphism for which the vector field X_μ, in the new coordinates system is topologically equivalent to $Y_\eta = \varphi \circ X_\mu$[2,33]; it has the form $Y_\eta = P(u, v)\, \frac{\partial}{\partial u} + Q(u, v) \frac{\partial}{\partial v}$ and the associated polynomial differential equations system with four parameters is

$$Y_\eta : \begin{cases} \frac{du}{d\tau} = \left((1 - u)(A + u^2) - Q\, v\right)\, u^2 \\ \frac{dv}{d\tau} = B\, (u - v)\, (A + u^2)v \end{cases} \tag{2}$$

where $A = (a/K^2) < 1$, $B = s/r$, $Q = qn/(rK)$ and $\eta = (A, B, Q) \in \nabla =]0, 1[\, \times \mathbb{R}_+^2$. The equilibrium points of Y_η in $\hat{\Omega}$ are: $O = (0, 0)$, $P_1 = (1, 0)$; moreover there exists those that are determined by the intersection of isoclinic

$$v = \frac{1}{Q}\, (1 - u)\, (A + u^2) \;\text{ and }\; v = u.$$

The abscise of this point at $\hat{\Omega}$, satisfy the third grade equation

$$h(u) = -u^3 + u^2 - (A + Q)\, u + A = 0 \tag{3}$$

By Descartes Rule, the polynomial $h(u)$, it can have a unique real positive root or three different real positive roots or two different real positive root, one of this with multiplicity two.

We write down by $u_e = H$, the real positive root that always exists for $h(u)$ when $A \in \,]0,1[$ and $Q \in \mathbb{R}$ and we denote by $P_e = (H,H)$, the equilibrium point that always exists at $\hat{\Omega}$. In this situation, we will see that

$$Q = \frac{1}{H}(1-H)\left(H^2+A\right).$$

The vector field Y_η becomes

$$Y_\lambda : \begin{cases} \frac{du}{d\tau} = \left((1-u)(A+u^2) - \frac{1}{H}(1-H)\left(H^2+A\right)\ v\right)\ u^2 \\ \frac{dv}{d\tau} = B\ (u-v)\ (A+u^2)v \end{cases} \tag{4}$$

with $\lambda = (H,A,B) \in \Delta = \,]0,1[\, \times \,]0,1[\, \times \mathbb{R}_+$, and the jacobian matrix (community matrix) is:

$$DY_\lambda(u,v) = \begin{pmatrix} Y(u,v)_{11} & -\frac{1}{H}(1-H)\left(H^2+A\right)u^2 \\ Bv\left(-2uv+A+3u^2\right) & B\left(-2v+u\right)\left(A+u^2\right) \end{pmatrix}$$

where

$$Y(u,v)_{11} = -3u^2A - 5u^4 + 4u^3 + 2uA - 2\frac{1}{H}(1-H)\left(H^2+A\right)uv.$$

3. Main Results

For system (2) or vector field Y_η, we have the following results:

Lemma 3.1.

(1) The set $\breve{\Gamma} = \left\{(u,v) \in \hat{\Omega} \ / \ 0 \le u \le 1,\ 0 \le v\right\}$ is an invariance region.

(2) The solution are bounded.

Lemma 3.2.

(1) For all $\eta = (A,B,Q) \in \nabla$ the equilibrium $(1,0)$ is a saddle point.

(2) The singularity $(0,0)$ is a nonhyperbolic saddle point, for all $\eta = (A,B,Q) \in \nabla$.

As the singularities $(0,0)$ and $(1,0)$ of vector field Y_η are saddle point, this results implies that, the prey and predator populations always coexists for any set of parameters in vector field X_μ. Also is important to prove that the solutions are bounded, since the strip $\breve{\Gamma}$, it is possible apply the Poincaré-Bendixon Theorem[2,3,8,17,22,33].

Now, let $Q = \frac{1}{4}(1-H)(H+1)^2$ and we define

$$T = H(1-H)^2 - 4A \text{ and } S^2 = HT.$$

Lemma 3.3.

(1) The third degree equation (3) has:

(a) A unique real positive root $u_e = H$, if and only if, $T < 0$.
(b) Three different real positives root, if and only if, $T > 0$.
(c) Two real positive root, some of which has multiplicity two, if and only if, $T = 0$, and then

$$A = \frac{H(1-H)^2}{4}.$$

(d) A unique real positive root of multiplicity three for $u_e = \frac{1}{3}$, if and only if, $A = \frac{1}{27}$ and $Q = \frac{8}{27}$.

(2) For system (3) or vector field Y_λ, we have:

(a) If $T < 0$, there exists a unique equilibrium point $P_e = (H,H)$ at interior of $\hat{\Omega}$.
(b) If $T = 0$, there exist two equilibrium points at interior of $\hat{\Omega}$ which are

$$P_e = (H,H) \text{ and } P_{e2} = \left(\frac{1-H}{2}, \frac{1-H}{2}\right).$$

(c) If $T > 0$, there exist three equilibrium points at interior of $\hat{\Omega}$ which are $P_e = (H,H)$, $P_{e3} = (L_1, L_1)$ and $P_{e4} = (L_2, L_2)$ with

$$L_1 = \frac{H(1-H)+S}{2H} \text{ and } L_2 = \frac{H(1-H)-S}{2H}.$$

(d) In particular, if $A = \frac{1}{27}$ and $Q = \frac{8}{27}$, then $P = \left(\frac{1}{3}, \frac{1}{3}\right)$ is the unique equilibrium point at interior of $\hat{\Omega}$.

We have that there exists a unique equilibrium point in the region of parameters values defined by

$$\Delta_1 = \left\{(H,A,B) \in \Delta \,|\, 0 < H < 1,\, 0 < A < 1,\, A > \frac{H(1-H)^2}{4}\right\}.$$

There are two positives equilibrium point at the region

$$\Delta_2 = \left\{(H, A, B) \in \Delta \,|\, 0 < H < 1, 0 < A < 1, \;\; A = \frac{H(1-H)^2}{4}\right\}.$$

There are three equilibrium point at the region

$$\Delta_3 = \left\{(H, A, B) \in \Delta \,|\, 0 < H < 1, 0 < A < 1, A < \frac{H(1-H)^2}{4}\right\}.$$

Also, there is a unique equilibrium by the collapse of the three equilibrium points when $H = \frac{1}{3}$, for $A = \frac{1}{27}$ and $Q = \frac{8}{27}$. In the following figures it is can see the above curves and region.

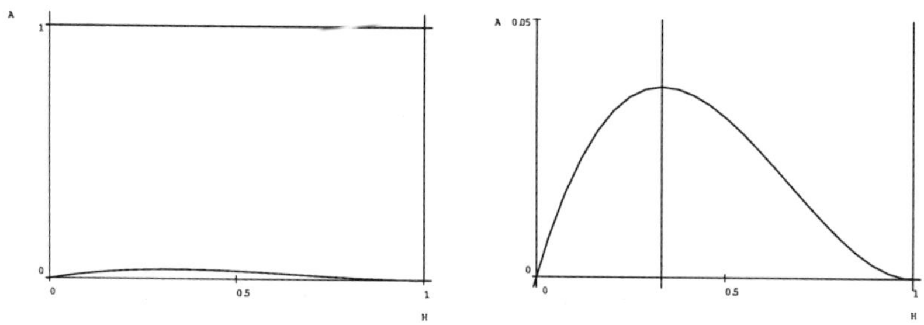

(a) (b)

Figure 2. (a) The parameter space for the existance of positive equilibrium points. (b) A zoom for the bifurcation curve $A = \frac{H(1-H)^2}{4}$ Above this curve there is a unique equilibrium point, below this curve there are Three equilibrium points. Over the curve two equilibrium points collapse.

We detach that the region Δ_1 is greatest that Δ_2, for which the probabilities of existence of three equilibrium points is lower than the existence of one equilibrium point.

3.1. *Stability of the positive equilibrium points*

Case a

For system (2) we have that

Theorem 3.1. *If $A = \frac{1}{27}$ and $Q = \frac{8}{27}$, then the point $P_e = \left(\frac{1}{3}, \frac{1}{3}\right)$ is the unique equilibrium point at the interior of the first quadrant, which is*

(1) an attractor node, if and only if, $B > \frac{2}{3}$.
(2) a repellor node, if and only if, $B < \frac{2}{3}$. *In this case. By the Poincaré-Bendixon Theorem there exists a noninfinitesimal limit cycle.*

We note that the existence of this non-infinitesimal limit cycle[5] that it persist below small perturbations on the parameter value. This has a great importance, because system (2) has a limit cycle for certain parameter value which is not obtained by Hopf bifurcation[2,8,34] as is usually in analytical system.

Case b

We suppose that $(H, A, B) \in \Delta_1$, then $T < 0$, and there exists a unique equilibrium point $P_e = (H, H)$ at $\hat{\Omega}$.

Theorem 3.2. *For system (4) we have that the nature of* P_e *depend of trace since*

$$\det DY_\eta(H, H) = H^4 B \left(A + 2H^3 - H^2\right)\left(A + H^2\right) > 0.$$

Then, (H, H) *is*

(1) a repellor focus, surrounded by a limit cycle, if and only if,

$$B < \frac{H\left(2H - 3H^2 - A\right)}{H^2 + A}.$$

(2) an attractor, if and only if,

$$B > \frac{H\left(2H - 3H^2 - A\right)}{H^2 + A}.$$

(3) a two order weak focus[8,26]*, if and only if,*

$$B = \frac{H\left(2H - 3H^2 - A\right)}{H^2 + A}.$$

Case c

Now, we assume that $(H, A, B) \in \Delta_2$, then $T = H\left(1 - H\right)^2 - 4A = 0$; we obtain that and $S = 0$, and there exist two equilibrium points at interior of $\hat{\Omega}$. The vector field Y_η it transforms to

$$Y_\theta : \begin{cases} \frac{du}{d\tau} = \left(\left(1 - u\right)\left(\frac{H(1-H)^2}{4} + u^2\right) - \frac{1}{4}\left(1 - H\right)\left(H + 1\right)^2 v\right) u^2 \\ \frac{dv}{d\tau} = B\,\left(u - v\right)\left(\frac{H(1-H)^2}{4} + u^2\right) v \end{cases} \tag{5}$$

where $\theta = (H, A, B) \in \Delta_2$. We define the following subset of Δ

$$R_1 = \left\{ (H, A, B) \in \Delta_2 \,|\, B > \frac{H\left(7 - 10H - H^2\right)}{\left(1 + H\right)^2} \text{ and } B > \frac{H + 1}{2} \right\}$$

$$R_2 = \left\{ (H, A, B) \in \Delta_2 |\, B > \frac{H\left(7 - 10H - H^2\right)}{(1+H)^2} \text{ and } B < \frac{H+1}{2} \right\}$$

$$R_3 = \left\{ (H, A, B) \in \Delta_2 \,|\, B < \frac{H\left(7 - 10H - H^2\right)}{(1+H)^2} \text{ and } B < \frac{H+1}{2} \right\}$$

$$R_4 = \left\{ (H, A, B) \in \Delta_2 \,|\, B < \frac{H\left(7 - 10H - H^2\right)}{(1+H)^2} \text{ and } B > \frac{H+1}{2} \right\}.$$

This regions are showed in Fig. 3.

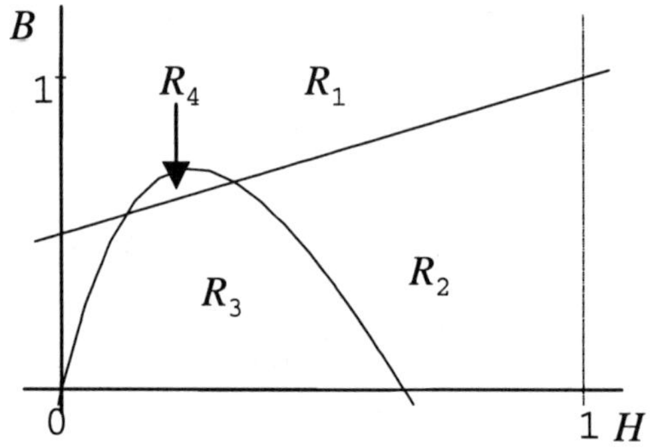

Figure 3. The subset of the parameter space Λ, indicating the nature of the two positive equilibrium points.

Theorem 3.3. *For system (5) we have that*

(1) If $(H, A, B) \in R_1$*, the singularity* $P_{e2} = \left(\frac{1-H}{2}, \frac{1-H}{2}\right)$ *is an attractor node and the singularity* $P_e = (H, H)$ *is:*

(a) an attractor focus if

$$B < \frac{-H^3 + 8H^2 - 5H + 2 + 2\,(H+1)\,|3H-1|\,\sqrt{(1-H)}}{(H+1)^2}$$

(b) an attractor node if

$$B > \frac{-H^3 + 8H^2 - 5H + 2 + 2\,(H+1)\,|3H-1|\,\sqrt{(1-H)}}{(H+1)^2}$$

Moreover, there exists a separtrix curve and depending on initial conditions the $\omega-$limit of trajectories will be the points P_e or P_{e2}.

(2) *If $(H,A,B) \in R_2$, the singularity $P_{e2} = (\frac{1-H}{2}, \frac{1-H}{2})$ is a repellor node and the singularity $P_e = (H,H)$ is:*

(a) *an attractor focus if*

$$B > \frac{-H^3 + 8H^2 - 5H + 2 - 2(H+1)|3H-1|\sqrt{(1-H)}}{(H+1)^2}$$

(b) *an attractor node if*

$$B < \frac{-H^3 + 8H^2 - 5H + 2 - 2(H+1)|3H-1|\sqrt{(1-H)}}{(H+1)^2}$$

(3) *If $(H,A,B) \in R_3$, the singularity $P_{e2} = \left(\frac{1-H}{2}, \frac{1-H}{2}\right)$ is a repellor node and the singularity $P_e = (H,H)$ is:*

(a) *a repellor focus if*

$$B > \frac{-H^3 + 8H^2 - 5H + 2 - 2(H+1)|3H-1|\sqrt{(1-H)}}{(H+1)^2}$$

(b) *a repellor node if*

$$B < \frac{-H^3 + 8H^2 - 5H + 2 - 2(H+1)|3H-1|\sqrt{(1-H)}}{(H+1)^2}$$

both surrounding by a noninfinitesimal limit cycle[5] *which is attractor (stable).*

(4) *If $(H,A,B) \in R_4$, the singularity $P_{e2} = \left(\frac{1-H}{2}, \frac{1-H}{2}\right)$ is an attractor node and the singularity $P_e = (H,H)$ is a repellor node. Moreover, there exist, an unique heteroclinic orbit*[17,22] *connecting them and an unique limit cycle surrounding them. According the initial conditions the $\omega-$limit of trajectories will be this limit cycle or the singularity P_{e2}.*

4. Proofs

Proof of Lemma 3.1.

(1) Clearly, the axis $u = 0$ and $v = 0$ are invariant set.

If $u = 1$, we have that $\frac{du}{d\tau} = -Q\, v < 0$, and any let be the sign of $\frac{dv}{d\tau} = B\,(1-v)\,(A+1)v$, the trajectories enter to the region.

(2) We study the behaviour of system at infinity by means of Poincaré compactification[8] given by the transformation $X = \frac{u}{v}$ and $Y = \frac{1}{v}$, thus[1]

$$\frac{dX}{d\tau} = \frac{1}{v^2}\left(v\frac{du}{d\tau} - u\frac{dv}{d\tau}\right) \text{ and } \frac{dY}{d\tau} = -\frac{1}{v^2}\frac{dv}{d\tau}. \ \square$$

Proof of Lemma 3.2

(1) As $DY_w(1,0) = \begin{pmatrix} -A-1 & -Q \\ 0 & B(A+1) \end{pmatrix}$, it has that

$$\det DY_w(1,0) = -B\,(A+1)^2 < 0\ ,$$

therefore $(1,0)$ is saddle point[3].

(2) As $DY_w(0,0) = \begin{pmatrix} 0\,0 \\ 0\,0 \end{pmatrix}$. To desingularize the origin we apply the Blowing-up method[11,30]

a) We consider the vertical Blowing-up given by the function

$$\Psi(p,q) = (pq, q) = (u,v),$$

thus

$$\frac{dp}{d\tau} = \frac{1}{q}\left(\frac{du}{d\tau} - p\,\frac{dq}{d\tau}\right) \text{ and } \frac{dq}{d\tau} = \frac{dv}{d\tau};$$

Substituting in system (2) we have that

$$Z_\omega : \begin{cases} \frac{dp}{d\tau} = pq\left(((1-pq)(A+(pq)^2) - Qq)\,p - B\,(p-1)\left(A+p^2q^2\right)\right) \\ \frac{dq}{d\tau} = Bq^2\,(p-1)\left(A+p^2q^2\right) \end{cases}$$

making a time rescaling given by $T = q\tau$ we become:

$$\frac{1}{q}Z_\omega = \overline{Z_\omega} : \begin{cases} \frac{dp}{dT} = p([(1-pq)(A+p^2q^2) - Qq]p - B\ (p-1)\ (A+p^2q^2)) \\ \frac{dq}{dT} = B\ q(p-1)\ (A+p^2q^2) \end{cases}$$

If $q = 0$ then $(dq/dT) = 0$; moreover,

$$\frac{dp}{dT} = (Ap - AB\ (p-1))\,p\ .$$

When $(dp/dT) = 0$ it has the solutions $p = 0$ or $p = B/(B-1)$. As $0 < p$, then $B/(B-1) > 0$ and $B > 1$. The jacobian matrix for vector field $\overline{Z_\omega}$ is,

$$D\overline{Z_\omega}(p,q) = \begin{pmatrix} \overline{Z}_{\omega 11} & -p^2\left(Ap + 3p^3q^2 - 2qp^2 + Q + 2Bqp^2 - 2Bqp\right) \\ Bq\left(3p^2q^2 - 2q^2p + A\right) & B\,(p-1)\left(3p^2q^2 + A\right) \end{pmatrix}$$

where

$$\overline{\tilde{Z}_{\omega 11}} = -3p^2qA - 5q^3p^4 + 4p^3q^2 + 2Ap - 2pQq - 2pAB - 4p^3q^2B + 3p^2q^2B + AB.$$

At $(0,0)$ the jacobian matrix is

$$D\overline{Z_\omega}(0,0) = \begin{pmatrix} AB & 0 \\ 0 & -AB \end{pmatrix}$$

therefore, $(0,0)$ at the vector field $\overline{Z_\omega}$ is a saddle point, for all parameters value.

At $\left(\frac{B}{B-1}, 0\right)$ we have that

$$D\overline{Z_\omega}\left(\frac{B}{B-1}, 0\right) = \begin{pmatrix} -AB & -\left(\frac{B}{B-1}\right)^2\left(\frac{AB}{B-1} + Q\right) \\ 0 & \frac{AB}{B-1} \end{pmatrix}$$

as

$$\det D\overline{Z_\omega}\left(\frac{B}{B-1}, 0\right) < 0,$$

the point $\left(\frac{B}{B-1}, 0\right)$ is also a saddle point for $B > 1$.

We note that $B < 1$, the singularity is in the second quadrant and it is not interesting situation.

(b) Now, we consider the horizontal blowing-up by means the function

$$\Psi(p,q) = (p, pq) = (u, v),$$

thus

$$\frac{dp}{d\tau} = \frac{du}{d\tau} \text{ and } \frac{dq}{d\tau} = \frac{1}{p}\left(\frac{dv}{d\tau} - q\,\frac{dp}{d\tau}\right);$$

Substituting in system (2) we have

$$Z_\omega = \begin{cases} \frac{dp}{dT} = ((1-p)(A+p^2) - Q\,pq)\ p^2 \\ \frac{dq}{dT} = pq\left(B\ (1-q)\ (A+p^2) - ((1-p)(A+p^2) - Q\,pq)\right) \end{cases}$$

making a time rescaling given by $T = p\tau$ we obtain:

$$\frac{1}{p}Z_\omega = \tilde{Z}_\omega : \begin{cases} \frac{dp}{dT} = ((1-p)(A+p^2) - Q\,pq)\ p \\ \frac{dq}{dT} = q\left(B\ (1-q)\ (A+p^2) - ((1-p)(A+p^2) - Q\,pq)\right) \end{cases}$$

Then, we have that the equilibrium points for vector field $\tilde{Z}_\omega$ are $(0,0)$ and $(0, \frac{B-1}{B})$ with $B > 1$. The jacobian matrix of $\tilde{Z}_\omega$ is

$$D\tilde{Z}_\omega(p,q) = \begin{pmatrix} -2pA - 4p^3 + 3p^2 - 2Qpq + A & -Qp^2 \\ q\left(2Bp - 2Bpq + A + 3p^2 - 2p + Qq\right) & \tilde{Z}_\omega 22 \end{pmatrix}$$

where

$$\tilde{Z}_\omega 22 = BA + Bp^2 - 2BqA - 2Bqp^2 - A - p^2 + pA + p^3 + 2Qpq.$$

Evaluating the matrix we have that:

$$D\tilde{Z}_\omega(0,0) = \begin{pmatrix} A & 0 \\ 0 & A(B-1) \end{pmatrix}.$$

As

$$\det D\tilde{Z}_\omega(0,0) = A^2(B-1) > 0$$

and

$$\mathrm{tr} D\tilde{Z}_\omega(0,0) = A + A(B-1) > 0$$

the point $(0,0)$ is a repellor.

$$D\tilde{Z}_\omega(0, \frac{B-1}{B}) = \begin{pmatrix} A & 0 \\ \dfrac{(B-1)(BA+QB-Q)}{B^2} & -(B-1)A \end{pmatrix}$$

As

$$\det D\tilde{Z}_\omega(0, \frac{B-1}{B}) = -A^2(B-1) < 0,$$

the singularity $(0, \frac{B-1}{B})$ for $B > 1$ is a saddle point. In similar form to the vertical blowing-up, if $B < 1$, the singularity has not interest.

Then the singularity $(0,0)$ in the system (2) is a nonhyperbolic saddle point. □

Proof of Lemma 3.3

The equilibrium points of system (2) satisfy the Eq. (3). According to Descartes Rule, the associated polynomial has three or one real positive roots or two some of this with multiplicity two. Let $u_e = H$, the solution that always exists for all $\eta = (A, B, Q) \in \Delta$; employing the synthetic division we obtain

$$h(u) = \left(-u^2 + (1-H)u - A - Q + H - H^2\right)(u - H)$$

and the rest is $A - HA - HQ + H^2 - H^3 = 0$, that is,

$$Q = \frac{1}{H}(1-H)\left(H^2 + A\right).$$

The roots of factor

$$h_1(u) = -u^2 + (1-H)\,u - A - Q + H - H^2$$
$$= -u^2 + (1-H)\,u - \frac{A}{H}$$

are

$$L_1 = \frac{H(1-H)+S}{2H} \text{ and } L_2 = \frac{H(1-H)-S}{2H}$$

with

$$S^2 = H\left(H\,(1-H)^2 - 4A\right).$$

Clearly,

(1) If $T = H\,(1-H)^2 - 4A < 0$, the equation $h_1(u) = 0$, has an unique root.
(2) If $T = H\,(1-H)^2 - 4A > 0$, the equation $h_1(u) = 0$, has three different real positive roots.
(3) If $T = H\,(1-H)^2 - 4A = 0$, then $L_1 = L_2 = L$ and the equation $h_1(u) = 0$, has two different real positive roots. Moreover $h_1(u) = (u-L)^2$, and in this case, $Q = \frac{1}{4}\,(1-H)\,(H+1)^2$.
(4) If $A = \frac{1}{27}$ and $Q = \frac{8}{27}$, we have the equation $h(u) = (u - \frac{1}{3})^3$ and there exists a unique real positive root.

Is immediate the second part of Lemma 3.3. □

Proof of Theorem 3.1

The jacobian matrix of system (4) evaluate at point $\left(\frac{1}{3}, \frac{1}{3}\right)$ is

$$DY_\eta\left(\frac{1}{3}, \frac{1}{3}\right) = \begin{pmatrix} -\frac{1}{9}A + \frac{1}{27} & -\frac{2}{9}A - \frac{2}{81} \\ \frac{1}{3}B\left(A + \frac{1}{9}\right) & -\frac{1}{3}B\left(A + \frac{1}{9}\right) \end{pmatrix}$$

if $A = \frac{1}{27}$, then

$$DY_\eta\left(\frac{1}{3}, \frac{1}{3}\right) = \begin{pmatrix} \frac{8}{243} & -\frac{8}{243} \\ \frac{4}{81}B & -\frac{4}{81}B \end{pmatrix},$$

and

$$\det DY_\eta\left(\frac{1}{3}, \frac{1}{3}\right) = 0,$$

the singularity is nonhyperbolic. As the

$$\operatorname{tr} DY_\eta\left(\frac{1}{3}, \frac{1}{3}\right) = \frac{8}{243} - \frac{4}{81}B,$$

we have the equilibrium point $\left(\frac{1}{3}, \frac{1}{3}\right)$ is

(1) a nonhyperbolic attractor node[29], if and only if, $B > \frac{2}{3}$.
(2) a nonhyperbolic repellor node[29], if and only if, $B < \frac{2}{3}$.
In this case, how the singularities are saddle point, by Poincaré-Bendixon Theorem must to exist a noninfitesimal limit cycle at the strip $\breve{\Gamma}$.
(3) $B = \frac{2}{3}$, the singularity has a behavior as a cusp point, surrounded by a non-infinitesimal limit cycle. □

We observe that this limit cycle remains under small perturbation of parameter values, and it explains why it can exist a non-infinitesimal limit cycle surrounding two or three equilibrium points, and it not originated by Hopf bifurcation.

Proof of Theorem 3.2

The jacobian matrix of system (4) evaluate at point (H, H) is

$$DY_\eta(H,H) = \begin{pmatrix} -H^2\left(A+3H^2-2H\right) & -(1-H)\left(A+H^2\right)H \\ BH\left(A+H^2\right) & -BH\left(A+H^2\right) \end{pmatrix},$$

and as $H(1-H)^2 - 4A < 0$, then

$$detDY_\eta(H,H) = BH^2\left(A+2H^3-H^2\right)\left(H^2+A\right) > 0,$$

since if it assume negative a contradiction appears. Then, the nature of this equilibrium point dependent of trace of jacobian matrix given by:

$$\mathrm{tr}DY_\eta(H,H) = -H\left(H\left(A+3H^2-2H\right)+B\left(H^2+A\right)\right),$$

and let

$$\begin{aligned} M &= \left(\mathrm{tr}DY_\eta(H,H)\right)^2 - 4\mathrm{det}DY_\eta(H,H) \\ &= \left(H^2\left(H\left(-3H^2+2H-A\right)-B\left(H^2+A\right)\right)\right)^2 - \\ &\quad -4H^4B\left(A+2H^3-H^2\right)\left(A+H^2\right) \\ &= \left(A+H^2\right)^2B^2 - 2\left(A+H^2\right)\left(2A-AH+H^3\right)B + \\ &\quad +H^2\left(A-2H+3H^2\right)^2 \end{aligned}$$

(1) If $H\left(A+3H^2-2H\right)+B\left(H^2+A\right) < 0$, the point (H, H) is a repellor and by Poincaré-Bendixon Theorem there exists a attractor limit cycle surrounding it.
   (a) If $M < 0$, is a repellor focus,
   (b) If $M > 0$, is a repellor node, and by the Poincaré-Bendixon Theorem it is surrounded by a non-infinitesimal limit cycle.

(2) If $H\left(A+3H^2-2H\right)+B\left(H^2+A\right)>0$, the point (H,H) is a global attractor.

(a) If $M<0$, is an attractor focus.
(b) If $M<0$, is an attractor node.

(3) Now, we assume that If $\mathrm{tr}DY_\eta(H,H)=0$, then

$$B=\frac{H\left(2H-3H^2-A\right)}{H^2+A}.$$

Let $u=U+H$ and $u=V+H$, then $\frac{du}{dT}=\frac{dU}{dT}$ and $\frac{dv}{dT}=\frac{dV}{dT}$ and we obtain the system translated to origin

$$Z_\nu:\begin{cases}\frac{dU}{dT}=(H(1-U-H)(A+(U+H)^2)-(1-H)\left(H^2+A\right)\ (V+H))\\ \qquad .\,(U+H)^2\\ \frac{dV}{dT}=\frac{H^2\left(2H-3H^2-A\right)}{H^2+A}\ (U-V)\ (A+(U+H)^2)(V+H)\end{cases}$$

A normal form[8,17] for this system is obtained making an adequate change of coordinates; for this, we use the canonical form employing the associated Jordan matrix given by[3]

$$J=\begin{pmatrix}0 & -W\\ W & 0\end{pmatrix},$$

where $W^2=H^5\left(-3H^2+2H-A\right)\left(A+2H^3-H^2\right)$.

The matrix for the change of coordinates is[3]

$$M=\begin{pmatrix}H^3\left(-3H^2+2H-A\right) & -W\\ H^3\left(-3H^2+2H-A\right) & 0\end{pmatrix}.$$

After a tedious algebraic calculus we obtain

$$\breve{Z}:\begin{cases}\frac{dx}{d\Upsilon}=-y-\frac{H^2\left(-3H^2+2H-A\right)\left(A+3H^2\right)}{(H^2+A)}xy+\frac{2HW}{(H^2+A)}y^2\\ \qquad -\frac{3H^6\left(-3H^2+2H-A\right)^2}{(H^2+A)}x^2y+H.O.T.\\ \frac{dy}{d\Upsilon}=x+\frac{H^2\left(-3H^2+2H-A\right)\left(2A-3H^2+7H^3\right)}{(A+2H^3-H^2)}x^2\\ \qquad -\frac{W\left(2A^2-2H^4+7H^5-4AH^2+A^2H+12AH^3\right)}{H(A+2H^3-H^2)(A+H^2)}xy\\ \qquad +\frac{\left(2A^2-5AH-H^3+3H^4+9AH^2\right)H^2}{(H^2+A)}y^2\\ \qquad +\frac{H^4\left(-3H^2+2H-A\right)^2\left(A-3H^2+9H^3\right)}{(A+2H^3-H^2)}x^3+H.O.T.\end{cases}$$

Using the Mathematica package[39], we obtain that the second Liapunov quantities[8] is:

$$\eta[2]=\frac{Hf(H,A,W)}{8\left(A+H^2\right)^2W^3}$$

where

$$
\begin{aligned}
f(H,A,W) = {} & H^{14}\left(1608H^3 - 376H^2 - 32H - 2225H^4 + 1023H^5 + 16\right) \\
& + 3H^3(H+1)A^7 + H^4\left(53H^2 - 6H + 9H^3 - 14\right)A^6 \\
& + H^5\left(6H - 247H^2 + 231H^3 + 156H^4 + 20\right)A^5 \\
& + H^6\left(556H^2 - 40H - 1004H^3 - 155H^4 + 1029H^5 - 8\right)A^4 \\
& + H^8\left(1052H^2 - 560H + 805H^3 - 4295H^4 + 3424H^5 + 56\right)A^3 \\
& + H^{10}\left(456H - 3206H^2 + 8458H^3 - 10\,921H^4 + 5523H^5 + 24\right)A^2 \\
& + H^{12}\left(1180H - 4098H^2 + 8239H^3 - 9059H^4 + 4004H^5 - 152\right)A \\
& + (2A^4 + 4H\left(3H + H^2 - 2\right)A^3 + 4H^2\left(7H^3 - 8H + 2\right)A^2 \\
& \quad + 4H\left(4H^3 - 23H^5 + 15H^6 + 2\right)A \\
& \quad + 6H^6(H-2)(2H-1)(3H-2))W
\end{aligned}
$$

It is possible to prove that $f(H,A,W)$ changes of sign, and the system has two limit cycles surrounding the unique equilibrium point (H,H), as is showing in Figure 4, the innermost unstable and the outermost stable. □

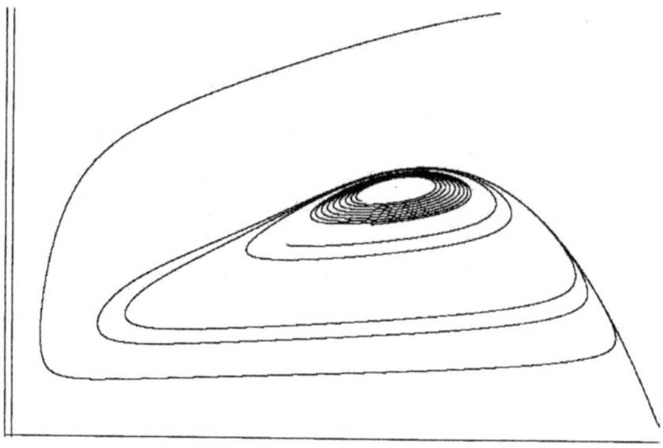

Figure 4. The unique equilibrium point surrounded by two limit cycles.

Proof of Theorem 3.3

When $T = H(1-H)^2 - 4A = 0$, then $A = \frac{H(1-H)^2}{4}$ and the jacobian matrix of vector field (5) in $P_{e2} = \left(\frac{1-H}{2}, \frac{1-H}{2}\right)$ is

$$
DY_\eta\left(\frac{1-H}{2}, \frac{1-H}{2}\right) = \begin{pmatrix} \frac{1}{16}(1-H)^3(H+1)^2 & -\frac{1}{16}(H+1)^2(1-H)^3 \\ B(H+1)\left(\frac{1-H}{2}\right)^3 & -\frac{1}{8}B(1-H)^3(H+1) \end{pmatrix}
$$

and

$$\det DY_w\left(\frac{1-H}{2},\frac{1-H}{2}\right)=0,$$

therefore is $\left(\frac{1-H}{2},\frac{1-H}{2}\right)$ a nonhyperbolic equilibrium points and its behavior of this point dependent of

$$\operatorname{tr} DY_w\left(\frac{1-H}{2},\frac{1-H}{2}\right)=\frac{1}{16}\left(H+1\right)\left(1-H\right)^3\left(H+1-2B\right),$$

and in particular of the factor

$$N=H+1-2B$$

(1) If $B<\frac{H+1}{2}$, the singularity $\left(\frac{1-H}{2},\frac{1-H}{2}\right)$ is a nonhyperbolic repellor node[29].
(2) If $B>\frac{H+1}{2}$, the singularity $\left(\frac{1-H}{2},\frac{1-H}{2}\right)$ is a nonhyperbolic attractor node[29].
(3) If $B=\frac{H+1}{2}$, the singularity $\left(\frac{1-H}{2},\frac{1-H}{2}\right)$ is a cusp point[3], and the system presents the Bogdanovich-Takens bifurcations[7,8,31,44] .

For other hand the Jacobian matrix of vector field of system (5) at $P_e=(H,H)$ is

$$DY_\eta(H,H)=\begin{pmatrix}\frac{H^3}{4}\left(-7+10H+H^2\right) & -\frac{1}{4}\left(1-H\right)\left(H+1\right)^2H^2\\ \frac{1}{4}BH^2\left(H+1\right)^2 & -\frac{1}{4}BH^2\left(H+1\right)^2\end{pmatrix},$$

and how

$$\det DY_\eta(H,H)=\frac{1}{16}H^4B\left(H+1\right)^2\left(3H-1\right)^2>0,$$

the sign of eigenvalues dependent of

$$\operatorname{tr} DY_\eta(H,H)=\frac{1}{4}H^2\left(H\left(7-10H-H^2\right)-\left(H+1\right)^2B\right),$$

in particular of the factor

$$L=H\left(7-10H-H^2\right)-\left(H+1\right)^2B.$$

Clearly,

(1) If

$$B<\frac{H\left(7-10H-H^2\right)}{\left(H+1\right)^2},$$

the singularity $P_e=(H,H)$ is a repellor.

(2) If

$$B > \frac{H\left(7 - 10H - H^2\right)}{\left(H+1\right)^2},$$

the singularity $P_e = (H, H)$ is an attractor.

For

$$\begin{aligned} M &= (\mathrm{tr} DY_\eta(H,H))^2 - 4\det DY_\eta(H,H) \\ &= \frac{1}{16} H^4 ((H+1)^4 B^2 + 2\left(H^3 - 8H^2 + 5H - 2\right)(H+1)^2 B + \\ &\quad + H^2 \left(H^2 + 10H - 7\right)^2) \end{aligned}$$

we have that the sign of M dependent of the factor

$$M_1 = (H+1)^4 B^2 + 2\left(H^3 - 8H^2 + 5H - 2\right)(H+1)^2 B + H^2\left(H^2 + 10H - 7\right)^2$$

determining the curves

$$B_1 = \frac{-H^3 + 8H^2 - 5H + 2 - 2\sqrt{(1-H)(H+1)^2(3H-1)^2}}{(H+1)^2}$$

and

$$B_2 = \frac{-H^3 + 8H^2 - 5H + 2 + 2\sqrt{(1-H)(H+1)^2(3H-1)^2}}{(H+1)^2}$$

Assuming that

$$B > \frac{H\left(7 - 10H - H^2\right)}{(H+1)^2},$$

the singularity $P = (H, H)$ is

(1) an attractor focus if

$$B < \frac{-H^3 + 8H^2 - 5H + 2 + 2(H+1)\left|3H-1\right|\sqrt{(1-H)}}{(H+1)^2}$$

(2) an attractor node if

$$B > \frac{-H^3 + 8H^2 - 5H + 2 + 2(H+1)\left|3H-1\right|\sqrt{(1-H)}}{(H+1)^2}.$$

If the equilibrium point $P_e = (H, H)$ is attractor and P_{e2} is a repellor, all trajectories has as $\omega-$limit the point P_e.

Assuming that

$$B < \frac{H\left(7 - 10H - H^2\right)}{\left(H+1\right)^2},$$

the singularity $P = (H, H)$ is

(1) a repellor focus if

$$B > \frac{-H^3 + 8H^2 - 5H - 2 + 2\left(H+1\right)\left|3H - 1\right|\sqrt{(1-H)}}{\left(H+1\right)^2}$$

(2) a repellor node if

$$B < \frac{-H^3 + 8H^2 - 5H + 2 - 2\left(H+1\right)\left|3H - 1\right|\sqrt{(1-H)}}{\left(H+1\right)^2}$$

If both equilibrium points are node attractor, there exists a separatrix curve that divide the behavior of trajectories of system according the position of initial conditions.

If the singularity $P = (H, H)$ is a repellor node then both equilibrium points are surrounded by a non-infinitesimal limit cycle by Poincaré-Bendixon Theorem (Figure 4) □

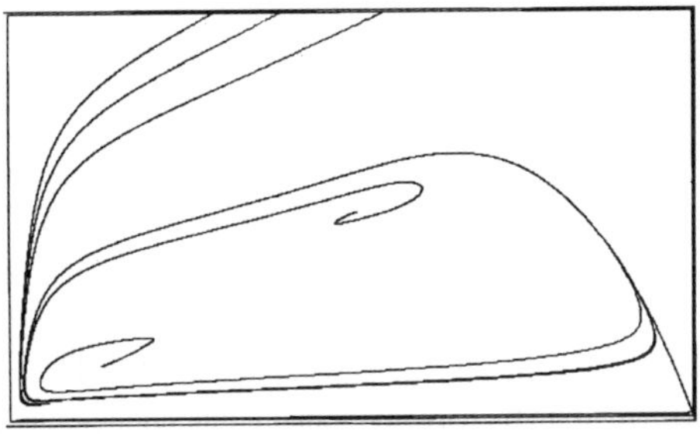

Figure 5. Two repellor equilibrium points and surrounded by a non-infinitesimal limite cycle.

In Figure 6 we give the bifurcation diagram of system (5).

5. Conclusions

In this work, the model proposed has been studied in two cases, when there exists an unique positive equilibrium point or if there exists two equilibrium

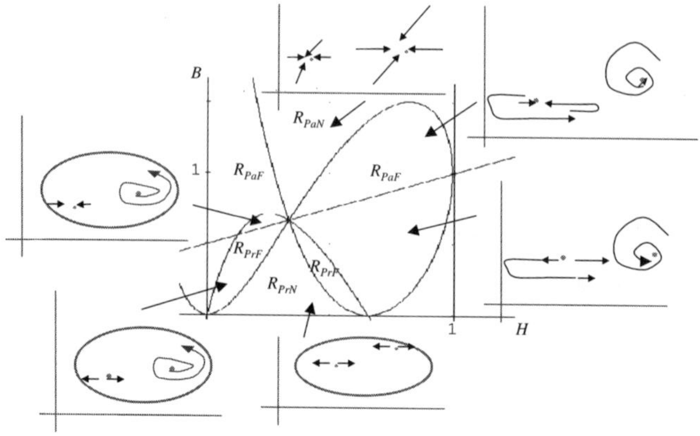

Figure 6. Bifurcation diagram for system (4).

points at the interior of first quadrant and we show that the system (2) has a reach dynamics.

We can see that the second case has more interesting due to mathematical complexities as are existence of separatrix, heteroclinic curves or non-infinitesimal limit cycles surrounding to two equilibrium points. Also is possible the existence of two limit cycles surrounding the equilibrium point $P_e = (H, H)$ that ever exists at interior of the first quadrant, where H is the positive root of a third grade polynomial equation. This singularity has a great importance upon whole behavior of system, anything be the value of H.

Taylor[35] hypothesizes that a type IV functional response should tend to destabilize a system and the results obtained would seem to support this because for a wide range of parameter bistability is prevalent.

With respect to the biological control of a pest by a predator, this results suggests that prey interference can lead to pest outbreaks earlier in the growing season[10]. Another example of the destabilizing nature for non-monotonic functional response is the presence of two attractor points for some values of the parameters, which illustrate the dynamical complexity possible in a relatively simple differential equations system based in a population interaction.

We show that the qualitatively behavior demonstrated above for the type IV functional response is not similar to the model of May-Holling-Tanner[3,32,38] because furthermore the exception of the number of the equilibrium point at interior of the first quadrant, which can be both attractor, system has heteroclinic curve separating the behavior of trajectories, some of which has as $\omega-$limit the stable equilibrium point and the other has a limit cycle surrounding both equilibrium points.

For the existence of two equilibrium points at the interior of first quadrant it can possible verify that, one of this only can a node (attractor repellor) meanwhile the other can be focus or node. Moreover, this model has sensitivity dependence on initial conditions[3], implying the high sensitivity of some trajectories of this deterministic system respect to initial size of both populations, because it presents the bistability phenomenon for certain set of parameter values. It is clear that can coexist:

(1) one stable equilibrium point surrounded by a stable limit cycle or,
(2) two stable equilibrium points.

Biologically, we can to say that this model is persistent, because $(0,0)$ and $(1,0)$ are always saddle point for any parameter values, then both populations coexist.

Acknowledgments

The authors wish to thank to the members of the Mathematical Ecology Group of the Institute of Mathematics on the Pontificia Universidad Católica of Valparaíso, for their valuable comments and suggestions.

References

1. A. Aguilera-Moya and E. González-Olivares, A Gause type model with a generalized class of nonmonotonic functional response, In R. Mondaini (Ed.) Proceedings of the Third Brazilian Symposium on Mathematical and Computational Biology, E-Papers Serviços Editoriais Ltda, Río de Janeiro, Vol 2 (2004) 206–217.
2. A. A. Andronov, E. A. Leontovich, I. Gordon and A. G. Maier, Qualitative theory of second-order dynamic systems, A Halsted Press Book, John Wiley & Sons, New York (1973).
3. D.K.Arrowsmith and C.M. Place, Dynamical System.Differential equations, maps and chaotic behaviour, Chapman and Hall, 1992.
4. M. A. Aziz-Alaoui and M. Daher Okiye, Boundedness and global stability for a predator-prey model with modified Leslie-Gower and Holling-type II schemes, Applied Mathematics Letters, 16 (2003), 1069–1075.
5. T. R. Blows and N. G. Lloyd, The number of limit cycles of certain polynomial differential equations, Proceedings of the Royal Society of Edimburgh, 98A (1984), 215–239.

6. A. A. Berryman, A. P. Gutierrez, and R. Arditi. Credible, Parsimonious and Useful Predator-Prey Models - A reply to Abrams, Gleeson, and Sarnelle. Ecology, 76 (1995), 1980–1985.
7. H. W. Broer, V. Naudot, R. Roussarie and K. Saleh, Bifurcations of a predator-prey model with non-monotonic response function, C. R. Acad. Sci. Paris, Ser. I, 341 (2005), 601–604.
8. C. Chicone, Ordinary differential equations with applications, Texts in Applied Mathematics 34, Springer, 1999.
9. C. S. Coleman, Hilbert's 16th. Problem: How Many Cycles? In: M. Braun, C. S. Coleman and D. Drew (Ed). Differential Equations Model. Springer Verlag, (1983) 279–297.
10. J. B. Collings, The effect of the functional response on the bifurcation behavior of a mite predator-prey interaction model, Journal of Mathematical Biology, 36 (1997) 149–168.
11. F. Dumortier, Singularities of vector fields, Monografías de Matemática Vol. 32, IMPA, Brazil, 1978.
12. H. I. Freedman, Deterministic Mathematical Model in Population Ecology, Marcel Dekker, New York, 1980.
13. H. I. Freedman and G. S. K. Wolkowicz, Predator-prey systems with group defence: The paradox of enrichment revisted. Bulletin of Mathematical Biology, 8 (1986) 493–508.
14. E. González-Olivares, A predator-prey model with a nonmonotonic consumption function, In R. Mondaini (Ed.) Proceedings of the Second Brazilian Symposium on Mathematical and Computational Biology, E-papers Serviços Editoriais Ltda. (2003) 23–39.
15. E. González-Olivares, B, González-Yañez, E Sáez and I. Szántó, On the number of limit cycles in a predator-prey model with nonmonotonic functional response, Discrete and Continuous Dynamical System, 6 (2006), 525–534.
16. B, González-Yañez. and E. González-Olivares, Consequences of Allee effect on a Gause type predator-prey model with nonmonotonic functional response, In R. Mondaini (Ed.) Proceedings of the Third Brazilian Symposium on Mathematical and Computational Biology, E-Papers Serviços Editoriais Ltda, Río de Janeiro, Vol 2 (2004), 358–373.
17. J. Guckenheimer and P. Holmes, Nonlinear Oscillations, Dynamical Systems, and Bifurcations of Vector Fields, Springer-Verlag, 1983.
18. I. Hanski, H. Hentonnen, E. Korpimaki, L Oksanen, and P. Turchin, Small-rodent dynamics and predation, Ecology, 82 (2001), 1505–1520.
19. J-C. Huang and D. Xiao, Analyses of bifurcations and stability in a predator-prey system with Holling Type-IV functional response, Acta Mathematicae Applicatae Sinica, 20 (2004) 167–178.
20. H. Koçak, Differential and difference equations through computer experiments, Springer-Verlag, 1989.
21. A. Korobeinikov, A Lyapunov function for Leslie-Gower predator-prey models, Applied Mathematical Letters, 14 (2001) 697–699.
22. Y. A. Kuznetsov, Elements of Applied Bifurcation, AMS 112, Springer, 1995.
23. P. H. Leslie, Some further notes on the use of matrices in population math-

ematics, Biometrica, 35 (1948) 213–245.

24. Y. Li and D. Xiao, Bifurcations of a predator–prey system of Holling and Leslie types, Chaos, Solitons and Fractals (2006), in press.
25. Z. Liu and R. Yuan, Bifurcations in predator-prey system with nonmonotonic functional response, Nonlinear Analysis, Real World Appications, 6 (2005), 187–205.
26. N. G. Lloyd, J. M. Pearson, E. Sáez and I. Szántó, Limit cycles of a cubic Kolmogorov system, Applied Mathematical Letter **9** (1996), 15–18.
27. R. M. May, Stability and complexity in model ecosystems, Princeton University Press, 1974.
28. P. Y. H. Pang and M. Wang, Non-constant positive steady states of a predator-prey system with non-monotonic funtional response and diffusion, Proceedings of London Mathematical Society 3, (2004), 135–157.
29. L. Perko, Differential equations and dynamical systems, Springer-Verlag, 1991
30. R. Roussarie, Bifurcations of planar vector fields and Hilbert's 16th problem, IMPA, 1995.
31. S. Ruan and D. Xiao, Global analysis in a predator-prey system with nonmonotonic functional response, SIAM Journal on Applied. Mathematics, 61 (2001), 1445–1472.
32. E. Sáez and E. González-Olivares, Dynamics on a Predator-prey Model, SIAM Journal on Applied Mathematics, 59 (1999) 1867–1878.
33. J. Sotomayor, Lições de Equações Diferenciais Ordinárias, Projeto Euclides IMPA, CNPq., 1979.
34. F. Takens, Unfoldings of certain singularities of vector fields. Generalized Hopf bifurcations, Journal Differential Equations, 14 (1973), 476–493.
35. R. J. Taylor, Predation, Chapman and Hall, 1984.
36. P. Turchin, Complex population dynamics. A theoretical/empirical synthesis, Monographs in Population Biology 35, Princeton University Press, 2003.
37. S. Véliz-Retamales and E. González-Olivares, Dynamics of a Gause type prey-predator model with a rational nonmonotonic consumption function, In R. Mondaini (Ed.) Proceedings of the Third Brazilian Symposium on Mathematical and Computational Biology, E-Papers Serviços Editoriais Ltda, Río de Janeiro, Vol. 2 (2004) 181–192.
38. J. J. Wollkind, J. B. Collings and J. A. Logan, Metastability in a temperature-dependent model system for a predator-prey mite outbreak interactions on fruit trees, Bulletin of Mathematical Biology, 5 (1988) 379–409.
39. Wolfram Research, Mathematica: A System for Doing Mathematics by Computer, 1988.
40. G. S. W. Wolkowicz, Bifurcation analysis of a predator–prey system involving group defense, SIAM Journal on Applied Mathematics, 48 (1988), 592–606.
41. D. Xiao and S. Ruan, Bifurcations in a predator-prey system with group defense, International Journal of Bifurcation and Chaos, 11 (2001), 2123–2131.
42. D. Xiao and S. Ruan, Bogdanov-Takens Bifurcations in predator-prey systems with constant rate harvesting, Field Institue Communications, 21

(1999), 493–506.
43. D. Xiao and H. Zhu, Multiple focus and Hopf bifurcations in a predator-prey system with nonmonotonic functional response SIAM Journal on Applied Mathematics, 66 (2006), 802–820.
44. H. Zhu, S.A. Campbell and G. S. K. Wolkowicz. Bifurcation analysis of a predator-prey system with nonmonotonic functional response. SIAM Journal on Applied Mathematics, 63 (2002), 636–682.

INDEX